# AUTOMATE THE BORING STUFF
# WITH PYTHON

# AUTOMATE THE BORING STUFF WITH PYTHON

## 2ND EDITION

## Practical Programming for Total Beginners

### by Al Sweigart

**no starch press**

San Francisco

**AUTOMATE THE BORING STUFF WITH PYTHON, 2ND EDITION.** Copyright © 2020 by Al Sweigart.

Printed in USA

First printing

24 23 22 21 20      1 2 3 4 5 6 7 8 9

ISBN-10: 1-59327-992-2
ISBN-13: 978-1-59327-992-9

Publisher: William Pollock
Production Editor: Laurel Chun
Cover Illustration: Josh Ellingson
Interior Design: Octopod Studios
Developmental Editors: Frances Saux and Jan Cash
Technical Reviewers: Ari Lacenski and Philip James
Copyeditors: Kim Wimpsett, Britt Bogan, and Paula L. Fleming
Compositors: Susan Glinert Stevens and Danielle Foster
Proofreaders: Lisa Devoto Farrell and Emelie Burnette

For information on distribution, translations, or bulk sales, please contact No Starch Press, Inc. directly:

No Starch Press, Inc.
245 8th Street, San Francisco, CA 94103
phone: 1.415.863.9900; info@nostarch.com
www.nostarch.com

The Library of Congress Control Number for the first edition is: 2014953114

For my nephew Jack

## About the Author

Al Sweigart is a software developer and tech book author. Python is his favorite programming language, and he is the developer of several open source modules for it. His other books are freely available under a Creative Commons license on his website *https://inventwithpython.com/*. His cat now weighs 11 pounds.

## About the Tech Reviewer

Philip James has been working in Python for over a decade and is a frequent speaker in the Python community. He speaks on topics ranging from Unix fundamentals to open source social networks. Philip is a Core Contributor to the BeeWare project and lives in the San Francisco Bay Area with his partner Nic and her cat River.

# BRIEF CONTENTS

# CONTENTS IN DETAIL

## 2
## FLOW CONTROL

## 3
## FUNCTIONS

# 6
# MANIPULATING STRINGS

# PART II: AUTOMATING TASKS

## 7
## PATTERN MATCHING WITH REGULAR EXPRESSIONS 161

## 8
## INPUT VALIDATION 187

# 9
# READING AND WRITING FILES
**201**

# 12
# WEB SCRAPING
**267**

## 13
## WORKING WITH EXCEL SPREADSHEETS      301

## 16
## WORKING WITH CSV FILES AND JSON DATA　　　　　371

## 17
## KEEPING TIME, SCHEDULING TASKS,
## AND LAUNCHING PROGRAMS　　　　　389

# 18
# SENDING EMAIL AND TEXT MESSAGES

**415**

# 19
# MANIPULATING IMAGES 447

## 20
## CONTROLLING THE KEYBOARD AND MOUSE
## WITH GUI AUTOMATION
**473**

# ACKNOWLEDGMENTS

It's misleading to have just my name on the cover. I couldn't have written a book like this without the help of a lot of people. I'd like to thank my publisher, Bill Pollock; my editors, Laurel Chun, Leslie Shen, Greg Poulos, Jennifer Griffith-Delgado, and Frances Saux; and the rest of the staff at No Starch Press for their invaluable help. Thanks to my tech reviewers, Ari Lacenski and Philip James, for great suggestions, edits, and support.

Many thanks to everyone at the Python Software Foundation for their great work. The Python community is the best one I've found in the tech industry.

Finally, I would like to thank my family, friends, and the gang at Shotwell's for not minding the busy life I've had while writing this book. Cheers!

# INTRODUCTION

"You've just done in two hours what it takes the three of us two days to do." My college roommate was working at a retail electronics store in the early 2000s. Occasionally, the store would receive a spreadsheet of thousands of product prices from other stores. A team of three employees would print the spreadsheet onto a thick stack of paper and split it among themselves. For each product price, they would look up their store's price and note all the products that their competitors sold for less. It usually took a couple of days.

"You know, I could write a program to do that if you have the original file for the printouts," my roommate told them, when he saw them sitting on the floor with papers scattered and stacked all around.

After a couple of hours, he had a short program that read a competitor's price from a file, found the product in the store's database, and noted whether the competitor was cheaper. He was still new to programming, so he spent most of his time looking up documentation in a programming book. The actual program took only a few seconds to run. My roommate and his co-workers took an extra-long lunch that day.

This is the power of computer programming. A computer is like a Swiss Army knife that you can configure for countless tasks. Many people spend hours clicking and typing to perform repetitive tasks, unaware that the machine they're using could do their job in seconds if they gave it the right instructions.

## Whom Is This Book For?

Software is at the core of so many of the tools we use today: nearly everyone uses social networks to communicate, many people have internet-connected computers in their phones, and most office jobs involve interacting with a computer to get work done. As a result, the demand for people who can code has skyrocketed. Countless books, interactive web tutorials, and developer boot camps promise to turn ambitious beginners into software engineers with six-figure salaries.

This book is not for those people. It's for everyone else.

On its own, this book won't turn you into a professional software developer any more than a few guitar lessons will turn you into a rock star. But if you're an office worker, administrator, academic, or anyone else who uses a computer for work or fun, you will learn the basics of programming so that you can automate simple tasks such as these:

- Moving and renaming thousands of files and sorting them into folders
- Filling out online forms—no typing required
- Downloading files or copying text from a website whenever it updates
- Having your computer text you custom notifications
- Updating or formatting Excel spreadsheets
- Checking your email and sending out prewritten responses

These tasks are simple but time-consuming for humans, and they're often so trivial or specific that there's no ready-made software to perform them. Armed with a little bit of programming knowledge, however, you can have your computer do these tasks for you.

## Conventions

This book is not designed as a reference manual; it's a guide for beginners. The coding style sometimes goes against best practices (for example, some programs use global variables), but that's a trade-off to make the code simpler to learn. This book is made for people to write throwaway code, so there's not much time spent on style and elegance. Sophisticated programming concepts—like object-oriented programming, list comprehensions, and generators—aren't covered because of the complexity they add. Veteran programmers may point out ways the code in this book could be changed to improve efficiency, but this book is mostly concerned with getting programs to work with the least amount of effort on your part.

## What Is Programming?

Television shows and films often show programmers furiously typing cryptic streams of 1s and 0s on glowing screens, but modern programming isn't that mysterious. *Programming* is simply the act of entering instructions for the computer to perform. These instructions might crunch some numbers, modify text, look up information in files, or communicate with other computers over the internet.

All programs use basic instructions as building blocks. Here are a few of the most common ones, in English:

- "Do this; then do that."
- "If this condition is true, perform this action; otherwise, do that action."
- "Do this action exactly 27 times."
- "Keep doing that until this condition is true."

You can combine these building blocks to implement more intricate decisions, too. For example, here are the programming instructions, called the *source code*, for a simple program written in the Python programming language. Starting at the top, the Python software runs each line of code (some lines are run only *if* a certain condition is true or *else* Python runs some other line) until it reaches the bottom.

```
❶ passwordFile = open('SecretPasswordFile.txt')
❷ secretPassword = passwordFile.read()
❸ print('Enter your password.')
  typedPassword = input()
❹ if typedPassword == secretPassword:
❺     print('Access granted')
❻     if typedPassword == '12345':
❼         print('That password is one that an idiot puts on their luggage.')
  else:
❽     print('Access denied')
```

You might not know anything about programming, but you could probably make a reasonable guess at what the previous code does just by reading it. First, the file *SecretPasswordFile.txt* is opened ❶, and the secret password in it is read ❷. Then, the user is prompted to input a password (from the keyboard) ❸. These two passwords are compared ❹, and if they're the same, the program prints *Access granted* to the screen ❺. Next, the program checks to see whether the password is *12345* ❻ and hints that this choice might not be the best for a password ❼. If the passwords are not the same, the program prints *Access denied* to the screen ❽.

## What Is Python?

*Python* is a programming language (with syntax rules for writing what is considered valid Python code) and the Python interpreter software that reads source code (written in the Python language) and performs its instructions. You can download the Python interpreter for free at *https://python.org/*, and there are versions for Linux, macOS, and Windows.

The name Python comes from the surreal British comedy group Monty Python, not from the snake. Python programmers are affectionately called Pythonistas, and both Monty Python and serpentine references usually pepper Python tutorials and documentation.

## Programmers Don't Need to Know Much Math

The most common anxiety I hear about learning to program is the notion that it requires a lot of math. Actually, most programming doesn't require math beyond basic arithmetic. In fact, being good at programming isn't that different from being good at solving Sudoku puzzles.

To solve a Sudoku puzzle, the numbers 1 through 9 must be filled in for each row, each column, and each 3×3 interior square of the full 9×9 board. Some numbers are provided to give you a start, and you find a solution by making deductions based on these numbers. In the puzzle shown in Figure 0-1, since 5 appears in the first and second rows, it cannot show up in these rows again. Therefore, in the upper-right grid, it must be in the third row. Since the last column also already has a 5 in it, the 5 cannot go to the right of the 6, so it must go to the left of the 6. Solving one row, column, or square will provide more clues to the rest of the puzzle, and as you fill in one group of numbers 1 to 9 and then another, you'll soon solve the entire grid.

| 5 | 3 |   |   | 7 |   |   |   |   |
|---|---|---|---|---|---|---|---|---|
| 6 |   |   | 1 | 9 | 5 |   |   |   |
|   | 9 | 8 |   |   |   |   | 6 |   |
| 8 |   |   |   | 6 |   |   |   | 3 |
| 4 |   |   | 8 |   | 3 |   |   | 1 |
| 7 |   |   |   | 2 |   |   |   | 6 |
|   | 6 |   |   |   |   | 2 | 8 |   |
|   |   |   | 4 | 1 | 9 |   |   | 5 |
|   |   |   |   | 8 |   |   | 7 | 9 |

| 5 | 3 | 4 | 6 | 7 | 8 | 9 | 1 | 2 |
|---|---|---|---|---|---|---|---|---|
| 6 | 7 | 2 | 1 | 9 | 5 | 3 | 4 | 8 |
| 1 | 9 | 8 | 3 | 4 | 2 | 5 | 6 | 7 |
| 8 | 5 | 9 | 7 | 6 | 1 | 4 | 2 | 3 |
| 4 | 2 | 6 | 8 | 5 | 3 | 7 | 9 | 1 |
| 7 | 1 | 3 | 9 | 2 | 4 | 8 | 5 | 6 |
| 9 | 6 | 1 | 5 | 3 | 7 | 2 | 8 | 4 |
| 2 | 8 | 7 | 4 | 1 | 9 | 6 | 3 | 5 |
| 3 | 4 | 5 | 2 | 8 | 6 | 1 | 7 | 9 |

*Figure 0-1: A new Sudoku puzzle (left) and its solution (right). Despite using numbers, Sudoku doesn't involve much math. (Images © Wikimedia Commons)*

Just because Sudoku involves numbers doesn't mean you have to be good at math to figure out the solution. The same is true of programming. Like solving a Sudoku puzzle, writing programs involves breaking down a problem into individual, detailed steps. Similarly, when *debugging* programs (that is, finding and fixing errors), you'll patiently observe what the program is doing and find the cause of the bugs. And like all skills, the more you program, the better you'll become.

### You Are Not Too Old to Learn Programming

The second most common anxiety I hear about learning to program is that people think they're too old to learn it. I read many internet comments from folks who think it's too late for them because they are already (gasp!) 23 years old. This is clearly not "too old" to learn to program: many people learn much later in life.

You don't need to have started as a child to become a capable programmer. But the image of programmers as whiz kids is a persistent one. Unfortunately, I contribute to this myth when I tell others that I was in grade school when I started programming.

However, programming is much easier to learn today than it was in the 1990s. Today, there are more books, better search engines, and many more online question-and-answer websites. On top of that, the programming languages themselves are far more user-friendly. For these reasons, **everything I learned about programming in the years between grade school and high school graduation could be learned today in about a dozen weekends**. My head start wasn't really much of a head start.

It's important to have a "growth mindset" about programming—in other words, understand that people develop programming skills through practice. They aren't just born as programmers, and being unskilled at programming now is not an indication that you can never become an expert.

### Programming Is a Creative Activity

Programming is a creative task, like painting, writing, knitting, or constructing LEGO castles. Like painting a blank canvas, making software has many constraints but endless possibilities.

The difference between programming and other creative activities is that when programming, you have all the raw materials you need in your computer; you don't need to buy any additional canvas, paint, film, yarn, LEGO bricks, or electronic components. A decade-old computer is more than powerful enough to write programs. Once your program is written, it can be copied perfectly an infinite number of times. A knit sweater can only be worn by one person at a time, but a useful program can easily be shared online with the entire world.

## About This Book

The first part of this book covers basic Python programming concepts, and the second part covers various tasks you can have your computer automate. Each chapter in the second part has project programs for you to study. Here's a brief rundown of what you'll find in each chapter.

**Part I: Python Programming Basics**

**Chapter 1: Python Basics**   Covers expressions, the most basic type of Python instruction, and how to use the Python interactive shell software to experiment with code.

**Chapter 2: Flow Control**   Explains how to make programs decide which instructions to execute so your code can intelligently respond to different conditions.

**Chapter 3: Functions**   Instructs you on how to define your own functions so that you can organize your code into more manageable chunks.

**Chapter 4: Lists**   Introduces the list data type and explains how to organize data.

**Chapter 5: Dictionaries and Structuring Data**   Introduces the dictionary data type and shows you more powerful ways to organize data.

**Chapter 6: Manipulating Strings**   Covers working with text data (called *strings* in Python).

**Part II: Automating Tasks**

**Chapter 7: Pattern Matching with Regular Expressions**   Covers how Python can manipulate strings and search for text patterns with regular expressions.

**Chapter 8: Input Validation**   Explains how your program can verify the information a user gives it, ensuring that the user's data arrives in a format that won't cause errors in the rest of the program.

**Chapter 9: Reading and Writing Files**   Explains how your program can read the contents of text files and save information to files on your hard drive.

**Chapter 10: Organizing Files**   Shows how Python can copy, move, rename, and delete large numbers of files much faster than a human user can. Also explains compressing and decompressing files.

**Chapter 11: Debugging**   Shows how to use Python's various bug-finding and bug-fixing tools.

**Chapter 12: Web Scraping**   Shows how to write programs that can automatically download web pages and parse them for information. This is called *web scraping*.

**Chapter 13: Working with Excel Spreadsheets**   Covers programmatically manipulating Excel spreadsheets so that you don't have to read them. This is helpful when the number of documents you have to analyze is in the hundreds or thousands.

**Chapter 14: Working with Google Sheets**   Covers how to read and update Google Sheets, a popular web-based spreadsheet application, using Python.

**Chapter 15: Working with PDF and Word Documents**   Covers programmatically reading Word and PDF documents.

**Chapter 16: Working with CSV Files and JSON Data**   Continues to explain how to programmatically manipulate documents, now discussing CSV and JSON files.

**Chapter 17: Keeping Time, Scheduling Tasks, and Launching Programs** Explains how Python programs handle time and dates and how to schedule your computer to perform tasks at certain times. Also shows how your Python programs can launch non-Python programs.

**Chapter 18: Sending Email and Text Messages**   Explains how to write programs that can send emails and text messages on your behalf.

**Chapter 19: Manipulating Images**   Explains how to programmatically manipulate images such as JPEG or PNG files.

**Chapter 20: Controlling the Keyboard and Mouse with GUI Automation** Explains how to programmatically control the mouse and keyboard to automate clicks and keypresses.

**Appendix A: Installing Third-Party Modules**   Shows you how to extend Python with useful additional modules.

**Appendix B: Running Programs**   Shows you how to run your Python programs on Windows, macOS, and Linux from outside of the code editor.

**Appendix C: Answers to the Practice Questions**   Provides answers and some additional context to the practice questions at the end of each chapter.

## Downloading and Installing Python

You can download Python for Windows, macOS, and Ubuntu for free at *https://python.org/downloads/*. If you download the latest version from the website's download page, all of the programs in this book should work.

**WARNING** *Be sure to download a version of Python 3 (such as 3.8.0). The programs in this book are written to run on Python 3 and may not run correctly, if at all, on Python 2.*

On the download page, you'll find Python installers for 64-bit and 32-bit computers for each operating system, so first figure out which installer you need. If you bought your computer in 2007 or later, it is most likely a 64-bit system. Otherwise, you have a 32-bit version, but here's how to find out for sure:

- On Windows, select **Start ▸ Control Panel ▸ System** and check whether System Type says 64-bit or 32-bit.

- On macOS, go the Apple menu, select **About This Mac ▸ More Info ▸ System Report ▸ Hardware**, and then look at the Processor Name field. If it says Intel Core Solo or Intel Core Duo, you have a 32-bit machine. If it says anything else (including Intel Core 2 Duo), you have a 64-bit machine.

- On Ubuntu Linux, open a Terminal and run the command `uname -m`. A response of `i686` means 32-bit, and `x86_64` means 64-bit.

On Windows, download the Python installer (the filename will end with *.msi*) and double-click it. Follow the instructions the installer displays on the screen to install Python, as listed here:

1. Select **Install for All Users** and click **Next**.
2. Accept the default options for the next several windows by clicking **Next**.

On macOS, download the *.dmg* file that's right for your version of macOS and double-click it. Follow the instructions the installer displays on the screen to install Python, as listed here:

1. When the DMG package opens in a new window, double-click the *Python.mpkg* file. You may have to enter the administrator password.
2. Accept the default options for the next several windows by clicking **Continue** and click **Agree** to accept the license.
3. On the final window, click **Install**.

If you're running Ubuntu, you can install Python from the Terminal by following these steps:

1. Open the Terminal window.
2. Enter `sudo apt-get install python3`.
3. Enter `sudo apt-get install idle3`.
4. Enter `sudo apt-get install python3-pip`.

## Downloading and Installing Mu

While the *Python interpreter* is the software that runs your Python programs, the *Mu editor software* is where you'll enter your programs, much the way you type in a word processor. You can download Mu from *https://codewith.mu/*.

On Windows and macOS, download the installer for your operating system and then run it by double-clicking the installer file. If you are on macOS, running the installer opens a window where you must drag the Mu icon to the Applications folder icon to continue the installation. If you are on Ubuntu, you'll need to install Mu as a Python package. In that case, click the Instructions button in the Python Package section of the download page.

## Starting Mu

Once it's installed, let's start Mu.

- On Windows 7 or later, click the Start icon in the lower-left corner of your screen, enter `Mu` in the search box, and select it.

- On macOS, open the Finder window, click **Applications**, and then click **mu-editor**.
- On Ubuntu, select **Applications ▶ Accessories ▶ Terminal** and then enter `python3 -m mu`.

The first time Mu runs, a Select Mode window will appear with options Adafruit CircuitPython, BBC micro:bit, Pygame Zero, and Python 3. Select **Python 3**. You can always change the mode later by clicking the Mode button at the top of the editor window.

**NOTE** *You'll need to download Mu version 1.10.0 or later in order to install the third-party modules featured in this book. As of this writing, 1.10.0 is an alpha release and is listed on the download page as a separate link from the main download links.*

## Starting IDLE

This book uses Mu as an editor and interactive shell. However, you can use any number of editors for writing Python code. The *Integrated Development and Learning Environment (IDLE)* software installs along with Python, and it can serve as a second editor if for some reason you can't get Mu installed or working. Let's start IDLE now.

- On Windows 7 or later, click the Start icon in the lower-left corner of your screen, enter `IDLE` in the search box, and select **IDLE (Python GUI)**.
- On macOS, open the Finder window, click **Applications**, click **Python 3.8**, and then click the IDLE icon.
- On Ubuntu, select **Applications ▶ Accessories ▶ Terminal** and then enter `idle3`. (You may also be able to click **Applications** at the top of the screen, select **Programming**, and then click **IDLE 3**.)

## The Interactive Shell

When you run Mu, the window that appears is called the *file editor* window. You can open the *interactive shell* by clicking the REPL button. A shell is a program that lets you type instructions into the computer, much like the Terminal or Command Prompt on macOS and Windows, respectively. Python's interactive shell lets you enter instructions for the Python interpreter software to run. The computer reads the instructions you enter and runs them immediately.

In Mu, the interactive shell is a pane in the lower half of the window with the following text:

```
Jupyter QtConsole 4.3.1
Python 3.6.3 (v3.6.3:2c5fed8, Oct  3 2017, 18:11:49) [MSC v.1900 64 bit
(AMD64)]
Type 'copyright', 'credits' or 'license' for more information
```

```
IPython 6.2.1 -- An enhanced Interactive Python. Type '?' for help.

In [1]:
```

If you run IDLE, the interactive shell is the window that first appears. It should be mostly blank except for text that looks something like this:

```
Python 3.8.0b1 (tags/v3.8.0b1:3b5deb0116, Jun  4 2019, 19:52:55) [MSC v.1916
64 bit (AMD64)] on win32
Type "help", "copyright", "credits" or "license" for more information.
>>>
```

In [1]: and >>> are called *prompts*. The examples in this book will use the >>> prompt for the interactive shell since it's more common. If you run Python from the Terminal or Command Prompt, they'll use the >>> prompt, as well. The In [1]: prompt was invented by Jupyter Notebook, another popular Python editor.

For example, enter the following into the interactive shell next to the prompt:

```
>>> print('Hello, world!')
```

After you type that line and press ENTER, the interactive shell should display this in response:

```
>>> print('Hello, world!')
Hello, world!
```

You've just given the computer an instruction, and it did what you told it to do!

## Installing Third-Party Modules

Some Python code requires your program to import modules. Some of these modules come with Python, but others are third-party modules created by developers outside of the Python core dev team. Appendix A has detailed instructions on how to use the pip program (on Windows) or pip3 program (on macOS and Linux) to install third-party modules. Consult Appendix A when this book instructs you to install a particular third-party module.

## How to Find Help

Programmers tend to learn by searching the internet for answers to their questions. This is quite different from the way many people are accustomed to learning—through an in-person teacher who lectures and can answer questions. What's great about using the internet as a schoolroom is that there are whole communities of folks who can answer your questions.

Indeed, your questions have probably already been answered, and the answers are waiting online for you to find them. If you encounter an error message or have trouble making your code work, you won't be the first person to have your problem, and finding a solution is easier than you might think.

For example, let's cause an error on purpose: enter `'42' + 3` into the interactive shell. You don't need to know what this instruction means right now, but the result should look like this:

```
>>> '42' + 3
❶ Traceback (most recent call last):
    File "<pyshell#0>", line 1, in <module>
      '42' + 3
❷ TypeError: Can't convert 'int' object to str implicitly
>>>
```

The error message ❷ appears because Python couldn't understand your instruction. The traceback part ❶ of the error message shows the specific instruction and line number that Python had trouble with. If you're not sure what to make of a particular error message, search for it online. Enter **"TypeError: Can't convert 'int' object to str implicitly"** (including the quotes) into your favorite search engine, and you should see tons of links explaining what the error message means and what causes it, as shown in Figure 0-2.

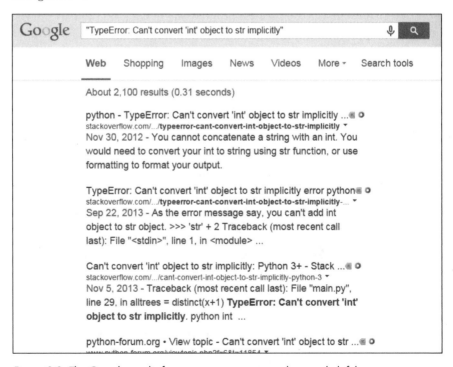

*Figure 0-2: The Google results for an error message can be very helpful.*

You'll often find that someone else had the same question as you and that some other helpful person has already answered it. No one person can know everything about programming, so an everyday part of any software developer's job is looking up answers to technical questions.

## Asking Smart Programming Questions

If you can't find the answer by searching online, try asking people in a web forum such as Stack Overflow (*https://stackoverflow.com/*) or the "learn programming" subreddit at *https://reddit.com/r/learnprogramming/*. But keep in mind there are smart ways to ask programming questions that help others help you. To begin with, be sure to read the FAQ sections at these websites about the proper way to post questions.

When asking programming questions, remember to do the following:

- Explain what you are trying to do, not just what you did. This lets your helper know if you are on the wrong track.
- Specify the point at which the error happens. Does it occur at the very start of the program or only after you do a certain action?
- Copy and paste the *entire* error message and your code to *https://pastebin .com/* or *https://gist.github.com/*.

    These websites make it easy to share large amounts of code with people online, without losing any text formatting. You can then put the URL of the posted code in your email or forum post. For example, here some pieces of code I've posted: *https://pastebin.com/SzP2DbFx/* and *https://gist.github.com/asweigart/6912168/*.

- Explain what you've already tried to do to solve your problem. This tells people you've already put in some work to figure things out on your own.
- List the version of Python you're using. (There are some key differences between version 2 Python interpreters and version 3 Python interpreters.) Also, say which operating system and version you're running.
- If the error came up after you made a change to your code, explain exactly what you changed.
- Say whether you're able to reproduce the error every time you run the program or whether it happens only after you perform certain actions. If the latter, then explain what those actions are.

Always follow good online etiquette as well. For example, don't post your questions in all caps or make unreasonable demands of the people trying to help you.

You can find more information on how to ask for programming help in the blog post at *https://autbor.com/help/*. You can find a list of frequently asked questions about programming at *https://www.reddit.com /r/learnprogramming/wiki/faq/* and a similar list about getting a job in software development at *https://www.reddit.com/r/cscareerquestions/wiki/ index/*.

I love helping people discover Python. I write programming tutorials on my blog at *https://inventwithpython.com/blog/*, and you can contact me with questions at *al@inventwithpython.com*. Although, you may get a faster response by posting your questions to *https://reddit.com/r/inventwithpython/*.

## Summary

For most people, their computer is just an appliance instead of a tool. But by learning how to program, you'll gain access to one of the most powerful tools of the modern world, and you'll have fun along the way. Programming isn't brain surgery—it's fine for amateurs to experiment and make mistakes.

This book assumes you have zero programming knowledge and will teach you quite a bit, but you may have questions beyond its scope. Remember that asking effective questions and knowing how to find answers are invaluable tools on your programming journey.

Let's begin!

# PART I

## PYTHON PROGRAMMING BASICS

# 1

## PYTHON BASICS

The Python programming language has a wide range of syntactical constructions, standard library functions, and interactive development environment features. Fortunately, you can ignore most of that; you just need to learn enough to write some handy little programs.

You will, however, have to learn some basic programming concepts before you can do anything. Like a wizard in training, you might think these concepts seem arcane and tedious, but with some knowledge and practice, you'll be able to command your computer like a magic wand and perform incredible feats.

This chapter has a few examples that encourage you to type into the *interactive shell*, also called the *REPL* (Read-Evaluate-Print Loop), which lets you run (or *execute*) Python instructions one at a time and instantly shows you the results. Using the interactive shell is great for learning what basic Python instructions do, so give it a try as you follow along. You'll remember the things you do much better than the things you only read.

## Entering Expressions into the Interactive Shell

You can run the interactive shell by launching the Mu editor, which you should have downloaded when going through the setup instructions in the Preface. On Windows, open the Start menu, type "Mu," and open the Mu app. On macOS, open your Applications folder and double-click **Mu**. Click the **New** button and save an empty file as *blank.py*. When you run this blank file by clicking the **Run** button or pressing F5, it will open the interactive shell, which will open as a new pane that opens at the bottom of the Mu editor's window. You should see a >>> prompt in the interactive shell.

Enter **2 + 2** at the prompt to have Python do some simple math. The Mu window should now look like this:

```
>>> 2 + 2
4
>>>
```

In Python, 2 + 2 is called an *expression*, which is the most basic kind of programming instruction in the language. Expressions consist of *values* (such as 2) and *operators* (such as +), and they can always *evaluate* (that is, reduce) down to a single value. That means you can use expressions anywhere in Python code that you could also use a value.

In the previous example, 2 + 2 is evaluated down to a single value, 4. A single value with no operators is also considered an expression, though it evaluates only to itself, as shown here:

```
>>> 2
2
```

---

### ERRORS ARE OKAY!

Programs will crash if they contain code the computer can't understand, which will cause Python to show an error message. An error message won't break your computer, though, so don't be afraid to make mistakes. A *crash* just means the program stopped running unexpectedly.

If you want to know more about an error, you can search for the exact error message text online for more information. You can also check out the resources at *https://nostarch.com/automatestuff2/* to see a list of common Python error messages and their meanings.

---

You can use plenty of other operators in Python expressions, too. For example, Table 1-1 lists all the math operators in Python.

**Table 1-1:** Math Operators from Highest to Lowest Precedence

| Operator | Operation | Example | Evaluates to . . . |
|----------|-----------|---------|--------------------|
| ** | Exponent | 2 ** 3 | 8 |
| % | Modulus/remainder | 22 % 8 | 6 |
| // | Integer division/floored quotient | 22 // 8 | 2 |
| / | Division | 22 / 8 | 2.75 |
| * | Multiplication | 3 * 5 | 15 |
| - | Subtraction | 5 - 2 | 3 |
| + | Addition | 2 + 2 | 4 |

The *order of operations* (also called *precedence*) of Python math operators is similar to that of mathematics. The ** operator is evaluated first; the *, /, //, and % operators are evaluated next, from left to right; and the + and - operators are evaluated last (also from left to right). You can use parentheses to override the usual precedence if you need to. Whitespace in between the operators and values doesn't matter for Python (except for the indentation at the beginning of the line), but a single space is convention. Enter the following expressions into the interactive shell:

```
>>> 2 + 3 * 6
20
>>> (2 + 3) * 6
30
>>> 48565878 * 578453
28093077826734
>>> 2 ** 8
256
>>> 23 / 7
3.2857142857142856
>>> 23 // 7
3
>>> 23 % 7
2
>>> 2      +         2
4
>>> (5 - 1) * ((7 + 1) / (3 - 1))
16.0
```

In each case, you as the programmer must enter the expression, but Python does the hard part of evaluating it down to a single value. Python will keep evaluating parts of the expression until it becomes a single value, as shown here:

```
(5 - 1) * ((7 + 1) / (3 - 1))

  4 * ((7 + 1) / (3 - 1))

  4 * (   8   ) / (3 - 1)

  4 * (   8   ) / (   2   )

  4 * 4.0

  16.0
```

These rules for putting operators and values together to form expressions are a fundamental part of Python as a programming language, just like the grammar rules that help us communicate. Here's an example:

**This is a grammatically correct English sentence.**

**This grammatically is sentence not English correct a.**

The second line is difficult to parse because it doesn't follow the rules of English. Similarly, if you enter a bad Python instruction, Python won't be able to understand it and will display a SyntaxError error message, as shown here:

```
>>> 5 +
  File "<stdin>", line 1
    5 +
      ^
SyntaxError: invalid syntax
>>> 42 + 5 + * 2
  File "<stdin>", line 1
    42 + 5 + * 2
            ^
SyntaxError: invalid syntax
```

You can always test to see whether an instruction works by entering it into the interactive shell. Don't worry about breaking the computer: the worst that could happen is that Python responds with an error message. Professional software developers get error messages while writing code all the time.

# The Integer, Floating-Point, and String Data Types

Remember that expressions are just values combined with operators, and they always evaluate down to a single value. A *data type* is a category for values, and every value belongs to exactly one data type. The most common data types in Python are listed in Table 1-2. The values -2 and 30, for example, are said to be *integer* values. The integer (or *int*) data type indicates values that are whole numbers. Numbers with a decimal point, such as 3.14, are called *floating-point numbers* (or *floats*). Note that even though the value 42 is an integer, the value 42.0 would be a floating-point number.

**Table 1-2:** Common Data Types

| Data type | Examples |
| --- | --- |
| Integers | -2, -1, 0, 1, 2, 3, 4, 5 |
| Floating-point numbers | -1.25, -1.0, -0.5, 0.0, 0.5, 1.0, 1.25 |
| Strings | 'a', 'aa', 'aaa', 'Hello!', '11 cats' |

Python programs can also have text values called *strings*, or *strs* (pronounced "stirs"). Always surround your string in single quote (') characters (as in 'Hello' or 'Goodbye cruel world!') so Python knows where the string begins and ends. You can even have a string with no characters in it, '', called a *blank string* or an *empty string*. Strings are explained in greater detail in Chapter 4.

If you ever see the error message SyntaxError: EOL while scanning string literal, you probably forgot the final single quote character at the end of the string, such as in this example:

```
>>> 'Hello, world!
SyntaxError: EOL while scanning string literal
```

# String Concatenation and Replication

The meaning of an operator may change based on the data types of the values next to it. For example, + is the addition operator when it operates on two integers or floating-point values. However, when + is used on two string values, it joins the strings as the *string concatenation* operator. Enter the following into the interactive shell:

```
>>> 'Alice' + 'Bob'
'AliceBob'
```

The expression evaluates down to a single, new string value that combines the text of the two strings. However, if you try to use the + operator on a string and an integer value, Python will not know how to handle this, and it will display an error message.

```
>>> 'Alice' + 42
Traceback (most recent call last):
  File "<pyshell#0>", line 1, in <module>
    'Alice' + 42
TypeError: can only concatenate str (not "int") to str
```

The error message can only concatenate str (not "int") to str means that Python thought you were trying to concatenate an integer to the string 'Alice'. Your code will have to explicitly convert the integer to a string because Python cannot do this automatically. (Converting data types will be explained in "Dissecting Your Program" on page 13 when we talk about the str(), int(), and float() functions.)

The * operator multiplies two integer or floating-point values. But when the * operator is used on one string value and one integer value, it becomes the *string replication* operator. Enter a string multiplied by a number into the interactive shell to see this in action.

```
>>> 'Alice' * 5
'AliceAliceAliceAliceAlice'
```

The expression evaluates down to a single string value that repeats the original string a number of times equal to the integer value. String replication is a useful trick, but it's not used as often as string concatenation.

The * operator can be used with only two numeric values (for multiplication), or one string value and one integer value (for string replication). Otherwise, Python will just display an error message, like the following:

```
>>> 'Alice' * 'Bob'
Traceback (most recent call last):
  File "<pyshell#32>", line 1, in <module>
    'Alice' * 'Bob'
TypeError: can't multiply sequence by non-int of type 'str'
>>> 'Alice' * 5.0
Traceback (most recent call last):
  File "<pyshell#33>", line 1, in <module>
    'Alice' * 5.0
TypeError: can't multiply sequence by non-int of type 'float'
```

It makes sense that Python wouldn't understand these expressions: you can't multiply two words, and it's hard to replicate an arbitrary string a fractional number of times.

## Storing Values in Variables

A *variable* is like a box in the computer's memory where you can store a single value. If you want to use the result of an evaluated expression later in your program, you can save it inside a variable.

### Assignment Statements

You'll store values in variables with an *assignment statement*. An assignment statement consists of a variable name, an equal sign (called the *assignment operator*), and the value to be stored. If you enter the assignment statement spam = 42, then a variable named spam will have the integer value 42 stored in it.

Think of a variable as a labeled box that a value is placed in, as in Figure 1-1.

Figure 1-1: spam = 42 is like telling the program, "The variable spam now has the integer value 42 in it."

For example, enter the following into the interactive shell:

```
❶ >>> spam = 40
   >>> spam
   40
   >>> eggs = 2
❷ >>> spam + eggs
   42
   >>> spam + eggs + spam
   82
❸ >>> spam = spam + 2
   >>> spam
   42
```

A variable is *initialized* (or created) the first time a value is stored in it ❶. After that, you can use it in expressions with other variables and values ❷. When a variable is assigned a new value ❸, the old value is forgotten, which is why spam evaluated to 42 instead of 40 at the end of the example. This is called *overwriting* the variable. Enter the following code into the interactive shell to try overwriting a string:

```
>>> spam = 'Hello'
>>> spam
'Hello'
>>> spam = 'Goodbye'
>>> spam
'Goodbye'
```

Just like the box in Figure 1-2, the spam variable in this example stores 'Hello' until you replace the string with 'Goodbye'.

Figure 1-2: When a new value is assigned to a variable, the old one is forgotten.

## Variable Names

A good variable name describes the data it contains. Imagine that you moved to a new house and labeled all of your moving boxes as *Stuff.* You'd never find anything! Most of this book's examples (and Python's documentation) use generic variable names like spam, eggs, and bacon, which come from the Monty Python "Spam" sketch. But in your programs, a descriptive name will help make your code more readable.

Though you can name your variables almost anything, Python does have some naming restrictions. Table 1-3 has examples of legal variable names. You can name a variable anything as long as it obeys the following three rules:

- It can be only one word with no spaces.
- It can use only letters, numbers, and the underscore (_) character.
- It can't begin with a number.

**Table 1-3:** Valid and Invalid Variable Names

| Valid variable names | Invalid variable names |
|---|---|
| current_balance | current-balance (hyphens are not allowed) |
| currentBalance | current balance (spaces are not allowed) |
| account4 | 4account (can't begin with a number) |
| _42 | 42 (can't begin with a number) |
| TOTAL_SUM | TOTAL_$UM (special characters like $ are not allowed) |
| hello | 'hello' (special characters like ' are not allowed) |

Variable names are case-sensitive, meaning that spam, SPAM, Spam, and sPaM are four different variables. Though Spam is a valid variable you can use in a program, it is a Python convention to start your variables with a lowercase letter.

This book uses *camelcase* for variable names instead of underscores; that is, variables lookLikeThis instead of looking_like_this. Some experienced programmers may point out that the official Python code style, PEP 8, says that underscores should be used. I unapologetically prefer camelcase and point to the "A Foolish Consistency Is the Hobgoblin of Little Minds" section in PEP 8 itself:

> Consistency with the style guide is important. But most importantly: know when to be inconsistent—sometimes the style guide just doesn't apply. When in doubt, use your best judgment.

## Your First Program

While the interactive shell is good for running Python instructions one at a time, to write entire Python programs, you'll type the instructions into the file editor. The *file editor* is similar to text editors such as Notepad or TextMate, but it has some features specifically for entering source code. To open a new file in Mu, click the **New** button on the top row.

The window that appears should contain a cursor awaiting your input, but it's different from the interactive shell, which runs Python instructions as soon as you press ENTER. The file editor lets you type in many instructions, save the file, and run the program. Here's how you can tell the difference between the two:

- The interactive shell window will always be the one with the >>> prompt.
- The file editor window will not have the >>> prompt.

Now it's time to create your first program! When the file editor window opens, enter the following into it:

```
❶ # This program says hello and asks for my name.

❷ print('Hello, world!')
  print('What is your name?')    # ask for their name
❸ myName = input()
❹ print('It is good to meet you, ' + myName)
❺ print('The length of your name is:')
  print(len(myName))
❻ print('What is your age?')    # ask for their age
  myAge = input()
  print('You will be ' + str(int(myAge) + 1) + ' in a year.')
```

Once you've entered your source code, save it so that you won't have to retype it each time you start Mu. Click the **Save** button, enter *hello.py* in the File Name field, and then click **Save**.

You should save your programs every once in a while as you type them. That way, if the computer crashes or you accidentally exit Mu, you won't lose the code. As a shortcut, you can press CTRL-S on Windows and Linux or ⌘-S on macOS to save your file.

Once you've saved, let's run our program. Press the **F5** key. Your program should run in the interactive shell window. Remember, you have to press **F5** from the file editor window, not the interactive shell window. Enter your name when your program asks for it. The program's output in the interactive shell should look something like this:

```
Python 3.7.0b4 (v3.7.0b4:eb96c37699, May  2 2018, 19:02:22) [MSC v.1913 64 bit
(AMD64)] on win32
Type "copyright", "credits" or "license()" for more information.
>>> ============================= RESTART =============================
>>>
Hello, world!
What is your name?
Al
It is good to meet you, Al
The length of your name is:
2
What is your age?
4
You will be 5 in a year.
>>>
```

When there are no more lines of code to execute, the Python program *terminates*; that is, it stops running. (You can also say that the Python program *exits*.)

You can close the file editor by clicking the X at the top of the window. To reload a saved program, select **File ▸ Open...** from the menu. Do that now, and in the window that appears, choose *hello.py* and click the **Open** button. Your previously saved *hello.py* program should open in the file editor window.

You can view the execution of a program using the Python Tutor visualization tool at *http://pythontutor.com/*. You can see the execution of this particular program at *https://autbor.com/hellopy/*. Click the forward button to move through each step of the program's execution. You'll be able to see how the variables' values and the output change.

## Dissecting Your Program

With your new program open in the file editor, let's take a quick tour of the Python instructions it uses by looking at what each line of code does.

### Comments

The following line is called a *comment*.

❶ # This program says hello and asks for my name.

Python ignores comments, and you can use them to write notes or remind yourself what the code is trying to do. Any text for the rest of the line following a hash mark (#) is part of a comment.

Sometimes, programmers will put a # in front of a line of code to temporarily remove it while testing a program. This is called *commenting out* code, and it can be useful when you're trying to figure out why a program isn't working. You can remove the # later when you are ready to put the line back in.

Python also ignores the blank line after the comment. You can add as many blank lines to your program as you want. This can make your code easier to read, like paragraphs in a book.

### The print() Function

The print() function displays the string value inside its parentheses on the screen.

❷ print('Hello, world!')
print('What is your name?') # ask for their name

The line print('Hello, world!') means "Print out the text in the string 'Hello, world!'." When Python executes this line, you say that Python is *calling* the print() function and the string value is being *passed* to the function. A value that is passed to a function call is an *argument*. Notice that the quotes are not printed to the screen. They just mark where the string begins and ends; they are not part of the string value.

*You can also use this function to put a blank line on the screen; just call* print() *with nothing in between the parentheses.*

When you write a function name, the opening and closing parentheses at the end identify it as the name of a function. This is why in this book, you'll see print() rather than print. Chapter 3 describes functions in more detail.

### The input() Function

The input() function waits for the user to type some text on the keyboard and press ENTER.

❸ myName = input()

This function call evaluates to a string equal to the user's text, and the line of code assigns the myName variable to this string value.

You can think of the input() function call as an expression that evaluates to whatever string the user typed in. If the user entered 'Al', then the expression would evaluate to myName = 'Al'.

If you call input() and see an error message, like NameError: name 'Al' is not defined, the problem is that you're running the code with Python 2 instead of Python 3.

### Printing the User's Name

The following call to print() actually contains the expression 'It is good to meet you, ' + myName between the parentheses.

❹ print('It is good to meet you, ' + myName)

Remember that expressions can always evaluate to a single value. If 'Al' is the value stored in myName on line ❸, then this expression evaluates to 'It is good to meet you, Al'. This single string value is then passed to print(), which prints it on the screen.

### The len() Function

You can pass the len() function a string value (or a variable containing a string), and the function evaluates to the integer value of the number of characters in that string.

❺ print('The length of your name is:')
print(len(myName))

Enter the following into the interactive shell to try this:

```
>>> len('hello')
5
>>> len('My very energetic monster just scarfed nachos.')
46
>>> len('')
0
```

Just like those examples, len(myName) evaluates to an integer. It is then passed to print() to be displayed on the screen. The print() function allows you to pass it either integer values or string values, but notice the error that shows up when you type the following into the interactive shell:

```
>>> print('I am ' + 29 + ' years old.')
Traceback (most recent call last):
  File "<pyshell#6>", line 1, in <module>
    print('I am ' + 29 + ' years old.')
TypeError: can only concatenate str (not "int") to str
```

The print() function isn't causing that error, but rather it's the expression you tried to pass to print(). You get the same error message if you type the expression into the interactive shell on its own.

```
>>> 'I am ' + 29 + ' years old.'
Traceback (most recent call last):
  File "<pyshell#7>", line 1, in <module>
    'I am ' + 29 + ' years old.'
TypeError: can only concatenate str (not "int") to str
```

Python gives an error because the + operator can only be used to add two integers together or concatenate two strings. You can't add an integer to a string, because this is ungrammatical in Python. You can fix this by using a string version of the integer instead, as explained in the next section.

## The str(), int(), and float() Functions

If you want to concatenate an integer such as 29 with a string to pass to print(), you'll need to get the value '29', which is the string form of 29. The str() function can be passed an integer value and will evaluate to a string value version of the integer, as follows:

```
>>> str(29)
'29'
>>> print('I am ' + str(29) + ' years old.')
I am 29 years old.
```

Because str(29) evaluates to '29', the expression 'I am ' + str(29) + ' years old.' evaluates to 'I am ' + '29' + ' years old.', which in turn evaluates to 'I am 29 years old.'. This is the value that is passed to the print() function.

The str(), int(), and float() functions will evaluate to the string, integer, and floating-point forms of the value you pass, respectively. Try converting some values in the interactive shell with these functions and watch what happens.

```
>>> str(0)
'0'
>>> str(-3.14)
'-3.14'
>>> int('42')
42
>>> int('-99')
-99
>>> int(1.25)
1
>>> int(1.99)
1
>>> float('3.14')
3.14
>>> float(10)
10.0
```

The previous examples call the str(), int(), and float() functions and pass them values of the other data types to obtain a string, integer, or floating-point form of those values.

The str() function is handy when you have an integer or float that you want to concatenate to a string. The int() function is also helpful if you have a number as a string value that you want to use in some mathematics. For example, the input() function always returns a string, even if the user enters a number. Enter **spam = input()** into the interactive shell and enter **101** when it waits for your text.

```
>>> spam = input()
101
>>> spam
'101'
```

The value stored inside spam isn't the integer 101 but the string '101'. If you want to do math using the value in spam, use the int() function to get the integer form of spam and then store this as the new value in spam.

```
>>> spam = int(spam)
>>> spam
101
```

Now you should be able to treat the spam variable as an integer instead of a string.

```
>>> spam * 10 / 5
202.0
```

Note that if you pass a value to int() that it cannot evaluate as an integer, Python will display an error message.

```
>>> int('99.99')
Traceback (most recent call last):
  File "<pyshell#18>", line 1, in <module>
    int('99.99')
ValueError: invalid literal for int() with base 10: '99.99'
>>> int('twelve')
Traceback (most recent call last):
  File "<pyshell#19>", line 1, in <module>
    int('twelve')
ValueError: invalid literal for int() with base 10: 'twelve'
```

The int() function is also useful if you need to round a floating-point number down.

```
>>> int(7.7)
7
>>> int(7.7) + 1
8
```

You used the int() and str() functions in the last three lines of your program to get a value of the appropriate data type for the code.

```
❻ print('What is your age?') # ask for their age
myAge = input()
print('You will be ' + str(int(myAge) + 1) + ' in a year.')
```

---

**TEXT AND NUMBER EQUIVALENCE**

Although the string value of a number is considered a completely different value from the integer or floating-point version, an integer can be equal to a floating point.

```
>>> 42 == '42'
False
>>> 42 == 42.0
True
>>> 42.0 == 0042.000
True
```

Python makes this distinction because strings are text, while integers and floats are both numbers.

The myAge variable contains the value returned from input(). Because the input() function always returns a string (even if the user typed in a number), you can use the int(myAge) code to return an integer value of the string in myAge. This integer value is then added to 1 in the expression int(myAge) + 1.

The result of this addition is passed to the str() function: str(int(myAge) + 1). The string value returned is then concatenated with the strings 'You will be ' and ' in a year.' to evaluate to one large string value. This large string is finally passed to print() to be displayed on the screen.

Let's say the user enters the string '4' for myAge. The string '4' is converted to an integer, so you can add one to it. The result is 5. The str() function converts the result back to a string, so you can concatenate it with the second string, 'in a year.', to create the final message. These evaluation steps would look something like the following:

```
print('You will be ' + str(int(myAge) + 1) + ' in a year.')

print('You will be ' + str(int( '4' ) + 1) + ' in a year.')

print('You will be ' + str(    4 + 1    ) + ' in a year.')

print('You will be ' + str(      5      ) + ' in a year.')

print('You will be ' +               '5'       + ' in a year.')

print('You will be 5'                          + ' in a year.')

print('You will be 5 in a year.')
```

## Summary

You can compute expressions with a calculator or enter string concatenations with a word processor. You can even do string replication easily by copying and pasting text. But expressions, and their component values—operators, variables, and function calls—are the basic building blocks that make programs. Once you know how to handle these elements, you will be able to instruct Python to operate on large amounts of data for you.

It is good to remember the different types of operators (+, -, *, /, //, %, and ** for math operations, and + and * for string operations) and the three data types (integers, floating-point numbers, and strings) introduced in this chapter.

I introduced a few different functions as well. The print() and input() functions handle simple text output (to the screen) and input (from the keyboard). The len() function takes a string and evaluates to an int of the number of characters in the string. The str(), int(), and float() functions will evaluate to the string, integer, or floating-point number form of the value they are passed.

In the next chapter, you'll learn how to tell Python to make intelligent decisions about what code to run, what code to skip, and what code to repeat based on the values it has. This is known as *flow control*, and it allows you to write programs that make intelligent decisions.

## Practice Questions

1. Which of the following are operators, and which are values?

```
*
'hello'
-88.8
-
/
+
5
```

2. Which of the following is a variable, and which is a string?

```
spam
'spam'
```

3. Name three data types.

4. What is an expression made up of? What do all expressions do?

5. This chapter introduced assignment statements, like spam = 10. What is the difference between an expression and a statement?

6. What does the variable bacon contain after the following code runs?

```
bacon = 20
bacon + 1
```

7. What should the following two expressions evaluate to?

```
'spam' + 'spamspam'
'spam' * 3
```

8. Why is eggs a valid variable name while 100 is invalid?

9. What three functions can be used to get the integer, floating-point number, or string version of a value?

10. Why does this expression cause an error? How can you fix it?

```
'I have eaten ' + 99 + ' burritos.'
```

**Extra credit:** Search online for the Python documentation for the `len()` function. It will be on a web page titled "Built-in Functions." Skim the list of other functions Python has, look up what the `round()` function does, and experiment with it in the interactive shell.

# 2

## FLOW CONTROL

So, you know the basics of individual instructions and that a program is just a series of instructions. But programming's real strength isn't just running one instruction after another like a weekend errand list. Based on how expressions evaluate, a program can decide to skip instructions, repeat them, or choose one of several instructions to run. In fact, you almost never want your programs to start from the first line of code and simply execute every line, straight to the end. *Flow control statements* can decide which Python instructions to execute under which conditions.

These flow control statements directly correspond to the symbols in a flowchart, so I'll provide flowchart versions of the code discussed in this chapter. Figure 2-1 shows a flowchart for what to do if it's raining. Follow the path made by the arrows from Start to End.

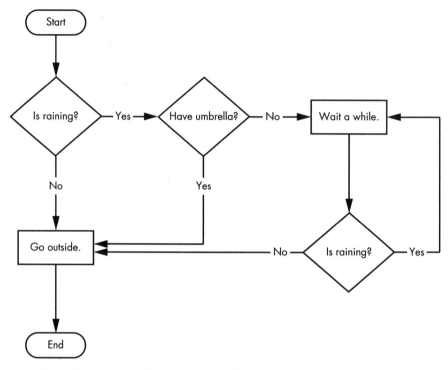

Figure 2-1: A flowchart to tell you what to do if it is raining

In a flowchart, there is usually more than one way to go from the start to the end. The same is true for lines of code in a computer program. Flowcharts represent these branching points with diamonds, while the other steps are represented with rectangles. The starting and ending steps are represented with rounded rectangles.

But before you learn about flow control statements, you first need to learn how to represent those *yes* and *no* options, and you need to understand how to write those branching points as Python code. To that end, let's explore Boolean values, comparison operators, and Boolean operators.

## Boolean Values

While the integer, floating-point, and string data types have an unlimited number of possible values, the *Boolean* data type has only two values: True and False. (Boolean is capitalized because the data type is named after mathematician George Boole.) When entered as Python code, the Boolean values True and False lack the quotes you place around strings, and they always start with a capital *T* or *F*, with the rest of the word in lowercase. Enter the following into the interactive shell. (Some of these instructions are intentionally incorrect, and they'll cause error messages to appear.)

```
❶ >>> spam = True
   >>> spam
   True
❷ >>> true
   Traceback (most recent call last):
     File "<pyshell#2>", line 1, in <module>
       true
   NameError: name 'true' is not defined
❸ >>> True = 2 + 2
   SyntaxError: can't assign to keyword
```

Like any other value, Boolean values are used in expressions and can be stored in variables ❶. If you don't use the proper case ❷ or you try to use True and False for variable names ❸, Python will give you an error message.

## Comparison Operators

*Comparison operators*, also called *relational operators*, compare two values and evaluate down to a single Boolean value. Table 2-1 lists the comparison operators.

**Table 2-1:** Comparison Operators

| Operator | Meaning |
| --- | --- |
| == | Equal to |
| != | Not equal to |
| < | Less than |
| > | Greater than |
| <= | Less than or equal to |
| >= | Greater than or equal to |

These operators evaluate to True or False depending on the values you give them. Let's try some operators now, starting with == and !=.

```
>>> 42 == 42
True
>>> 42 == 99
False
>>> 2 != 3
True
>>> 2 != 2
False
```

As you might expect, == (equal to) evaluates to True when the values on both sides are the same, and != (not equal to) evaluates to True when the two values are different. The == and != operators can actually work with values of any data type.

```
>>> 'hello' == 'hello'
True
>>> 'hello' == 'Hello'
False
>>> 'dog' != 'cat'
True
>>> True == True
True
>>> True != False
True
>>> 42 == 42.0
True
❶ >>> 42 == '42'
False
```

Note that an integer or floating-point value will always be unequal to a string value. The expression 42 == '42' ❶ evaluates to False because Python considers the integer 42 to be different from the string '42'.

The <, >, <=, and >= operators, on the other hand, work properly only with integer and floating-point values.

```
>>> 42 < 100
True
>>> 42 > 100
False
>>> 42 < 42
False
>>> eggCount = 42
❶ >>> eggCount <= 42
True
>>> myAge = 29
❷ >>> myAge >= 10
True
```

**THE DIFFERENCE BETWEEN THE == AND = OPERATORS**

You might have noticed that the == operator (equal to) has two equal signs, while the = operator (assignment) has just one equal sign. It's easy to confuse these two operators with each other. Just remember these points:

- The == operator (equal to) asks whether two values are the same as each other.

- The = operator (assignment) puts the value on the right into the variable on the left.

To help remember which is which, notice that the == operator (equal to) consists of two characters, just like the != operator (not equal to) consists of two characters.

You'll often use comparison operators to compare a variable's value to some other value, like in the eggCount <= 42 ❶ and myAge >= 10 ❷ examples. (After all, instead of entering 'dog' != 'cat' in your code, you could have just entered True.) You'll see more examples of this later when you learn about flow control statements.

# Boolean Operators

The three Boolean operators (and, or, and not) are used to compare Boolean values. Like comparison operators, they evaluate these expressions down to a Boolean value. Let's explore these operators in detail, starting with the and operator.

## Binary Boolean Operators

The and and or operators always take two Boolean values (or expressions), so they're considered *binary* operators. The and operator evaluates an expression to True if *both* Boolean values are True; otherwise, it evaluates to False. Enter some expressions using and into the interactive shell to see it in action.

```
>>> True and True
True
>>> True and False
False
```

A *truth table* shows every possible result of a Boolean operator. Table 2-2 is the truth table for the and operator.

**Table 2-2:** The and Operator's Truth Table

| Expression | Evaluates to . . . |
| --- | --- |
| True and True | True |
| True and False | False |
| False and True | False |
| False and False | False |

On the other hand, the or operator evaluates an expression to True if *either* of the two Boolean values is True. If both are False, it evaluates to False.

```
>>> False or True
True
>>> False or False
False
```

You can see every possible outcome of the or operator in its truth table, shown in Table 2-3.

**Table 2-3:** The or Operator's Truth Table

| Expression | Evaluates to . . . |
|---|---|
| True or True | True |
| True or False | True |
| False or True | True |
| False or False | False |

### The not Operator

Unlike and and or, the not operator operates on only one Boolean value (or expression). This makes it a *unary* operator. The not operator simply evaluates to the opposite Boolean value.

```
>>> not True
False
❶ >>> not not not not True
True
```

Much like using double negatives in speech and writing, you can nest not operators ❶, though there's never not no reason to do this in real programs. Table 2-4 shows the truth table for not.

**Table 2-4:** The not Operator's Truth Table

| Expression | Evaluates to . . . |
|---|---|
| not True | False |
| not False | True |

## Mixing Boolean and Comparison Operators

Since the comparison operators evaluate to Boolean values, you can use them in expressions with the Boolean operators.

Recall that the and, or, and not operators are called Boolean operators because they always operate on the Boolean values True and False. While expressions like 4 < 5 aren't Boolean values, they are expressions that evaluate down to Boolean values. Try entering some Boolean expressions that use comparison operators into the interactive shell.

```
>>> (4 < 5) and (5 < 6)
True
>>> (4 < 5) and (9 < 6)
False
>>> (1 == 2) or (2 == 2)
True
```

The computer will evaluate the left expression first, and then it will evaluate the right expression. When it knows the Boolean value for each, it will then evaluate the whole expression down to one Boolean value. You can think of the computer's evaluation process for (4 < 5) and (5 < 6) as the following:

```
(4 < 5) and (5 < 6)
        ↓
True and (5 < 6)
        ↓
  True and True
        ↓
       True
```

You can also use multiple Boolean operators in an expression, along with the comparison operators:

```
>>> 2 + 2 == 4 and not 2 + 2 == 5 and 2 * 2 == 2 + 2
True
```

The Boolean operators have an order of operations just like the math operators do. After any math and comparison operators evaluate, Python evaluates the not operators first, then the and operators, and then the or operators.

# Elements of Flow Control

Flow control statements often start with a part called the *condition* and are always followed by a block of code called the *clause*. Before you learn about Python's specific flow control statements, I'll cover what a condition and a block are.

## Conditions

The Boolean expressions you've seen so far could all be considered conditions, which are the same thing as expressions; *condition* is just a more specific name in the context of flow control statements. Conditions always evaluate down to a Boolean value, True or False. A flow control statement decides what to do based on whether its condition is True or False, and almost every flow control statement uses a condition.

## Blocks of Code

Lines of Python code can be grouped together in *blocks*. You can tell when a block begins and ends from the indentation of the lines of code. There are three rules for blocks.

- Blocks begin when the indentation increases.
- Blocks can contain other blocks.
- Blocks end when the indentation decreases to zero or to a containing block's indentation.

Blocks are easier to understand by looking at some indented code, so let's find the blocks in part of a small game program, shown here:

```
name = 'Mary'
password = 'swordfish'
if name == 'Mary':
❶ print('Hello, Mary')
    if password == 'swordfish':
      ❷ print('Access granted.')
    else:
      ❸ print('Wrong password.')
```

You can view the execution of this program at *https://autbor.com/blocks/*. The first block of code ❶ starts at the line print('Hello, Mary') and contains all the lines after it. Inside this block is another block ❷, which has only a single line in it: print('Access Granted.'). The third block ❸ is also one line long: print('Wrong password.').

## Program Execution

In the previous chapter's *hello.py* program, Python started executing instructions at the top of the program going down, one after another. The *program execution* (or simply, *execution*) is a term for the current instruction being executed. If you print the source code on paper and put your finger on each line as it is executed, you can think of your finger as the program execution.

Not all programs execute by simply going straight down, however. If you use your finger to trace through a program with flow control statements, you'll likely find yourself jumping around the source code based on conditions, and you'll probably skip entire clauses.

## Flow Control Statements

Now, let's explore the most important piece of flow control: the statements themselves. The statements represent the diamonds you saw in the flowchart in Figure 2-1, and they are the actual decisions your programs will make.

### if Statements

The most common type of flow control statement is the if statement. An if statement's clause (that is, the block following the if statement) will execute if the statement's condition is True. The clause is skipped if the condition is False.

In plain English, an if statement could be read as, "If this condition is true, execute the code in the clause." In Python, an if statement consists of the following:

- The if keyword
- A condition (that is, an expression that evaluates to True or False)

- A colon
- Starting on the next line, an indented block of code (called the if clause)

For example, let's say you have some code that checks to see whether someone's name is Alice. (Pretend name was assigned some value earlier.)

```
if name == 'Alice':
    print('Hi, Alice.')
```

All flow control statements end with a colon and are followed by a new block of code (the clause). This if statement's clause is the block with print('Hi, Alice.'). Figure 2-2 shows what a flowchart of this code would look like.

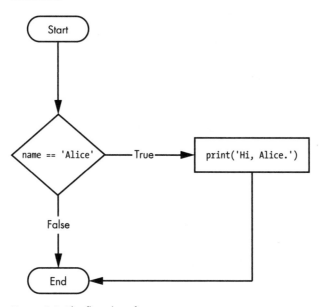

Figure 2-2: The flowchart for an if statement

### else Statements

An if clause can optionally be followed by an else statement. The else clause is executed only when the if statement's condition is False. In plain English, an else statement could be read as, "If this condition is true, execute this code. Or else, execute that code." An else statement doesn't have a condition, and in code, an else statement always consists of the following:

- The else keyword
- A colon
- Starting on the next line, an indented block of code (called the else clause)

Returning to the Alice example, let's look at some code that uses an else statement to offer a different greeting if the person's name isn't Alice.

```
if name == 'Alice':
    print('Hi, Alice.')
else:
    print('Hello, stranger.')
```

Figure 2-3 shows what a flowchart of this code would look like.

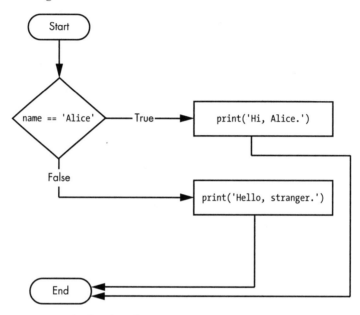

*Figure 2-3: The flowchart for an else statement*

## elif Statements

While only one of the if or else clauses will execute, you may have a case where you want one of *many* possible clauses to execute. The elif statement is an "else if" statement that always follows an if or another elif statement. It provides another condition that is checked only if all of the previous conditions were False. In code, an elif statement always consists of the following:

- The elif keyword
- A condition (that is, an expression that evaluates to True or False)
- A colon
- Starting on the next line, an indented block of code (called the elif clause)

Let's add an elif to the name checker to see this statement in action.

```
if name == 'Alice':
    print('Hi, Alice.')
```

```
elif age < 12:
    print('You are not Alice, kiddo.')
```

This time, you check the person's age, and the program will tell them something different if they're younger than 12. You can see the flowchart for this in Figure 2-4.

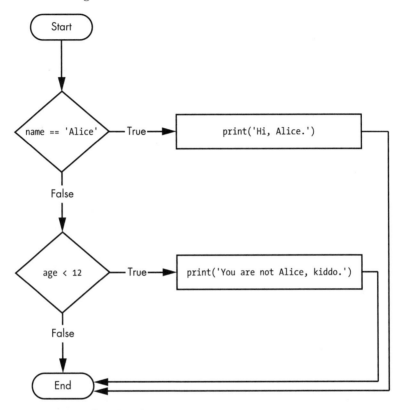

Figure 2-4: The flowchart for an elif statement

The elif clause executes if age < 12 is True and name == 'Alice' is False. However, if both of the conditions are False, then both of the clauses are skipped. It is *not* guaranteed that at least one of the clauses will be executed. When there is a chain of elif statements, only one or none of the clauses will be executed. Once one of the statements' conditions is found to be True, the rest of the elif clauses are automatically skipped. For example, open a new file editor window and enter the following code, saving it as *vampire.py*:

```
name = 'Carol'
age = 3000
if name == 'Alice':
    print('Hi, Alice.')
elif age < 12:
    print('You are not Alice, kiddo.')
elif age > 2000:
```

```
    print('Unlike you, Alice is not an undead, immortal vampire.')
elif age > 100:
    print('You are not Alice, grannie.')
```

You can view the execution of this program at *https://autbor.com/vampire/*. Here, I've added two more elif statements to make the name checker greet a person with different answers based on age. Figure 2-5 shows the flowchart for this.

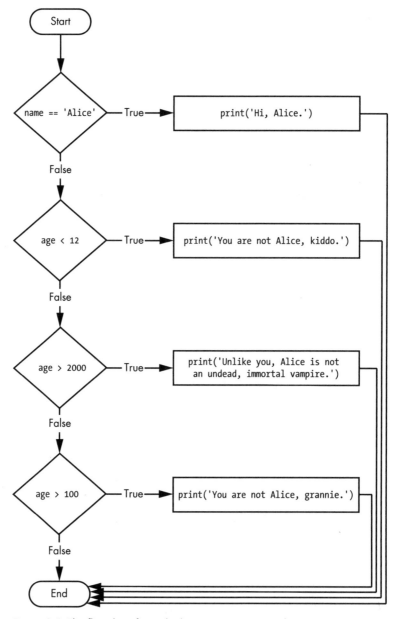

*Figure 2-5: The flowchart for multiple* elif *statements in the* vampire.py *program*

The order of the elif statements does matter, however. Let's rearrange them to introduce a bug. Remember that the rest of the elif clauses are automatically skipped once a True condition has been found, so if you swap around some of the clauses in *vampire.py*, you run into a problem. Change the code to look like the following, and save it as *vampire2.py*:

```
name = 'Carol'
age = 3000
if name == 'Alice':
    print('Hi, Alice.')
elif age < 12:
    print('You are not Alice, kiddo.')
❶ elif age > 100:
    print('You are not Alice, grannie.')
elif age > 2000:
    print('Unlike you, Alice is not an undead, immortal vampire.')
```

You can view the execution of this program at *https://autbor.com/vampire2/*. Say the age variable contains the value 3000 before this code is executed. You might expect the code to print the string 'Unlike you, Alice is not an undead, immortal vampire.'. However, because the age > 100 condition is True (after all, 3,000 *is* greater than 100) ❶, the string 'You are not Alice, grannie.' is printed, and the rest of the elif statements are automatically skipped. Remember that at most only one of the clauses will be executed, and for elif statements, the order matters!

Figure 2-6 shows the flowchart for the previous code. Notice how the diamonds for age > 100 and age > 2000 are swapped.

Optionally, you can have an else statement after the last elif statement. In that case, it *is* guaranteed that at least one (and only one) of the clauses will be executed. If the conditions in every if and elif statement are False, then the else clause is executed. For example, let's re-create the Alice program to use if, elif, and else clauses.

```
name = 'Carol'
age = 3000
if name == 'Alice':
    print('Hi, Alice.')
elif age < 12:
    print('You are not Alice, kiddo.')
else:
    print('You are neither Alice nor a little kid.')
```

You can view the execution of this program at *https://autbor.com /littlekid/*. Figure 2-7 shows the flowchart for this new code, which we'll save as *littleKid.py*.

In plain English, this type of flow control structure would be "If the first condition is true, do this. Else, if the second condition is true, do that. Otherwise, do something else." When you use if, elif, and else statements together, remember these rules about how to order them to avoid bugs like the one in Figure 2-6. First, there is always exactly one if statement. Any elif statements you need should follow the if statement. Second, if you want to be sure that at least one clause is executed, close the structure with an else statement.

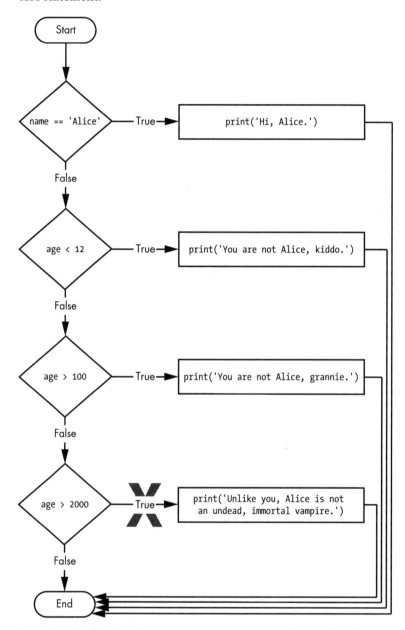

Figure 2-6: The flowchart for the vampire2.py program. The X path will logically never happen, because if age were greater than 2000, it would have already been greater than 100.

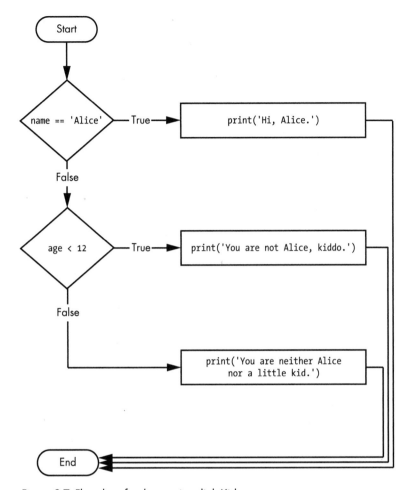

*Figure 2-7: Flowchart for the previous* littleKid.py *program*

## while Loop Statements

You can make a block of code execute over and over again using a while statement. The code in a while clause will be executed as long as the while statement's condition is True. In code, a while statement always consists of the following:

- The while keyword
- A condition (that is, an expression that evaluates to True or False)
- A colon
- Starting on the next line, an indented block of code (called the while clause)

You can see that a while statement looks similar to an if statement. The difference is in how they behave. At the end of an if clause, the program execution continues after the if statement. But at the end of a while clause, the program execution jumps back to the start of the while statement. The while clause is often called the *while loop* or just the *loop*.

Let's look at an if statement and a while loop that use the same condition and take the same actions based on that condition. Here is the code with an if statement:

```
spam = 0
if spam < 5:
    print('Hello, world.')
    spam = spam + 1
```

Here is the code with a while statement:

```
spam = 0
while spam < 5:
    print('Hello, world.')
    spam = spam + 1
```

These statements are similar—both if and while check the value of spam, and if it's less than 5, they print a message. But when you run these two code snippets, something very different happens for each one. For the if statement, the output is simply "Hello, world.". But for the while statement, it's "Hello, world." repeated five times! Take a look at the flowcharts for these two pieces of code, Figures 2-8 and 2-9, to see why this happens.

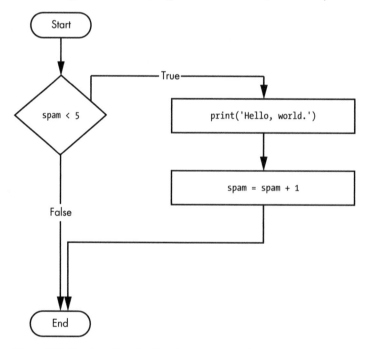

Figure 2-8: The flowchart for the if statement code

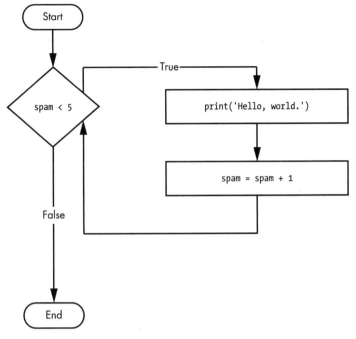

*Figure 2-9: The flowchart for the* while *statement code*

The code with the if statement checks the condition, and it prints Hello, world. only once if that condition is true. The code with the while loop, on the other hand, will print it five times. The loop stops after five prints because the integer in spam increases by one at the end of each loop iteration, which means that the loop will execute five times before spam < 5 is False.

In the while loop, the condition is always checked at the start of each *iteration* (that is, each time the loop is executed). If the condition is True, then the clause is executed, and afterward, the condition is checked again. The first time the condition is found to be False, the while clause is skipped.

### An Annoying while Loop

Here's a small example program that will keep asking you to type, literally, your name. Select **File ▶ New** to open a new file editor window, enter the following code, and save the file as *yourName.py*:

```
❶ name = ''
❷ while name != 'your name':
      print('Please type your name.')
   ❸ name = input()
❹ print('Thank you!')
```

You can view the execution of this program at *https://autbor.com/yourname/*. First, the program sets the name variable ❶ to an empty string. This is so that the name != 'your name' condition will evaluate to True and the program execution will enter the while loop's clause ❷.

The code inside this clause asks the user to type their name, which is assigned to the name variable ❸. Since this is the last line of the block, the execution moves back to the start of the while loop and reevaluates the condition. If the value in name is *not equal* to the string 'your name', then the condition is True, and the execution enters the while clause again.

But once the user types your name, the condition of the while loop will be 'your name' != 'your name', which evaluates to False. The condition is now False, and instead of the program execution reentering the while loop's clause, Python skips past it and continues running the rest of the program ❹. Figure 2-10 shows a flowchart for the *yourName.py* program.

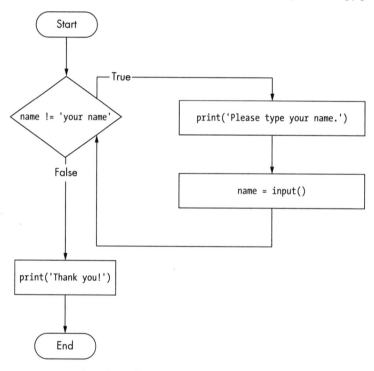

Figure 2-10: A flowchart of the yourName.py program

Now, let's see *yourName.py* in action. Press **F5** to run it, and enter something other than your name a few times before you give the program what it wants.

```
Please type your name.
Al
Please type your name.
Albert
Please type your name.
%#@#%*(^&!!!
```

```
Please type your name.
your name
Thank you!
```

If you never enter your name, then the while loop's condition will never be False, and the program will just keep asking forever. Here, the input() call lets the user enter the right string to make the program move on. In other programs, the condition might never actually change, and that can be a problem. Let's look at how you can break out of a while loop.

### break Statements

There is a shortcut to getting the program execution to break out of a while loop's clause early. If the execution reaches a break statement, it immediately exits the while loop's clause. In code, a break statement simply contains the break keyword.

Pretty simple, right? Here's a program that does the same thing as the previous program, but it uses a break statement to escape the loop. Enter the following code, and save the file as *yourName2.py*:

```
❶ while True:
      print('Please type your name.')
   ❷ name = input()
   ❸ if name == 'your name':
      ❹ break
❺ print('Thank you!')
```

You can view the execution of this program at *https://autbor.com/ yourname2/*. The first line ❶ creates an *infinite loop*; it is a while loop whose condition is always True. (The expression True, after all, always evaluates down to the value True.) After the program execution enters this loop, it will exit the loop only when a break statement is executed. (An infinite loop that *never* exits is a common programming bug.)

Just like before, this program asks the user to enter your name ❷. Now, however, while the execution is still inside the while loop, an if statement checks ❸ whether name is equal to 'your name'. If this condition is True, the break statement is run ❹, and the execution moves out of the loop to print('Thank you!') ❺. Otherwise, the if statement's clause that contains the break statement is skipped, which puts the execution at the end of the while loop. At this point, the program execution jumps back to the start of the while statement ❶ to recheck the condition. Since this condition is merely the True Boolean value, the execution enters the loop to ask the user to type your name again. See Figure 2-11 for this program's flowchart.

Run *yourName2.py*, and enter the same text you entered for *yourName.py*. The rewritten program should respond in the same way as the original.

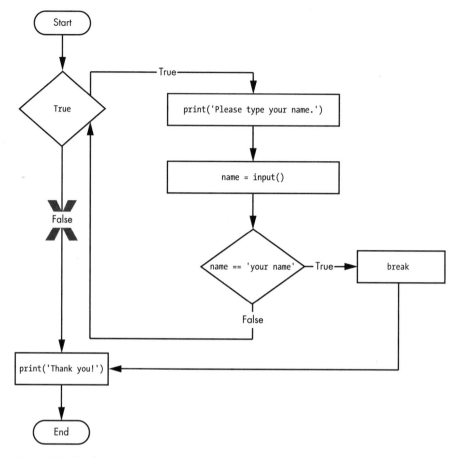

Figure 2-11: The flowchart for the yourName2.py program with an infinite loop. Note that the X path will logically never happen, because the loop condition is always True.

### continue Statements

Like break statements, continue statements are used inside loops. When the program execution reaches a continue statement, the program execution immediately jumps back to the start of the loop and reevaluates the loop's condition. (This is also what happens when the execution reaches the end of the loop.)

Let's use continue to write a program that asks for a name and password. Enter the following code into a new file editor window and save the program as *swordfish.py*.

```python
while True:
    print('Who are you?')
    name = input()
❶  if name != 'Joe':
❷      continue
    print('Hello, Joe. What is the password? (It is a fish.)')
❸  password = input()
    if password == 'swordfish':
❹      break
❺ print('Access granted.')
```

If the user enters any name besides Joe ❶, the continue statement ❷ causes the program execution to jump back to the start of the loop. When the program reevaluates the condition, the execution will always enter the loop, since the condition is simply the value True. Once the user makes it past that if statement, they are asked for a password ❸. If the password entered is swordfish, then the break statement ❹ is run, and the execution jumps out of the while loop to print Access granted ❺. Otherwise, the execution continues to the end of the while loop, where it then jumps back to the start of the loop. See Figure 2-12 for this program's flowchart.

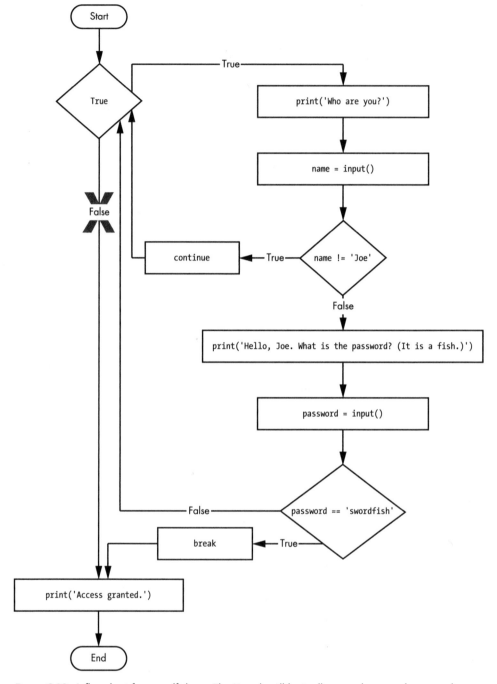

Figure 2-12: A flowchart for swordfish.py. The X path will logically never happen, because the loop condition is always True.

Conditions will consider some values in other data types equivalent to True and False. When used in conditions, 0, 0.0, and '' (the empty string) are considered False, while all other values are considered True. For example, look at the following program:

```
name = ''
❶ while not name:
    print('Enter your name:')
    name = input()
print('How many guests will you have?')
numOfGuests = int(input())
❷ if numOfGuests:
    ❸ print('Be sure to have enough room for all your guests.')
print('Done')
```

You can view the execution of this program at *https://autbor.com /howmanyguests/*. If the user enters a blank string for name, then the while statement's condition will be True ❶, and the program continues to ask for a name. If the value for numOfGuests is not 0 ❷, then the condition is considered to be True, and the program will print a reminder for the user ❸.

You could have entered not name != '' instead of not name, and numOfGuests != 0 instead of numOfGuests, but using the truthy and falsey values can make your code easier to read.

Run this program and give it some input. Until you claim to be Joe, the program shouldn't ask for a password, and once you enter the correct password, it should exit.

```
Who are you?
I'm fine, thanks. Who are you?
Who are you?
Joe
Hello, Joe. What is the password? (It is a fish.)
Mary
Who are you?
Joe
Hello, Joe. What is the password? (It is a fish.)
swordfish
Access granted.
```

You can view the execution of this program at *https://autbor.com/hellojoe/*.

### for Loops and the range() Function

The while loop keeps looping while its condition is True (which is the reason for its name), but what if you want to execute a block of code only a certain number of times? You can do this with a for loop statement and the range() function.

In code, a for statement looks something like for i in range(5): and includes the following:

- The for keyword
- A variable name
- The in keyword
- A call to the range() method with up to three integers passed to it
- A colon
- Starting on the next line, an indented block of code (called the for clause)

Let's create a new program called *fiveTimes.py* to help you see a for loop in action.

```
print('My name is')
for i in range(5):
    print('Jimmy Five Times (' + str(i) + ')')
```

You can view the execution of this program at *https://autbor.com/fivetimesfor/*. The code in the for loop's clause is run five times. The first time it is run, the variable i is set to 0. The print() call in the clause will print Jimmy Five Times (0). After Python finishes an iteration through all the code inside the for loop's clause, the execution goes back to the top of the loop, and the for statement increments i by one. This is why range(5) results in five iterations through the clause, with i being set to 0, then 1, then 2, then 3, and then 4. The variable i will go up to, but will not include, the integer passed to range(). Figure 2-13 shows a flowchart for the *fiveTimes.py* program.

When you run this program, it should print Jimmy Five Times followed by the value of i five times before leaving the for loop.

```
My name is
Jimmy Five Times (0)
Jimmy Five Times (1)
Jimmy Five Times (2)
Jimmy Five Times (3)
Jimmy Five Times (4)
```

**NOTE**    *You can use break and continue statements inside for loops as well. The continue statement will continue to the next value of the for loop's counter, as if the program execution had reached the end of the loop and returned to the start. In fact, you can use continue and break statements only inside while and for loops. If you try to use these statements elsewhere, Python will give you an error.*

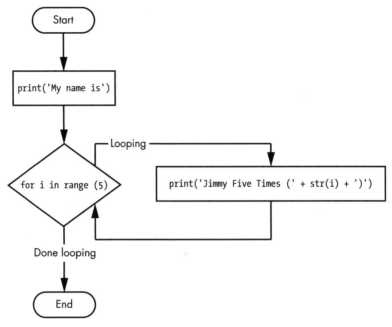

*Figure 2-13: The flowchart for* fiveTimes.py

As another for loop example, consider this story about the mathematician Carl Friedrich Gauss. When Gauss was a boy, a teacher wanted to give the class some busywork. The teacher told them to add up all the numbers from 0 to 100. Young Gauss came up with a clever trick to figure out the answer in a few seconds, but you can write a Python program with a for loop to do this calculation for you.

```
❶ total = 0
❷ for num in range(101):
    ❸ total = total + num
❹ print(total)
```

The result should be 5,050. When the program first starts, the total variable is set to 0 ❶. The for loop ❷ then executes total = total + num ❸ 100 times. By the time the loop has finished all of its 100 iterations, every integer from 0 to 100 will have been added to total. At this point, total is printed to the screen ❹. Even on the slowest computers, this program takes less than a second to complete.

(Young Gauss figured out a way to solve the problem in seconds. There are 50 pairs of numbers that add up to 101: 1 + 100, 2 + 99, 3 + 98, and so on, until 50 + 51. Since 50 × 101 is 5,050, the sum of all the numbers from 0 to 100 is 5,050. Clever kid!)

### An Equivalent while Loop

You can actually use a while loop to do the same thing as a for loop; for loops are just more concise. Let's rewrite *fiveTimes.py* to use a while loop equivalent of a for loop.

```
print('My name is')
i = 0
while i < 5:
    print('Jimmy Five Times (' + str(i) + ')')
    i = i + 1
```

You can view the execution of this program at *https://autbor.com /fivetimeswhile/*. If you run this program, the output should look the same as the *fiveTimes.py* program, which uses a for loop.

### The Starting, Stopping, and Stepping Arguments to range()

Some functions can be called with multiple arguments separated by a comma, and range() is one of them. This lets you change the integer passed to range() to follow any sequence of integers, including starting at a number other than zero.

```
for i in range(12, 16):
    print(i)
```

The first argument will be where the for loop's variable starts, and the second argument will be up to, but not including, the number to stop at.

```
12
13
14
15
```

The range() function can also be called with three arguments. The first two arguments will be the start and stop values, and the third will be the *step argument*. The step is the amount that the variable is increased by after each iteration.

```
for i in range(0, 10, 2):
    print(i)
```

So calling range(0, 10, 2) will count from zero to eight by intervals of two.

```
0
2
4
6
8
```

The range() function is flexible in the sequence of numbers it produces for for loops. *For* example (I never apologize for my puns), you can even use a negative number for the step argument to make the for loop count down instead of up.

```
for i in range(5, -1, -1):
    print(i)
```

This for loop would have the following output:

```
5
4
3
2
1
0
```

Running a for loop to print i with range(5, -1, -1) should print from five down to zero.

## Importing Modules

All Python programs can call a basic set of functions called *built-in functions*, including the print(), input(), and len() functions you've seen before. Python also comes with a set of modules called the *standard library*. Each module is a Python program that contains a related group of functions that can be embedded in your programs. For example, the math module has mathematics-related functions, the random module has random number-related functions, and so on.

Before you can use the functions in a module, you must import the module with an import statement. In code, an import statement consists of the following:

- The import keyword
- The name of the module
- Optionally, more module names, as long as they are separated by commas

Once you import a module, you can use all the cool functions of that module. Let's give it a try with the random module, which will give us access to the random.randint() function.

Enter this code into the file editor, and save it as *printRandom.py*:

```
import random
for i in range(5):
    print(random.randint(1, 10))
```

When you run this program, the output will look something like this:

```
4
1
8
4
1
```

You can view the execution of this program at *https://autbor.com/printrandom/*. The random.randint() function call evaluates to a random integer value between the two integers that you pass it. Since randint() is in the random module, you must first type **random.** in front of the function name to tell Python to look for this function inside the random module.

Here's an example of an import statement that imports four different modules:

```
import random, sys, os, math
```

Now we can use any of the functions in these four modules. We'll learn more about them later in the book.

### from import Statements

An alternative form of the import statement is composed of the from keyword, followed by the module name, the import keyword, and a star; for example, from random import *.

With this form of import statement, calls to functions in random will not need the random. prefix. However, using the full name makes for more readable code, so it is better to use the import random form of the statement.

## Ending a Program Early with the sys.exit() Function

The last flow control concept to cover is how to terminate the program. Programs always terminate if the program execution reaches the bottom of the instructions. However, you can cause the program to terminate, or exit, before the last instruction by calling the sys.exit() function. Since this function is in the sys module, you have to import sys before your program can use it.

Open a file editor window and enter the following code, saving it as *exitExample.py*:

```
import sys

while True:
    print('Type exit to exit.')
    response = input()
    if response == 'exit':
        sys.exit()
    print('You typed ' + response + '.')
```

Run this program in IDLE. This program has an infinite loop with no break statement inside. The only way this program will end is if the execution reaches the sys.exit() call. When response is equal to exit, the line containing the sys.exit() call is executed. Since the response variable is set by the input() function, the user must enter exit in order to stop the program.

## A Short Program: Guess the Number

The examples I've shown you so far are useful for introducing basic concepts, but now let's see how everything you've learned comes together in a more complete program. In this section, I'll show you a simple "guess the number" game. When you run this program, the output will look something like this:

```
I am thinking of a number between 1 and 20.
Take a guess.
10
Your guess is too low.
Take a guess.
15
Your guess is too low.
Take a guess.
17
Your guess is too high.
Take a guess.
16
Good job! You guessed my number in 4 guesses!
```

Enter the following source code into the file editor, and save the file as *guessTheNumber.py*:

```
# This is a guess the number game.
import random
secretNumber = random.randint(1, 20)
print('I am thinking of a number between 1 and 20.')

# Ask the player to guess 6 times.
for guessesTaken in range(1, 7):
    print('Take a guess.')
    guess = int(input())

    if guess < secretNumber:
        print('Your guess is too low.')
    elif guess > secretNumber:
        print('Your guess is too high.')
    else:
        break    # This condition is the correct guess!

if guess == secretNumber:
    print('Good job! You guessed my number in ' + str(guessesTaken) + '
guesses!')
else:
    print('Nope. The number I was thinking of was ' + str(secretNumber))
```

You can view the execution of this program at *https://autbor.com /guessthenumber/*. Let's look at this code line by line, starting at the top.

```
# This is a guess the number game.
import random
secretNumber = random.randint(1, 20)
```

First, a comment at the top of the code explains what the program does. Then, the program imports the random module so that it can use the random.randint() function to generate a number for the user to guess. The return value, a random integer between 1 and 20, is stored in the variable secretNumber.

```
print('I am thinking of a number between 1 and 20.')

# Ask the player to guess 6 times.
for guessesTaken in range(1, 7):
    print('Take a guess.')
    guess = int(input())
```

The program tells the player that it has come up with a secret number and will give the player six chances to guess it. The code that lets the player enter a guess and checks that guess is in a for loop that will loop at most six times. The first thing that happens in the loop is that the player types in a guess. Since input() returns a string, its return value is passed straight into int(), which translates the string into an integer value. This gets stored in a variable named guess.

```
if guess < secretNumber:
    print('Your guess is too low.')
elif guess > secretNumber:
    print('Your guess is too high.')
```

These few lines of code check to see whether the guess is less than or greater than the secret number. In either case, a hint is printed to the screen.

```
else:
    break    # This condition is the correct guess!
```

If the guess is neither higher nor lower than the secret number, then it must be equal to the secret number—in which case, you want the program execution to break out of the for loop.

```
if guess == secretNumber:
    print('Good job! You guessed my number in ' + str(guessesTaken) + ' guesses!')
else:
    print('Nope. The number I was thinking of was ' + str(secretNumber))
```

After the for loop, the previous if...else statement checks whether the player has correctly guessed the number and then prints an appropriate message to the screen. In both cases, the program displays a variable that contains an integer value (guessesTaken and secretNumber). Since it must concatenate these integer values to strings, it passes these variables to the str() function, which returns the string value form of these integers. Now these strings can be concatenated with the + operators before finally being passed to the print() function call.

## A Short Program: Rock, Paper, Scissors

Let's use the programming concepts we've learned so far to create a simple rock, paper, scissors game. The output will look like this:

```
ROCK, PAPER, SCISSORS
0 Wins, 0 Losses, 0 Ties
Enter your move: (r)ock (p)aper (s)cissors or (q)uit
p
PAPER versus...
PAPER
It is a tie!
0 Wins, 1 Losses, 1 Ties
Enter your move: (r)ock (p)aper (s)cissors or (q)uit
s
SCISSORS versus...
PAPER
You win!
```

```
1 Wins, 1 Losses, 1 Ties
Enter your move: (r)ock (p)aper (s)cissors or (q)uit
q
```

Type the following source code into the file editor, and save the file as *rpsGame.py*:

```python
import random, sys

print('ROCK, PAPER, SCISSORS')

# These variables keep track of the number of wins, losses, and ties.
wins = 0
losses = 0
ties = 0

while True: # The main game loop.
    print('%s Wins, %s Losses, %s Ties' % (wins, losses, ties))
    while True: # The player input loop.
        print('Enter your move: (r)ock (p)aper (s)cissors or (q)uit')
        playerMove = input()
        if playerMove == 'q':
            sys.exit() # Quit the program.
        if playerMove == 'r' or playerMove == 'p' or playerMove == 's':
            break # Break out of the player input loop.
        print('Type one of r, p, s, or q.')

    # Display what the player chose:
    if playerMove == 'r':
        print('ROCK versus...')
    elif playerMove == 'p':
        print('PAPER versus...')
    elif playerMove == 's':
        print('SCISSORS versus...')

    # Display what the computer chose:
    randomNumber = random.randint(1, 3)
    if randomNumber == 1:
        computerMove = 'r'
        print('ROCK')
    elif randomNumber == 2:
        computerMove = 'p'
        print('PAPER')
    elif randomNumber == 3:
        computerMove = 's'
        print('SCISSORS')

    # Display and record the win/loss/tie:
    if playerMove == computerMove:
        print('It is a tie!')
        ties = ties + 1
    elif playerMove == 'r' and computerMove == 's':
        print('You win!')
        wins = wins + 1
```

```
    elif playerMove == 'p' and computerMove == 'r':
        print('You win!')
        wins = wins + 1
    elif playerMove == 's' and computerMove == 'p':
        print('You win!')
        wins = wins + 1
    elif playerMove == 'r' and computerMove == 'p':
        print('You lose!')
        losses = losses + 1
    elif playerMove == 'p' and computerMove == 's':
        print('You lose!')
        losses = losses + 1
    elif playerMove == 's' and computerMove == 'r':
        print('You lose!')
        losses = losses + 1
```

Let's look at this code line by line, starting at the top.

```
import random, sys

print('ROCK, PAPER, SCISSORS')

# These variables keep track of the number of wins, losses, and ties.
wins = 0
losses = 0
ties = 0
```

First, we import the random and sys module so that our program can call the random.randint() and sys.exit() functions. We also set up three variables to keep track of how many wins, losses, and ties the player has had.

```
while True: # The main game loop.
    print('%s Wins, %s Losses, %s Ties' % (wins, losses, ties))
    while True: # The player input loop.
        print('Enter your move: (r)ock (p)aper (s)cissors or (q)uit')
        playerMove = input()
        if playerMove == 'q':
            sys.exit() # Quit the program.
        if playerMove == 'r' or playerMove == 'p' or playerMove == 's':
            break # Break out of the player input loop.
        print('Type one of r, p, s, or q.')
```

This program uses a while loop inside of another while loop. The first loop is the main game loop, and a single game of rock, paper, scissors is player on each iteration through this loop. The second loop asks for input from the player, and keeps looping until the player has entered an r, p, s, or q for their move. The r, p, and s correspond to rock, paper, and scissors, respectively, while the q means the player intends to quit. In that case, sys.exit() is called and the program exits. If the player has entered r, p, or s, the execution breaks out of the loop. Otherwise, the program reminds the player to enter r, p, s, or q and goes back to the start of the loop.

```
# Display what the player chose:
if playerMove == 'r':
    print('ROCK versus...')
elif playerMove == 'p':
    print('PAPER versus...')
elif playerMove == 's':
    print('SCISSORS versus...')
```

The player's move is displayed on the screen.

```
# Display what the computer chose:
randomNumber = random.randint(1, 3)
if randomNumber == 1:
    computerMove = 'r'
    print('ROCK')
elif randomNumber == 2:
    computerMove = 'p'
    print('PAPER')
elif randomNumber == 3:
    computerMove = 's'
    print('SCISSORS')
```

Next, the computer's move is randomly selected. Since random.randint() can only return a random number, the 1, 2, or 3 integer value it returns is stored in a variable named randomNumber. The program stores a 'r', 'p', or 's' string in computerMove based on the integer in randomNumber, as well as displays the computer's move.

```
# Display and record the win/loss/tie:
if playerMove == computerMove:
    print('It is a tie!')
    ties = ties + 1
elif playerMove == 'r' and computerMove == 's':
    print('You win!')
    wins = wins + 1
elif playerMove == 'p' and computerMove == 'r':
    print('You win!')
    wins = wins + 1
elif playerMove == 's' and computerMove == 'p':
    print('You win!')
    wins = wins + 1
elif playerMove == 'r' and computerMove == 'p':
    print('You lose!')
    losses = losses + 1
elif playerMove == 'p' and computerMove == 's':
    print('You lose!')
    losses = losses + 1
elif playerMove == 's' and computerMove == 'r':
    print('You lose!')
    losses = losses + 1
```

Finally, the program compares the strings in playerMove and computerMove, and displays the results on the screen. It also increments the wins, losses, or ties variable appropriately. Once the execution reaches the end, it jumps back to the start of the main program loop to begin another game.

## Summary

By using expressions that evaluate to True or False (also called conditions), you can write programs that make decisions on what code to execute and what code to skip. You can also execute code over and over again in a loop while a certain condition evaluates to True. The break and continue statements are useful if you need to exit a loop or jump back to the loop's start.

These flow control statements will let you write more intelligent programs. You can also use another type of flow control by writing your own functions, which is the topic of the next chapter.

## Practice Questions

1. What are the two values of the Boolean data type? How do you write them?

2. What are the three Boolean operators?

3. Write out the truth tables of each Boolean operator (that is, every possible combination of Boolean values for the operator and what they evaluate to).

4. What do the following expressions evaluate to?

```
(5 > 4) and (3 == 5)
not (5 > 4)
(5 > 4) or (3 == 5)
not ((5 > 4) or (3 == 5))
(True and True) and (True == False)
(not False) or (not True)
```

5. What are the six comparison operators?

6. What is the difference between the equal to operator and the assignment operator?

7. Explain what a condition is and where you would use one.

8. Identify the three blocks in this code:

```
spam = 0
if spam == 10:
    print('eggs')
    if spam > 5:
        print('bacon')
```

```
        else:
            print('ham')
        print('spam')
    print('spam')
```

9. Write code that prints Hello if 1 is stored in spam, prints Howdy if 2 is stored in spam, and prints Greetings! if anything else is stored in spam.

10. What keys can you press if your program is stuck in an infinite loop?

11. What is the difference between break and continue?

12. What is the difference between range(10), range(0, 10), and range(0, 10, 1) in a for loop?

13. Write a short program that prints the numbers 1 to 10 using a for loop. Then write an equivalent program that prints the numbers 1 to 10 using a while loop.

14. If you had a function named bacon() inside a module named spam, how would you call it after importing spam?

    **Extra credit:** Look up the round() and abs() functions on the internet, and find out what they do. Experiment with them in the interactive shell.

# 3

## FUNCTIONS

You're already familiar with the print(),
input(), and len() functions from the previous chapters. Python provides several built-
in functions like these, but you can also write
your own functions. A *function* is like a miniprogram
within a program.

To better understand how functions work, let's create one. Enter this
program into the file editor and save it as *helloFunc.py*:

```
❶ def hello():
    ❷ print('Howdy!')
       print('Howdy!!!')
       print('Hello there.')

❸ hello()
   hello()
   hello()
```

You can view the execution of this program at *https://autbor.com /hellofunc/*. The first line is a def statement ❶, which defines a function named hello(). The code in the block that follows the def statement ❷ is the body of the function. This code is executed when the function is called, not when the function is first defined.

The hello() lines after the function ❸ are function calls. In code, a function call is just the function's name followed by parentheses, possibly with some number of arguments in between the parentheses. When the program execution reaches these calls, it will jump to the top line in the function and begin executing the code there. When it reaches the end of the function, the execution returns to the line that called the function and continues moving through the code as before.

Since this program calls hello() three times, the code in the hello() function is executed three times. When you run this program, the output looks like this:

```
Howdy!
Howdy!!!
Hello there.
Howdy!
Howdy!!!
Hello there.
Howdy!
Howdy!!!
Hello there.
```

A major purpose of functions is to group code that gets executed multiple times. Without a function defined, you would have to copy and paste this code each time, and the program would look like this:

```
print('Howdy!')
print('Howdy!!!')
print('Hello there.')
print('Howdy!')
print('Howdy!!!')
print('Hello there.')
print('Howdy!')
print('Howdy!!!')
print('Hello there.')
```

In general, you always want to avoid duplicating code because if you ever decide to update the code—if, for example, you find a bug you need to fix—you'll have to remember to change the code everywhere you copied it.

As you get more programming experience, you'll often find yourself *deduplicating* code, which means getting rid of duplicated or copy-and-pasted code. Deduplication makes your programs shorter, easier to read, and easier to update.

# def Statements with Parameters

When you call the print() or len() function, you pass them values, called *arguments,* by typing them between the parentheses. You can also define your own functions that accept arguments. Type this example into the file editor and save it as *helloFunc2.py*:

```
❶ def hello(name):
    ❷ print('Hello, ' + name)

❸ hello('Alice')
  hello('Bob')
```

When you run this program, the output looks like this:

```
Hello, Alice
Hello, Bob
```

You can view the execution of this program at *https://autbor.com /hellofunc2/.* The definition of the hello() function in this program has a parameter called name ❶. *Parameters* are variables that contain arguments. When a function is called with arguments, the arguments are stored in the parameters. The first time the hello() function is called, it is passed the argument 'Alice' ❸. The program execution enters the function, and the parameter name is automatically set to 'Alice', which is what gets printed by the print() statement ❷.

One special thing to note about parameters is that the value stored in a parameter is forgotten when the function returns. For example, if you added print(name) after hello('Bob') in the previous program, the program would give you a NameError because there is no variable named name. This variable is destroyed after the function call hello('Bob') returns, so print(name) would refer to a name variable that does not exist.

This is similar to how a program's variables are forgotten when the program terminates. I'll talk more about why that happens later in the chapter, when I discuss what a function's local scope is.

## Define, Call, Pass, Argument, Parameter

The terms *define, call, pass, argument,* and *parameter* can be confusing. Let's look at a code example to review these terms:

```
❶ def sayHello(name):
      print('Hello, ' + name)
❷ sayHello('Al')
```

To *define* a function is to create it, just like an assignment statement like spam = 42 creates the spam variable. The def statement defines the sayHello() function ❶. The sayHello('Al') line ❷ *calls* the now-created function, sending the execution to the top of the function's code. This function call is also known as *passing* the string value 'Al' to the function. A value being passed to a function in a function call is an *argument*. The argument 'Al' is assigned to a local variable named name. Variables that have arguments assigned to them are *parameters*.

It's easy to mix up these terms, but keeping them straight will ensure that you know precisely what the text in this chapter means.

## Return Values and return Statements

When you call the len() function and pass it an argument such as 'Hello', the function call evaluates to the integer value 5, which is the length of the string you passed it. In general, the value that a function call evaluates to is called the *return value* of the function.

When creating a function using the def statement, you can specify what the return value should be with a return statement. A return statement consists of the following:

- The return keyword
- The value or expression that the function should return

When an expression is used with a return statement, the return value is what this expression evaluates to. For example, the following program defines a function that returns a different string depending on what number it is passed as an argument. Enter this code into the file editor and save it as *magic8Ball.py*:

```
❶ import random

❷ def getAnswer(answerNumber):
    ❸ if answerNumber == 1:
           return 'It is certain'
       elif answerNumber == 2:
           return 'It is decidedly so'
       elif answerNumber == 3:
           return 'Yes'
       elif answerNumber == 4:
           return 'Reply hazy try again'
       elif answerNumber == 5:
           return 'Ask again later'
       elif answerNumber == 6:
           return 'Concentrate and ask again'
       elif answerNumber == 7:
           return 'My reply is no'
       elif answerNumber == 8:
           return 'Outlook not so good'
       elif answerNumber == 9:
           return 'Very doubtful'

❹ r = random.randint(1, 9)
❺ fortune = getAnswer(r)
❻ print(fortune)
```

You can view the execution of this program at *https://autbor.com /magic8ball/*. When this program starts, Python first imports the random module ❶. Then the getAnswer() function is defined ❷. Because the function is being defined (and not called), the execution skips over the code in it. Next, the random.randint() function is called with two arguments: 1 and 9 ❹. It evaluates to a random integer between 1 and 9 (including 1 and 9 themselves), and this value is stored in a variable named r.

The getAnswer() function is called with r as the argument ❺. The program execution moves to the top of the getAnswer() function ❸, and the value r is stored in a parameter named answerNumber. Then, depending on the value in answerNumber, the function returns one of many possible string values. The program execution returns to the line at the bottom of the program that originally called getAnswer() ❺. The returned string is assigned to a variable named fortune, which then gets passed to a print() call ❻ and is printed to the screen.

Note that since you can pass return values as an argument to another function call, you could shorten these three lines:

```
r = random.randint(1, 9)
fortune = getAnswer(r)
print(fortune)
```

to this single equivalent line:

```
print(getAnswer(random.randint(1, 9)))
```

Remember, expressions are composed of values and operators. A function call can be used in an expression because the call evaluates to its return value.

## The None Value

In Python, there is a value called None, which represents the absence of a value. The None value is the only value of the NoneType data type. (Other programming languages might call this value null, nil, or undefined.) Just like the Boolean True and False values, None must be typed with a capital *N*.

This value-without-a-value can be helpful when you need to store something that won't be confused for a real value in a variable. One place where None is used is as the return value of print(). The print() function displays text on the screen, but it doesn't need to return anything in the same way len() or input() does. But since all function calls need to evaluate to a return value, print() returns None. To see this in action, enter the following into the interactive shell:

```
>>> spam = print('Hello!')
Hello!
>>> None == spam
True
```

Behind the scenes, Python adds return None to the end of any function definition with no return statement. This is similar to how a while or for loop implicitly ends with a continue statement. Also, if you use a return statement without a value (that is, just the return keyword by itself), then None is returned.

## Keyword Arguments and the print() Function

Most arguments are identified by their position in the function call. For example, random.randint(1, 10) is different from random.randint(10, 1). The function call random.randint(1, 10) will return a random integer between 1 and 10 because the first argument is the low end of the range and the second argument is the high end (while random.randint(10, 1) causes an error).

However, rather than through their position, *keyword arguments* are identified by the keyword put before them in the function call. Keyword arguments are often used for *optional parameters*. For example, the print() function has the optional parameters end and sep to specify what should be printed at the end of its arguments and between its arguments (separating them), respectively.

If you ran a program with the following code:

```
print('Hello')
print('World')
```

the output would look like this:

```
Hello
World
```

The two outputted strings appear on separate lines because the print() function automatically adds a newline character to the end of the string it is passed. However, you can set the end keyword argument to change the newline character to a different string. For example, if the code were this:

```
print('Hello', end='')
print('World')
```

the output would look like this:

```
HelloWorld
```

The output is printed on a single line because there is no longer a newline printed after 'Hello'. Instead, the blank string is printed. This is useful if you need to disable the newline that gets added to the end of every print() function call.

Similarly, when you pass multiple string values to print(), the function will automatically separate them with a single space. Enter the following into the interactive shell:

```
>>> print('cats', 'dogs', 'mice')
cats dogs mice
```

But you could replace the default separating string by passing the sep keyword argument a different string. Enter the following into the interactive shell:

```
>>> print('cats', 'dogs', 'mice', sep=',')
cats,dogs,mice
```

You can add keyword arguments to the functions you write as well, but first you'll have to learn about the list and dictionary data types in the next two chapters. For now, just know that some functions have optional keyword arguments that can be specified when the function is called.

## The Call Stack

Imagine that you have a meandering conversation with someone. You talk about your friend Alice, which then reminds you of a story about your coworker Bob, but first you have to explain something about your cousin Carol. You finish you story about Carol and go back to talking about Bob, and when you finish your story about Bob, you go back to talking about Alice. But then you are reminded about your brother David, so you tell a story about him, and then get back to finishing your original story about Alice. Your conversation followed a *stack*-like structure, like in Figure 3-1. The conversation is stack-like because the current topic is always at the top of the stack.

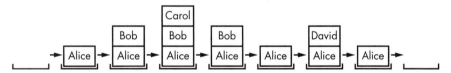

Figure 3-1: Your meandering conversation stack

Similar to our meandering conversation, calling a function doesn't send the execution on a one-way trip to the top of a function. Python will remember which line of code called the function so that the execution can return there when it encounters a return statement. If that original function called other functions, the execution would return to *those* function calls first, before returning from the original function call.

Open a file editor window and enter the following code, saving it as *abcdCallStack.py*:

```
def a():
    print('a() starts')
❶  b()
❷  d()
    print('a() returns')

def b():
    print('b() starts')
❸  c()
    print('b() returns')

def c():
❹  print('c() starts')
    print('c() returns')

def d():
    print('d() starts')
    print('d() returns')

❺ a()
```

If you run this program, the output will look like this:

```
a() starts
b() starts
c() starts
c() returns
b() returns
d() starts
d() returns
a() returns
```

You can view the execution of this program at *https://autbor.com /abcdcallstack/*. When a() is called ❺, it calls b() ❶, which in turn calls c() ❸. The c() function doesn't call anything; it just displays c() starts ❹ and c() returns before returning to the line in b() that called it ❸. Once execution returns to the code in b() that called c(), it returns to the line in a() that called b() ❶. The execution continues to the next line in the b() function ❷, which is a call to d(). Like the c() function, the d() function also doesn't call anything. It just displays d() starts and d() returns before returning to the line in b() that called it. Since b() contains no other code, the execution returns to the line in a() that called b() ❷. The last line in a() displays a() returns before returning to the original a() call at the end of the program ❺.

The *call stack* is how Python remembers where to return the execution after each function call. The call stack isn't stored in a variable in your program; rather, Python handles it behind the scenes. When your program calls a function, Python creates a *frame object* on the top of the call stack. Frame objects store the line number of the original function call so that Python can remember where to return. If another function call is made, Python puts another frame object on the call stack above the other one.

When a function call returns, Python removes a frame object from the top of the stack and moves the execution to the line number stored in it. Note that frame objects are always added and removed from the top of the stack and not from any other place. Figure 3-2 illustrates the state of the call stack in *abcdCallStack.py* as each function is called and returns.

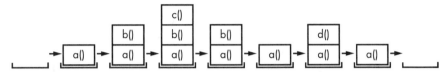

*Figure 3-2: The frame objects of the call stack as abcdCallStack.py calls and returns from functions*

The top of the call stack is which function the execution is currently in. When the call stack is empty, the execution is on a line outside of all functions.

The call stack is a technical detail that you don't strictly need to know about to write programs. It's enough to understand that function calls return to the line number they were called from. However, understanding call stacks makes it easier to understand local and global scopes, described in the next section.

## Local and Global Scope

Parameters and variables that are assigned in a called function are said to exist in that function's *local scope*. Variables that are assigned outside all functions are said to exist in the *global scope*. A variable that exists in a local scope is called a *local variable*, while a variable that exists in the global scope is called a *global variable*. A variable must be one or the other; it cannot be both local and global.

Think of a *scope* as a container for variables. When a scope is destroyed, all the values stored in the scope's variables are forgotten. There is only one global scope, and it is created when your program begins. When your program terminates, the global scope is destroyed, and all its variables are forgotten. Otherwise, the next time you ran a program, the variables would remember their values from the last time you ran it.

A local scope is created whenever a function is called. Any variables assigned in the function exist within the function's local scope. When the function returns, the local scope is destroyed, and these variables are forgotten. The next time you call the function, the local variables will not remember the values stored in them from the last time the function was called. Local variables are also stored in frame objects on the call stack.

Scopes matter for several reasons:

- Code in the global scope, outside of all functions, cannot use any local variables.
- However, code in a local scope can access global variables.
- Code in a function's local scope cannot use variables in any other local scope.
- You can use the same name for different variables if they are in different scopes. That is, there can be a local variable named spam and a global variable also named spam.

The reason Python has different scopes instead of just making everything a global variable is so that when variables are modified by the code in a particular call to a function, the function interacts with the rest of the program only through its parameters and the return value. This narrows down the number of lines of code that may be causing a bug. If your program contained nothing but global variables and had a bug because of a variable being set to a bad value, then it would be hard to track down where this bad value was set. It could have been set from anywhere in the program, and your program could be hundreds or thousands of lines long! But if the bug is caused by a local variable with a bad value, you know that only the code in that one function could have set it incorrectly.

While using global variables in small programs is fine, it is a bad habit to rely on global variables as your programs get larger and larger.

### Local Variables Cannot Be Used in the Global Scope

Consider this program, which will cause an error when you run it:

```
def spam():
❶ eggs = 31337
spam()
print(eggs)
```

If you run this program, the output will look like this:

```
Traceback (most recent call last):
  File "C:/test1.py", line 4, in <module>
    print(eggs)
NameError: name 'eggs' is not defined
```

The error happens because the eggs variable exists only in the local scope created when spam() is called ❶. Once the program execution returns from spam, that local scope is destroyed, and there is no longer a variable named eggs. So when your program tries to run print(eggs), Python gives you an error saying that eggs is not defined. This makes sense if you think about it; when the program execution is in the global scope, no local scopes exist, so there can't be any local variables. This is why only global variables can be used in the global scope.

## Local Scopes Cannot Use Variables in Other Local Scopes

A new local scope is created whenever a function is called, including when a function is called from another function. Consider this program:

```
def spam():
❶  eggs = 99
❷  bacon()
❸  print(eggs)

def bacon():
    ham = 101
❹  eggs = 0

❺ spam()
```

You can view the execution of this program at *https://autbor.com /otherlocalscopes/*. When the program starts, the spam() function is called ❺, and a local scope is created. The local variable eggs ❶ is set to 99. Then the bacon() function is called ❷, and a second local scope is created. Multiple local scopes can exist at the same time. In this new local scope, the local variable ham is set to 101, and a local variable eggs—which is different from the one in spam()'s local scope—is also created ❹ and set to 0.

When bacon() returns, the local scope for that call is destroyed, including its eggs variable. The program execution continues in the spam() function to print the value of eggs ❸. Since the local scope for the call to spam() still exists, the only eggs variable is the spam() function's eggs variable, which was set to 99. This is what the program prints.

The upshot is that local variables in one function are completely separate from the local variables in another function.

## Global Variables Can Be Read from a Local Scope

Consider the following program:

```
def spam():
    print(eggs)
eggs = 42
spam()
print(eggs)
```

You can view the execution of this program at *https://autbor.com /readglobal/*. Since there is no parameter named eggs or any code that assigns eggs a value in the spam() function, when eggs is used in spam(), Python considers it a reference to the global variable eggs. This is why 42 is printed when the previous program is run.

### Local and Global Variables with the Same Name

Technically, it's perfectly acceptable to use the same variable name for a global variable and local variables in different scopes in Python. But, to simplify your life, avoid doing this. To see what happens, enter the following code into the file editor and save it as *localGlobalSameName.py*:

```
def spam():
❶  eggs = 'spam local'
    print(eggs)     # prints 'spam local'

def bacon():
❷  eggs = 'bacon local'
    print(eggs)     # prints 'bacon local'
    spam()
    print(eggs)     # prints 'bacon local'

❸ eggs = 'global'
bacon()
print(eggs)         # prints 'global'
```

When you run this program, it outputs the following:

```
bacon local
spam local
bacon local
global
```

You can view the execution of this program at *https://autbor.com /localglobalsamename/*. There are actually three different variables in this program, but confusingly they are all named eggs. The variables are as follows:

❶  A variable named eggs that exists in a local scope when spam() is called.
❷  A variable named eggs that exists in a local scope when bacon() is called.
❸  A variable named eggs that exists in the global scope.

Since these three separate variables all have the same name, it can be confusing to keep track of which one is being used at any given time. This is why you should avoid using the same variable name in different scopes.

## The global Statement

If you need to modify a global variable from within a function, use the global statement. If you have a line such as global eggs at the top of a function, it tells Python, "In this function, eggs refers to the global variable, so don't create a local variable with this name." For example, enter the following code into the file editor and save it as *globalStatement.py*:

```
def spam():
❶  global eggs
❷  eggs = 'spam'
```

```
eggs = 'global'
spam()
print(eggs)
```

When you run this program, the final print() call will output this:

```
spam
```

You can view the execution of this program at *https://autbor.com /globalstatement/*. Because eggs is declared global at the top of spam() ❶, when eggs is set to 'spam' ❷, this assignment is done to the globally scoped eggs. No local eggs variable is created.

There are four rules to tell whether a variable is in a local scope or global scope:

- If a variable is being used in the global scope (that is, outside of all functions), then it is always a global variable.
- If there is a global statement for that variable in a function, it is a global variable.
- Otherwise, if the variable is used in an assignment statement in the function, it is a local variable.
- But if the variable is not used in an assignment statement, it is a global variable.

To get a better feel for these rules, here's an example program. Enter the following code into the file editor and save it as *sameNameLocalGlobal.py*:

```
def spam():
❶   global eggs
     eggs = 'spam' # this is the global

def bacon():
❷   eggs = 'bacon' # this is a local

def ham():
❸   print(eggs) # this is the global

eggs = 42 # this is the global
spam()
print(eggs)
```

In the spam() function, eggs is the global eggs variable because there's a global statement for eggs at the beginning of the function ❶. In bacon(), eggs is a local variable because there's an assignment statement for it in that function ❷. In ham() ❸, eggs is the global variable because there is no assignment statement or global statement for it in that function. If you run *sameNameLocalGlobal.py*, the output will look like this:

```
spam
```

You can view the execution of this program at *https://autbor.com/sameNameLocalGlobal/*. In a function, a variable will either always be global or always be local. The code in a function can't use a local variable named eggs and then use the global eggs variable later in that same function.

**NOTE** *If you ever want to modify the value stored in a global variable from in a function, you must use a global statement on that variable.*

If you try to use a local variable in a function before you assign a value to it, as in the following program, Python will give you an error. To see this, enter the following into the file editor and save it as *sameNameError.py*:

```
def spam():
    print(eggs) # ERROR!
 ❶ eggs = 'spam local'

❷ eggs = 'global'
spam()
```

If you run the previous program, it produces an error message.

```
Traceback (most recent call last):
  File "C:/sameNameError.py", line 6, in <module>
    spam()
  File "C:/sameNameError.py", line 2, in spam
    print(eggs) # ERROR!
UnboundLocalError: local variable 'eggs' referenced before assignment
```

You can view the execution of this program at *https://autbor.com/sameNameError/*. This error happens because Python sees that there is an assignment statement for eggs in the spam() function ❶ and, therefore, considers eggs to be local. But because print(eggs) is executed before eggs is assigned anything, the local variable eggs doesn't exist. Python will *not* fall back to using the global eggs variable ❷.

---

### FUNCTIONS AS "BLACK BOXES"

Often, all you need to know about a function are its inputs (the parameters) and output value; you don't always have to burden yourself with how the function's code actually works. When you think about functions in this high-level way, it's common to say that you're treating a function as a "black box."

This idea is fundamental to modern programming. Later chapters in this book will show you several modules with functions that were written by other people. While you can take a peek at the source code if you're curious, you don't need to know how these functions work in order to use them. And because writing functions without global variables is encouraged, you usually don't have to worry about the function's code interacting with the rest of your program.

## Exception Handling

Right now, getting an error, or *exception*, in your Python program means the entire program will crash. You don't want this to happen in real-world programs. Instead, you want the program to detect errors, handle them, and then continue to run.

For example, consider the following program, which has a divide-by-zero error. Open a file editor window and enter the following code, saving it as *zeroDivide.py*:

```
def spam(divideBy):
    return 42 / divideBy

print(spam(2))
print(spam(12))
print(spam(0))
print(spam(1))
```

We've defined a function called spam, given it a parameter, and then printed the value of that function with various parameters to see what happens. This is the output you get when you run the previous code:

```
21.0
3.5
Traceback (most recent call last):
  File "C:/zeroDivide.py", line 6, in <module>
    print(spam(0))
  File "C:/zeroDivide.py", line 2, in spam
    return 42 / divideBy
ZeroDivisionError: division by zero
```

You can view the execution of this program at *https://autbor.com /zerodivide/*. A ZeroDivisionError happens whenever you try to divide a number by zero. From the line number given in the error message, you know that the return statement in spam() is causing an error.

Errors can be handled with try and except statements. The code that could potentially have an error is put in a try clause. The program execution moves to the start of a following except clause if an error happens.

You can put the previous divide-by-zero code in a try clause and have an except clause contain code to handle what happens when this error occurs.

```
def spam(divideBy):
    try:
        return 42 / divideBy
    except ZeroDivisionError:
        print('Error: Invalid argument.')

print(spam(2))
print(spam(12))
print(spam(0))
print(spam(1))
```

When code in a try clause causes an error, the program execution immediately moves to the code in the except clause. After running that code, the execution continues as normal. The output of the previous program is as follows:

```
21.0
3.5
Error: Invalid argument.
None
42.0
```

You can view the execution of this program at *https://autbor.com /tryexceptzerodivide/*. Note that any errors that occur in function calls in a try block will also be caught. Consider the following program, which instead has the spam() calls in the try block:

```
def spam(divideBy):
    return 42 / divideBy

try:
    print(spam(2))
    print(spam(12))
    print(spam(0))
    print(spam(1))
except ZeroDivisionError:
    print('Error: Invalid argument.')
```

When this program is run, the output looks like this:

```
21.0
3.5
Error: Invalid argument.
```

You can view the execution of this program at *https://autbor.com /spamintry/*. The reason print(spam(1)) is never executed is because once the execution jumps to the code in the except clause, it does not return to the try clause. Instead, it just continues moving down the program as normal.

# A Short Program: Zigzag

Let's use the programming concepts you've learned so far to create a small animation program. This program will create a back-and-forth, zigzag pattern until the user stops it by pressing the Mu editor's Stop button or by pressing CTRL-C. When you run this program, the output will look something like this:

```
********
 ********
  ********
```

```
*******
*******
 *******
  *******
   *******
    *******
```

Type the following source code into the file editor, and save the file as *zigzag.py*:

```
import time, sys
indent = 0 # How many spaces to indent.
indentIncreasing = True # Whether the indentation is increasing or not.

try:
    while True: # The main program loop.
        print(' ' * indent, end='')
        print('********')
        time.sleep(0.1) # Pause for 1/10 of a second.

        if indentIncreasing:
            # Increase the number of spaces:
            indent = indent + 1
            if indent == 20:
                # Change direction:
                indentIncreasing = False
        else:
            # Decrease the number of spaces:
            indent = indent - 1
            if indent == 0:
                # Change direction:
                indentIncreasing = True
except KeyboardInterrupt:
    sys.exit()
```

Let's look at this code line by line, starting at the top.

```
import time, sys
indent = 0 # How many spaces to indent.
indentIncreasing = True # Whether the indentation is increasing or not.
```

First, we'll import the time and sys modules. Our program uses two variables: the indent variable keeps track of how many spaces of indentation are before the band of eight asterisks and indentIncreasing contains a Boolean value to determine if the amount of indentation is increasing or decreasing.

```
try:
    while True: # The main program loop.
        print(' ' * indent, end='')
        print('********')
        time.sleep(0.1) # Pause for 1/10 of a second.
```

Next, we place the rest of the program inside a try statement. When the user presses CTRL-C while a Python program is running, Python raises the KeyboardInterrupt exception. If there is no try-except statement to catch this exception, the program crashes with an ugly error message. However, for our program, we want it to cleanly handle the KeyboardInterrupt exception by calling sys.exit(). (The code for this is in the except statement at the end of the program.)

The while True: infinite loop will repeat the instructions in our program forever. This involves using ' ' * indent to print the correct amount of spaces of indentation. We don't want to automatically print a newline after these spaces, so we also pass end='' to the first print() call. A second print() call prints the band of asterisks. The time.sleep() function hasn't been covered yet, but suffice it to say that it introduces a one-tenth-second pause in our program at this point.

```
if indentIncreasing:
    # Increase the number of spaces:
    indent = indent + 1
    if indent == 20:
        indentIncreasing = False # Change direction.
```

Next, we want to adjust the amount of indentation for the next time we print asterisks. If indentIncreasing is True, then we want to add one to indent. But once indent reaches 20, we want the indentation to decrease.

```
else:
    # Decrease the number of spaces:
    indent = indent - 1
    if indent == 0:
        indentIncreasing = True # Change direction.
```

Meanwhile, if indentIncreasing was False, we want to subtract one from indent. Once indent reaches 0, we want the indentation to increase once again. Either way, the program execution will jump back to the start of the main program loop to print the asterisks again.

```
except KeyboardInterrupt:
    sys.exit()
```

If the user presses CTRL-C at any point that the program execution is in the try block, the KeyboardInterrrupt exception is raised and handled by this except statement. The program execution moves inside the except block, which runs sys.exit() and quits the program. This way, even though the main program loop is an infinite loop, the user has a way to shut down the program.

# Summary

Functions are the primary way to compartmentalize your code into logical groups. Since the variables in functions exist in their own local scopes, the code in one function cannot directly affect the values of variables in other functions. This limits what code could be changing the values of your variables, which can be helpful when it comes to debugging your code.

Functions are a great tool to help you organize your code. You can think of them as black boxes: they have inputs in the form of parameters and outputs in the form of return values, and the code in them doesn't affect variables in other functions.

In previous chapters, a single error could cause your programs to crash. In this chapter, you learned about try and except statements, which can run code when an error has been detected. This can make your programs more resilient to common error cases.

# Practice Questions

1.  Why are functions advantageous to have in your programs?

2.  When does the code in a function execute: when the function is defined or when the function is called?

3.  What statement creates a function?

4.  What is the difference between a function and a function call?

5.  How many global scopes are there in a Python program? How many local scopes?

6.  What happens to variables in a local scope when the function call returns?

7.  What is a return value? Can a return value be part of an expression?

8.  If a function does not have a return statement, what is the return value of a call to that function?

9.  How can you force a variable in a function to refer to the global variable?

10.  What is the data type of None?

11.  What does the import areallyourpetsnamederic statement do?

12.  If you had a function named bacon() in a module named spam, how would you call it after importing spam?

13.  How can you prevent a program from crashing when it gets an error?

14.  What goes in the try clause? What goes in the except clause?

## Practice Projects

For practice, write programs to do the following tasks.

### The Collatz Sequence

Write a function named collatz() that has one parameter named number. If number is even, then collatz() should print number // 2 and return this value. If number is odd, then collatz() should print and return 3 * number + 1.

Then write a program that lets the user type in an integer and that keeps calling collatz() on that number until the function returns the value 1. (Amazingly enough, this sequence actually works for any integer—sooner or later, using this sequence, you'll arrive at 1! Even mathematicians aren't sure why. Your program is exploring what's called the *Collatz sequence*, sometimes called "the simplest impossible math problem.")

Remember to convert the return value from input() to an integer with the int() function; otherwise, it will be a string value.

Hint: An integer number is even if number % 2 == 0, and it's odd if number % 2 == 1.

The output of this program could look something like this:

```
Enter number:
3
10
5
16
8
4
2
1
```

### Input Validation

Add try and except statements to the previous project to detect whether the user types in a noninteger string. Normally, the int() function will raise a ValueError error if it is passed a noninteger string, as in int('puppy'). In the except clause, print a message to the user saying they must enter an integer.

# 4

## LISTS

One more topic you'll need to understand before you can begin writing programs in earnest is the list data type and its cousin, the tuple. Lists and tuples can contain multiple values, which makes writing programs that handle large amounts of data easier. And since lists themselves can contain other lists, you can use them to arrange data into hierarchical structures.

In this chapter, I'll discuss the basics of lists. I'll also teach you about methods, which are functions that are tied to values of a certain data type. Then I'll briefly cover the sequence data types (lists, tuples, and strings) and show how they compare with each other. In the next chapter, I'll introduce you to the dictionary data type.

# The List Data Type

A *list* is a value that contains multiple values in an ordered sequence. The term *list value* refers to the list itself (which is a value that can be stored in a variable or passed to a function like any other value), not the values inside the list value. A list value looks like this: ['cat', 'bat', 'rat', 'elephant']. Just as string values are typed with quote characters to mark where the string begins and ends, a list begins with an opening square bracket and ends with a closing square bracket, []. Values inside the list are also called *items*. Items are separated with commas (that is, they are *comma-delimited*). For example, enter the following into the interactive shell:

```
>>> [1, 2, 3]
[1, 2, 3]
>>> ['cat', 'bat', 'rat', 'elephant']
['cat', 'bat', 'rat', 'elephant']
>>> ['hello', 3.1415, True, None, 42]
['hello', 3.1415, True, None, 42]
❶ >>> spam = ['cat', 'bat', 'rat', 'elephant']
>>> spam
['cat', 'bat', 'rat', 'elephant']
```

The spam variable ❶ is still assigned only one value: the list value. But the list value itself contains other values. The value [] is an empty list that contains no values, similar to '', the empty string.

## Getting Individual Values in a List with Indexes

Say you have the list ['cat', 'bat', 'rat', 'elephant'] stored in a variable named spam. The Python code spam[0] would evaluate to 'cat', and spam[1] would evaluate to 'bat', and so on. The integer inside the square brackets that follows the list is called an *index*. The first value in the list is at index 0, the second value is at index 1, the third value is at index 2, and so on. Figure 4-1 shows a list value assigned to spam, along with what the index expressions would evaluate to. Note that because the first index is 0, the last index is one less than the size of the list; a list of four items has 3 as its last index.

```
spam = ["cat", "bat", "rat", "elephant"]
        /        /        \          \
    spam[0]  spam[1]    spam[2]    spam[3]
```

*Figure 4-1: A list value stored in the variable spam, showing which value each index refers to*

For example, enter the following expressions into the interactive shell. Start by assigning a list to the variable spam.

```
>>> spam = ['cat', 'bat', 'rat', 'elephant']
>>> spam[0]
'cat'
```

```
>>> spam[1]
'bat'
>>> spam[2]
'rat'
>>> spam[3]
'elephant'
>>> ['cat', 'bat', 'rat', 'elephant'][3]
'elephant'
❶ >>> 'Hello, ' + spam[0]
❷ 'Hello, cat'
>>> 'The ' + spam[1] + ' ate the ' + spam[0] + '.'
'The bat ate the cat.'
```

Notice that the expression 'Hello, ' + spam[0] ❶ evaluates to 'Hello, ' + 'cat' because spam[0] evaluates to the string 'cat'. This expression in turn evaluates to the string value 'Hello, cat' ❷.

Python will give you an IndexError error message if you use an index that exceeds the number of values in your list value.

```
>>> spam = ['cat', 'bat', 'rat', 'elephant']
>>> spam[10000]
Traceback (most recent call last):
  File "<pyshell#9>", line 1, in <module>
    spam[10000]
IndexError: list index out of range
```

Indexes can be only integer values, not floats. The following example will cause a TypeError error:

```
>>> spam = ['cat', 'bat', 'rat', 'elephant']
>>> spam[1]
'bat'
>>> spam[1.0]
Traceback (most recent call last):
  File "<pyshell#13>", line 1, in <module>
    spam[1.0]
TypeError: list indices must be integers or slices, not float
>>> spam[int(1.0)]
'bat'
```

Lists can also contain other list values. The values in these lists of lists can be accessed using multiple indexes, like so:

```
>>> spam = [['cat', 'bat'], [10, 20, 30, 40, 50]]
>>> spam[0]
['cat', 'bat']
>>> spam[0][1]
'bat'
>>> spam[1][4]
50
```

The first index dictates which list value to use, and the second indicates the value within the list value. For example, spam[0][1] prints 'bat', the second value in the first list. If you only use one index, the program will print the full list value at that index.

## Negative Indexes

While indexes start at 0 and go up, you can also use negative integers for the index. The integer value -1 refers to the last index in a list, the value -2 refers to the second-to-last index in a list, and so on. Enter the following into the interactive shell:

```
>>> spam = ['cat', 'bat', 'rat', 'elephant']
>>> spam[-1]
'elephant'
>>> spam[-3]
'bat'
>>> 'The ' + spam[-1] + ' is afraid of the ' + spam[-3] + '.'
'The elephant is afraid of the bat.'
```

## Getting a List from Another List with Slices

Just as an index can get a single value from a list, a *slice* can get several values from a list, in the form of a new list. A slice is typed between square brackets, like an index, but it has two integers separated by a colon. Notice the difference between indexes and slices.

- spam[2] is a list with an index (one integer).
- spam[1:4] is a list with a slice (two integers).

In a slice, the first integer is the index where the slice starts. The second integer is the index where the slice ends. A slice goes up to, but will not include, the value at the second index. A slice evaluates to a new list value. Enter the following into the interactive shell:

```
>>> spam = ['cat', 'bat', 'rat', 'elephant']
>>> spam[0:4]
['cat', 'bat', 'rat', 'elephant']
>>> spam[1:3]
['bat', 'rat']
>>> spam[0:-1]
['cat', 'bat', 'rat']
```

As a shortcut, you can leave out one or both of the indexes on either side of the colon in the slice. Leaving out the first index is the same as using 0, or the beginning of the list. Leaving out the second index is the same as using the length of the list, which will slice to the end of the list. Enter the following into the interactive shell:

```
>>> spam = ['cat', 'bat', 'rat', 'elephant']
>>> spam[:2]
```

```
['cat', 'bat']
>>> spam[1:]
['bat', 'rat', 'elephant']
>>> spam[:]
['cat', 'bat', 'rat', 'elephant']
```

## Getting a List's Length with the len() Function

The len() function will return the number of values that are in a list value passed to it, just like it can count the number of characters in a string value. Enter the following into the interactive shell:

```
>>> spam = ['cat', 'dog', 'moose']
>>> len(spam)
3
```

## Changing Values in a List with Indexes

Normally, a variable name goes on the left side of an assignment statement, like spam = 42. However, you can also use an index of a list to change the value at that index. For example, spam[1] = 'aardvark' means "Assign the value at index 1 in the list spam to the string 'aardvark'." Enter the following into the interactive shell:

```
>>> spam = ['cat', 'bat', 'rat', 'elephant']
>>> spam[1] = 'aardvark'
>>> spam
['cat', 'aardvark', 'rat', 'elephant']
>>> spam[2] = spam[1]
>>> spam
['cat', 'aardvark', 'aardvark', 'elephant']
>>> spam[-1] = 12345
>>> spam
['cat', 'aardvark', 'aardvark', 12345]
```

## List Concatenation and List Replication

Lists can be concatenated and replicated just like strings. The + operator combines two lists to create a new list value and the * operator can be used with a list and an integer value to replicate the list. Enter the following into the interactive shell:

```
>>> [1, 2, 3] + ['A', 'B', 'C']
[1, 2, 3, 'A', 'B', 'C']
>>> ['X', 'Y', 'Z'] * 3
['X', 'Y', 'Z', 'X', 'Y', 'Z', 'X', 'Y', 'Z']
>>> spam = [1, 2, 3]
>>> spam = spam + ['A', 'B', 'C']
>>> spam
[1, 2, 3, 'A', 'B', 'C']
```

### Removing Values from Lists with del Statements

The del statement will delete values at an index in a list. All of the values in the list after the deleted value will be moved up one index. For example, enter the following into the interactive shell:

```
>>> spam = ['cat', 'bat', 'rat', 'elephant']
>>> del spam[2]
>>> spam
['cat', 'bat', 'elephant']
>>> del spam[2]
>>> spam
['cat', 'bat']
```

The del statement can also be used on a simple variable to delete it, as if it were an "unassignment" statement. If you try to use the variable after deleting it, you will get a NameError error because the variable no longer exists. In practice, you almost never need to delete simple variables. The del statement is mostly used to delete values from lists.

## Working with Lists

When you first begin writing programs, it's tempting to create many individual variables to store a group of similar values. For example, if I wanted to store the names of my cats, I might be tempted to write code like this:

```
catName1 = 'Zophie'
catName2 = 'Pooka'
catName3 = 'Simon'
catName4 = 'Lady Macbeth'
catName5 = 'Fat-tail'
catName6 = 'Miss Cleo'
```

It turns out that this is a bad way to write code. (Also, I don't actually own this many cats, I swear.) For one thing, if the number of cats changes, your program will never be able to store more cats than you have variables. These types of programs also have a lot of duplicate or nearly identical code in them. Consider how much duplicate code is in the following program, which you should enter into the file editor and save as *allMyCats1.py*:

```
print('Enter the name of cat 1:')
catName1 = input()
print('Enter the name of cat 2:')
catName2 = input()
print('Enter the name of cat 3:')
catName3 = input()
print('Enter the name of cat 4:')
catName4 = input()
print('Enter the name of cat 5:')
catName5 = input()
print('Enter the name of cat 6:')
```

```
catName6 = input()
print('The cat names are:')
print(catName1 + ' ' + catName2 + ' ' + catName3 + ' ' + catName4 + ' ' +
catName5 + ' ' + catName6)
```

Instead of using multiple, repetitive variables, you can use a single variable that contains a list value. For example, here's a new and improved version of the *allMyCats1.py* program. This new version uses a single list and can store any number of cats that the user types in. In a new file editor window, enter the following source code and save it as *allMyCats2.py*:

```
catNames = []
while True:
    print('Enter the name of cat ' + str(len(catNames) + 1) +
      ' (Or enter nothing to stop.):')
    name = input()
    if name == '':
        break
    catNames = catNames + [name]  # list concatenation
print('The cat names are:')
for name in catNames:
    print('  ' + name)
```

When you run this program, the output will look something like this:

```
Enter the name of cat 1 (Or enter nothing to stop.):
Zophie
Enter the name of cat 2 (Or enter nothing to stop.):
Pooka
Enter the name of cat 3 (Or enter nothing to stop.):
Simon
Enter the name of cat 4 (Or enter nothing to stop.):
Lady Macbeth
Enter the name of cat 5 (Or enter nothing to stop.):
Fat-tail
Enter the name of cat 6 (Or enter nothing to stop.):
Miss Cleo
Enter the name of cat 7 (Or enter nothing to stop.):

The cat names are:
  Zophie
  Pooka
  Simon
  Lady Macbeth
  Fat-tail
  Miss Cleo
```

You can view the execution of these programs at *https://autbor.com /allmycats1/* and *https://autbor.com/allmycats2/*. The benefit of using a list is that your data is now in a structure, so your program is much more flexible in processing the data than it would be with several repetitive variables.

## Using for Loops with Lists

In Chapter 2, you learned about using for loops to execute a block of code a certain number of times. Technically, a for loop repeats the code block once for each item in a list value. For example, if you ran this code:

```
for i in range(4):
    print(i)
```

the output of this program would be as follows:

```
0
1
2
3
```

This is because the return value from range(4) is a sequence value that Python considers similar to [0, 1, 2, 3]. (Sequences are described in "Sequence Data Types" on page 93.) The following program has the same output as the previous one:

```
for i in [0, 1, 2, 3]:
    print(i)
```

The previous for loop actually loops through its clause with the variable i set to a successive value in the [0, 1, 2, 3] list in each iteration.

A common Python technique is to use range(len(*someList*)) with a for loop to iterate over the indexes of a list. For example, enter the following into the interactive shell:

```
>>> supplies = ['pens', 'staplers', 'flamethrowers', 'binders']
>>> for i in range(len(supplies)):
...     print('Index ' + str(i) + ' in supplies is: ' + supplies[i])

Index 0 in supplies is: pens
Index 1 in supplies is: staplers
Index 2 in supplies is: flamethrowers
Index 3 in supplies is: binders
```

Using range(len(supplies)) in the previously shown for loop is handy because the code in the loop can access the index (as the variable i) and the value at that index (as supplies[i]). Best of all, range(len(supplies)) will iterate through all the indexes of supplies, no matter how many items it contains.

## The in and not in Operators

You can determine whether a value is or isn't in a list with the in and not in operators. Like other operators, in and not in are used in expressions and connect two values: a value to look for in a list and the list where it may be found. These expressions will evaluate to a Boolean value. Enter the following into the interactive shell:

```
>>> 'howdy' in ['hello', 'hi', 'howdy', 'heyas']
True
>>> spam = ['hello', 'hi', 'howdy', 'heyas']
>>> 'cat' in spam
False
>>> 'howdy' not in spam
False
>>> 'cat' not in spam
True
```

For example, the following program lets the user type in a pet name and then checks to see whether the name is in a list of pets. Open a new file editor window, enter the following code, and save it as *myPets.py*:

```
myPets = ['Zophie', 'Pooka', 'Fat-tail']
print('Enter a pet name:')
name = input()
if name not in myPets:
    print('I do not have a pet named ' + name)
else:
    print(name + ' is my pet.')
```

The output may look something like this:

```
Enter a pet name:
Footfoot
I do not have a pet named Footfoot
```

You can view the execution of this program at *https://autbor.com/mypets/*.

### The Multiple Assignment Trick

The *multiple assignment trick* (technically called *tuple unpacking*) is a shortcut that lets you assign multiple variables with the values in a list in one line of code. So instead of doing this:

```
>>> cat = ['fat', 'gray', 'loud']
>>> size = cat[0]
>>> color = cat[1]
>>> disposition = cat[2]
```

you could type this line of code:

```
>>> cat = ['fat', 'gray', 'loud']
>>> size, color, disposition = cat
```

The number of variables and the length of the list must be exactly equal, or Python will give you a ValueError:

```
>>> cat = ['fat', 'gray', 'loud']
>>> size, color, disposition, name = cat
Traceback (most recent call last):
  File "<pyshell#84>", line 1, in <module>
    size, color, disposition, name = cat
ValueError: not enough values to unpack (expected 4, got 3)
```

## Using the enumerate() Function with Lists

Instead of using the range(len(someList)) technique with a for loop to obtain the integer index of the items in the list, you can call the enumerate() function instead. On each iteration of the loop, enumerate() will return two values: the index of the item in the list, and the item in the list itself. For example, this code is equivalent to the code in the "Using for Loops with Lists" on page 84:

```
>>> supplies = ['pens', 'staplers', 'flamethrowers', 'binders']
>>> for index, item in enumerate(supplies):
...     print('Index ' + str(index) + ' in supplies is: ' + item)

Index 0 in supplies is: pens
Index 1 in supplies is: staplers
Index 2 in supplies is: flamethrowers
Index 3 in supplies is: binders
```

The enumerate() function is useful if you need both the item and the item's index in the loop's block.

## Using the random.choice() and random.shuffle() Functions with Lists

The random module has a couple functions that accept lists for arguments. The random.choice() function will return a randomly selected item from the list. Enter the following into the interactive shell:

```
>>> import random
>>> pets = ['Dog', 'Cat', 'Moose']
>>> random.choice(pets)
'Dog'
>>> random.choice(pets)
'Cat'
>>> random.choice(pets)
'Cat'
```

You can consider random.choice(someList) to be a shorter form of someList[random.randint(0, len(someList) - 1)].

The `random.shuffle()` function will reorder the items in a list. This function modifies the list in place, rather than returning a new list. Enter the following into the interactive shell:

```
>>> import random
>>> people = ['Alice', 'Bob', 'Carol', 'David']
>>> random.shuffle(people)
>>> people
['Carol', 'David', 'Alice', 'Bob']
>>> random.shuffle(people)
>>> people
['Alice', 'David', 'Bob', 'Carol']
```

## Augmented Assignment Operators

When assigning a value to a variable, you will frequently use the variable itself. For example, after assigning 42 to the variable spam, you would increase the value in spam by 1 with the following code:

```
>>> spam = 42
>>> spam = spam + 1
>>> spam
43
```

As a shortcut, you can use the augmented assignment operator += to do the same thing:

```
>>> spam = 42
>>> spam += 1
>>> spam
43
```

There are augmented assignment operators for the +, -, *, /, and % operators, described in Table 4-1.

**Table 4-1:** The Augmented Assignment Operators

| Augmented assignment statement | Equivalent assignment statement |
|---|---|
| spam += 1 | spam = spam + 1 |
| spam -= 1 | spam = spam - 1 |
| spam *= 1 | spam = spam * 1 |
| spam /= 1 | spam = spam / 1 |
| spam %= 1 | spam = spam % 1 |

The += operator can also do string and list concatenation, and the
*= operator can do string and list replication. Enter the following into the
interactive shell:

```
>>> spam = 'Hello,'
>>> spam += ' world!'
>>> spam
'Hello world!'
>>> bacon = ['Zophie']
>>> bacon *= 3
>>> bacon
['Zophie', 'Zophie', 'Zophie']
```

# Methods

A *method* is the same thing as a function, except it is "called on" a value. For
example, if a list value were stored in spam, you would call the index() list
method (which I'll explain shortly) on that list like so: spam.index('hello').
The method part comes after the value, separated by a period.

Each data type has its own set of methods. The list data type, for
example, has several useful methods for finding, adding, removing, and
otherwise manipulating values in a list.

## Finding a Value in a List with the index() Method

List values have an index() method that can be passed a value, and if that
value exists in the list, the index of the value is returned. If the value isn't
in the list, then Python produces a ValueError error. Enter the following into
the interactive shell:

```
>>> spam = ['hello', 'hi', 'howdy', 'heyas']
>>> spam.index('hello')
0
>>> spam.index('heyas')
3
>>> spam.index('howdy howdy howdy')
Traceback (most recent call last):
  File "<pyshell#31>", line 1, in <module>
    spam.index('howdy howdy howdy')
ValueError: 'howdy howdy howdy' is not in list
```

When there are duplicates of the value in the list, the index of its first
appearance is returned. Enter the following into the interactive shell, and
notice that index() returns 1, not 3:

```
>>> spam = ['Zophie', 'Pooka', 'Fat-tail', 'Pooka']
>>> spam.index('Pooka')
1
```

## Adding Values to Lists with the append() and insert() Methods

To add new values to a list, use the append() and insert() methods. Enter the following into the interactive shell to call the append() method on a list value stored in the variable spam:

```
>>> spam = ['cat', 'dog', 'bat']
>>> spam.append('moose')
>>> spam
['cat', 'dog', 'bat', 'moose']
```

The previous append() method call adds the argument to the end of the list. The insert() method can insert a value at any index in the list. The first argument to insert() is the index for the new value, and the second argument is the new value to be inserted. Enter the following into the interactive shell:

```
>>> spam = ['cat', 'dog', 'bat']
>>> spam.insert(1, 'chicken')
>>> spam
['cat', 'chicken', 'dog', 'bat']
```

Notice that the code is spam.append('moose') and spam.insert(1, 'chicken'), not spam = spam.append('moose') and spam = spam.insert(1, 'chicken'). Neither append() nor insert() gives the new value of spam as its return value. (In fact, the return value of append() and insert() is None, so you definitely wouldn't want to store this as the new variable value.) Rather, the list is modified *in place*. Modifying a list in place is covered in more detail later in "Mutable and Immutable Data Types" on page 94.

Methods belong to a single data type. The append() and insert() methods are list methods and can be called only on list values, not on other values such as strings or integers. Enter the following into the interactive shell, and note the AttributeError error messages that show up:

```
>>> eggs = 'hello'
>>> eggs.append('world')
Traceback (most recent call last):
  File "<pyshell#19>", line 1, in <module>
    eggs.append('world')
AttributeError: 'str' object has no attribute 'append'
>>> bacon = 42
>>> bacon.insert(1, 'world')
Traceback (most recent call last):
  File "<pyshell#22>", line 1, in <module>
    bacon.insert(1, 'world')
AttributeError: 'int' object has no attribute 'insert'
```

### Removing Values from Lists with the remove() Method

The remove() method is passed the value to be removed from the list it is called on. Enter the following into the interactive shell:

```
>>> spam = ['cat', 'bat', 'rat', 'elephant']
>>> spam.remove('bat')
>>> spam
['cat', 'rat', 'elephant']
```

Attempting to delete a value that does not exist in the list will result in a ValueError error. For example, enter the following into the interactive shell and notice the error that is displayed:

```
>>> spam = ['cat', 'bat', 'rat', 'elephant']
>>> spam.remove('chicken')
Traceback (most recent call last):
  File "<pyshell#11>", line 1, in <module>
    spam.remove('chicken')
ValueError: list.remove(x): x not in list
```

If the value appears multiple times in the list, only the first instance of the value will be removed. Enter the following into the interactive shell:

```
>>> spam = ['cat', 'bat', 'rat', 'cat', 'hat', 'cat']
>>> spam.remove('cat')
>>> spam
['bat', 'rat', 'cat', 'hat', 'cat']
```

The del statement is good to use when you know the index of the value you want to remove from the list. The remove() method is useful when you know the value you want to remove from the list.

### Sorting the Values in a List with the sort() Method

Lists of number values or lists of strings can be sorted with the sort() method. For example, enter the following into the interactive shell:

```
>>> spam = [2, 5, 3.14, 1, -7]
>>> spam.sort()
>>> spam
[-7, 1, 2, 3.14, 5]
>>> spam = ['ants', 'cats', 'dogs', 'badgers', 'elephants']
>>> spam.sort()
>>> spam
['ants', 'badgers', 'cats', 'dogs', 'elephants']
```

You can also pass True for the reverse keyword argument to have sort() sort the values in reverse order. Enter the following into the interactive shell:

```
>>> spam.sort(reverse=True)
>>> spam
['elephants', 'dogs', 'cats', 'badgers', 'ants']
```

There are three things you should note about the sort() method. First, the sort() method sorts the list in place; don't try to capture the return value by writing code like spam = spam.sort().

Second, you cannot sort lists that have both number values *and* string values in them, since Python doesn't know how to compare these values. Enter the following into the interactive shell and notice the TypeError error:

```
>>> spam = [1, 3, 2, 4, 'Alice', 'Bob']
>>> spam.sort()
Traceback (most recent call last):
  File "<pyshell#70>", line 1, in <module>
    spam.sort()
TypeError: '<' not supported between instances of 'str' and 'int'
```

Third, sort() uses "ASCIIbetical order" rather than actual alphabetical order for sorting strings. This means uppercase letters come before lowercase letters. Therefore, the lowercase *a* is sorted so that it comes *after* the uppercase *Z*. For an example, enter the following into the interactive shell:

```
>>> spam = ['Alice', 'ants', 'Bob', 'badgers', 'Carol', 'cats']
>>> spam.sort()
>>> spam
['Alice', 'Bob', 'Carol', 'ants', 'badgers', 'cats']
```

If you need to sort the values in regular alphabetical order, pass str.lower for the key keyword argument in the sort() method call.

```
>>> spam = ['a', 'z', 'A', 'Z']
>>> spam.sort(key=str.lower)
>>> spam
['a', 'A', 'z', 'Z']
```

This causes the sort() function to treat all the items in the list as if they were lowercase without actually changing the values in the list.

### Reversing the Values in a List with the reverse() Method

If you need to quickly reverse the order of the items in a list, you can call the reverse() list method. Enter the following into the interactive shell:

```
>>> spam = ['cat', 'dog', 'moose']
>>> spam.reverse()
>>> spam
['moose', 'dog', 'cat']
```

Like the sort() list method, reverse() doesn't return a list. This is why you write spam.reverse(), instead of spam = spam.reverse().

## Example Program: Magic 8 Ball with a List

Using lists, you can write a much more elegant version of the previous chapter's Magic 8 Ball program. Instead of several lines of nearly identical elif statements, you can create a single list that the code works with. Open a new file editor window and enter the following code. Save it as *magic8Ball2.py*.

```
import random

messages = ['It is certain',
    'It is decidedly so',
    'Yes definitely',
    'Reply hazy try again',
    'Ask again later',
    'Concentrate and ask again',
    'My reply is no',
```

```
'Outlook not so good',
'Very doubtful']

print(messages[random.randint(0, len(messages) - 1)])
```

You can view the execution of this program at *https://autbor.com/magic8ball2/*.

When you run this program, you'll see that it works the same as the previous *magic8Ball.py* program.

Notice the expression you use as the index for messages: random.randint (0, len(messages) - 1). This produces a random number to use for the index, regardless of the size of messages. That is, you'll get a random number between 0 and the value of len(messages) - 1. The benefit of this approach is that you can easily add and remove strings to the messages list without changing other lines of code. If you later update your code, there will be fewer lines you have to change and fewer chances for you to introduce bugs.

## Sequence Data Types

Lists aren't the only data types that represent ordered sequences of values. For example, strings and lists are actually similar if you consider a string to be a "list" of single text characters. The Python sequence data types include lists, strings, range objects returned by range(), and tuples (explained in the "The Tuple Data Type" on page 96). Many of the things you can do with lists can also be done with strings and other values of sequence types: indexing; slicing; and using them with for loops, with len(), and with the in and not in operators. To see this, enter the following into the interactive shell:

```
>>> name = 'Zophie'
>>> name[0]
'Z'
>>> name[-2]
'i'
>>> name[0:4]
'Zoph'
>>> 'Zo' in name
True
>>> 'z' in name
False
>>> 'p' not in name
False
>>> for i in name:
...     print('* * * ' + i + ' * * *')

* * * Z * * *
* * * o * * *
* * * p * * *
* * * h * * *
* * * i * * *
* * * e * * *
```

## Mutable and Immutable Data Types

But lists and strings are different in an important way. A list value is a *mutable* data type: it can have values added, removed, or changed. However, a string is *immutable*: it cannot be changed. Trying to reassign a single character in a string results in a TypeError error, as you can see by entering the following into the interactive shell:

```
>>> name = 'Zophie a cat'
>>> name[7] = 'the'
Traceback (most recent call last):
  File "<pyshell#50>", line 1, in <module>
    name[7] = 'the'
TypeError: 'str' object does not support item assignment
```

The proper way to "mutate" a string is to use slicing and concatenation to build a *new* string by copying from parts of the old string. Enter the following into the interactive shell:

```
>>> name = 'Zophie a cat'
>>> newName = name[0:7] + 'the' + name[8:12]
>>> name
'Zophie a cat'
>>> newName
'Zophie the cat'
```

We used [0:7] and [8:12] to refer to the characters that we don't wish to replace. Notice that the original 'Zophie a cat' string is not modified, because strings are immutable.

Although a list value *is* mutable, the second line in the following code does not modify the list eggs:

```
>>> eggs = [1, 2, 3]
>>> eggs = [4, 5, 6]
>>> eggs
[4, 5, 6]
```

The list value in eggs isn't being changed here; rather, an entirely new and different list value ([4, 5, 6]) is overwriting the old list value ([1, 2, 3]). This is depicted in Figure 4-2.

If you wanted to actually modify the original list in eggs to contain [4, 5, 6], you would have to do something like this:

```
>>> eggs = [1, 2, 3]
>>> del eggs[2]
>>> del eggs[1]
>>> del eggs[0]
>>> eggs.append(4)
>>> eggs.append(5)
>>> eggs.append(6)
>>> eggs
[4, 5, 6]
```

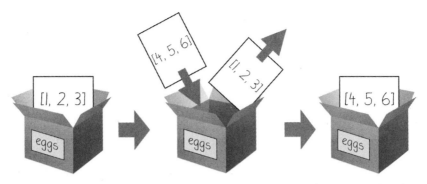

Figure 4-2: When eggs = [4, 5, 6] is executed, the contents of eggs are replaced with a new list value.

In the first example, the list value that eggs ends up with is the same list value it started with. It's just that this list has been changed, rather than overwritten. Figure 4-3 depicts the seven changes made by the first seven lines in the previous interactive shell example.

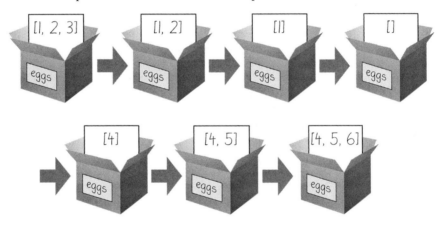

Figure 4-3: The del statement and the append() method modify the same list value in place.

Changing a value of a mutable data type (like what the del statement and append() method do in the previous example) changes the value in place, since the variable's value is not replaced with a new list value.

Mutable versus immutable types may seem like a meaningless distinction, but "Passing References" on page 100 will explain the different behavior when calling functions with mutable arguments versus immutable arguments. But first, let's find out about the tuple data type, which is an immutable form of the list data type.

## The Tuple Data Type

The *tuple* data type is almost identical to the list data type, except in two ways. First, tuples are typed with parentheses, ( and ), instead of square brackets, [ and ]. For example, enter the following into the interactive shell:

```
>>> eggs = ('hello', 42, 0.5)
>>> eggs[0]
'hello'
>>> eggs[1:3]
(42, 0.5)
>>> len(eggs)
3
```

But the main way that tuples are different from lists is that tuples, like strings, are immutable. Tuples cannot have their values modified, appended, or removed. Enter the following into the interactive shell, and look at the TypeError error message:

```
>>> eggs = ('hello', 42, 0.5)
>>> eggs[1] = 99
Traceback (most recent call last):
  File "<pyshell#5>", line 1, in <module>
    eggs[1] = 99
TypeError: 'tuple' object does not support item assignment
```

If you have only one value in your tuple, you can indicate this by placing a trailing comma after the value inside the parentheses. Otherwise, Python will think you've just typed a value inside regular parentheses. The comma is what lets Python know this is a tuple value. (Unlike some other programming languages, it's fine to have a trailing comma after the last item in a list or tuple in Python.) Enter the following type() function calls into the interactive shell to see the distinction:

```
>>> type(('hello',))
<class 'tuple'>
>>> type(('hello'))
<class 'str'>
```

You can use tuples to convey to anyone reading your code that you don't intend for that sequence of values to change. If you need an ordered sequence of values that never changes, use a tuple. A second benefit of using tuples instead of lists is that, because they are immutable and their contents don't change, Python can implement some optimizations that make code using tuples slightly faster than code using lists.

### Converting Types with the list() and tuple() Functions

Just like how str(42) will return '42', the string representation of the integer 42, the functions list() and tuple() will return list and tuple versions of the values passed to them. Enter the following into the interactive shell, and notice that the return value is of a different data type than the value passed:

```
>>> tuple(['cat', 'dog', 5])
('cat', 'dog', 5)
>>> list(('cat', 'dog', 5))
['cat', 'dog', 5]
>>> list('hello')
['h', 'e', 'l', 'l', 'o']
```

Converting a tuple to a list is handy if you need a mutable version of a tuple value.

## References

As you've seen, variables "store" strings and integer values. However, this explanation is a simplification of what Python is actually doing. Technically, variables are storing references to the computer memory locations where the values are stored. Enter the following into the interactive shell:

```
>>> spam = 42
>>> cheese = spam
>>> spam = 100
>>> spam
100
>>> cheese
42
```

When you assign 42 to the spam variable, you are actually creating the 42 value in the computer's memory and storing a *reference* to it in the spam variable. When you copy the value in spam and assign it to the variable cheese, you are actually copying the reference. Both the spam and cheese variables refer to the 42 value in the computer's memory. When you later change the value in spam to 100, you're creating a new 100 value and storing a reference to it in spam. This doesn't affect the value in cheese. Integers are *immutable* values that don't change; changing the *spam* variable is actually making it refer to a completely different value in memory.

But lists don't work this way, because list values can change; that is, lists are *mutable*. Here is some code that will make this distinction easier to understand. Enter this into the interactive shell:

```
❶ >>> spam = [0, 1, 2, 3, 4, 5]
❷ >>> cheese = spam # The reference is being copied, not the list.
❸ >>> cheese[1] = 'Hello!' # This changes the list value.
  >>> spam
```

```
[0, 'Hello!', 2, 3, 4, 5]
>>> cheese # The cheese variable refers to the same list.
[0, 'Hello!', 2, 3, 4, 5]
```

This might look odd to you. The code touched only the cheese list, but it seems that both the cheese and spam lists have changed.

When you create the list ❶, you assign a reference to it in the spam variable. But the next line ❷ copies only the list reference in spam to cheese, not the list value itself. This means the values stored in spam and cheese now both refer to the same list. There is only one underlying list because the list itself was never actually copied. So when you modify the first element of cheese ❸, you are modifying the same list that spam refers to.

Remember that variables are like boxes that contain values. The previous figures in this chapter show that lists in boxes aren't exactly accurate, because list variables don't actually contain lists—they contain *references* to lists. (These references will have ID numbers that Python uses internally, but you can ignore them.) Using boxes as a metaphor for variables, Figure 4-4 shows what happens when a list is assigned to the spam variable.

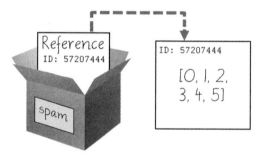

Figure 4-4: spam = [0, 1, 2, 3, 4, 5] stores a reference to a list, not the actual list.

Then, in Figure 4-5, the reference in spam is copied to cheese. Only a new reference was created and stored in cheese, not a new list. Note how both references refer to the same list.

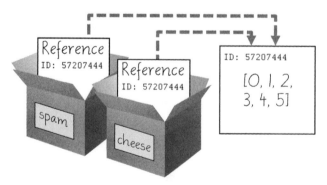

Figure 4-5: spam = cheese copies the reference, not the list.

When you alter the list that cheese refers to, the list that spam refers to is also changed, because both cheese and spam refer to the same list. You can see this in Figure 4-6.

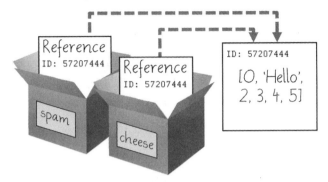

Figure 4-6: cheese[1] = 'Hello!' modifies the list that both variables refer to.

Although Python variables technically contain references to values, people often casually say that the variable contains the value.

## Identity and the id() Function

You may be wondering why the weird behavior with mutable lists in the previous section doesn't happen with immutable values like integers or strings. We can use Python's id() function to understand this. All values in Python have a unique identity that can be obtained with the id() function. Enter the following into the interactive shell:

```
>>> id('Howdy') # The returned number will be different on your machine.
44491136
```

When Python runs id('Howdy'), it creates the 'Howdy' string in the computer's memory. The numeric memory address where the string is stored is returned by the id() function. Python picks this address based on which memory bytes happen to be free on your computer at the time, so it'll be different each time you run this code.

Like all strings, 'Howdy' is immutable and cannot be changed. If you "change" the string in a variable, a new string object is being made at a different place in memory, and the variable refers to this new string. For example, enter the following into the interactive shell and see how the identity of the string referred to by bacon changes:

```
>>> bacon = 'Hello'
>>> id(bacon)
44491136
>>> bacon += ' world!' # A new string is made from 'Hello' and ' world!'.
>>> id(bacon) # bacon now refers to a completely different string.
44609712
```

However, lists can be modified because they are mutable objects. The append() method doesn't create a new list object; it changes the existing list object. We call this "modifying the object *in-place*."

```
>>> eggs = ['cat', 'dog'] # This creates a new list.
>>> id(eggs)
35152584
>>> eggs.append('moose') # append() modifies the list "in place".
>>> id(eggs) # eggs still refers to the same list as before.
35152584
>>> eggs = ['bat', 'rat', 'cow'] # This creates a new list, which has a new
identity.
>>> id(eggs) # eggs now refers to a completely different list.
44409800
```

If two variables refer to the same list (like spam and cheese in the previous section) and the list value itself changes, both variables are affected because they both refer to the same list. The append(), extend(), remove(), sort(), reverse(), and other list methods modify their lists in place.

Python's *automatic garbage collector* deletes any values not being referred to by any variables to free up memory. You don't need to worry about how the garbage collector works, which is a good thing: manual memory management in other programming languages is a common source of bugs.

## Passing References

References are particularly important for understanding how arguments get passed to functions. When a function is called, the values of the arguments are copied to the parameter variables. For lists (and dictionaries, which I'll describe in the next chapter), this means a copy of the reference is used for the parameter. To see the consequences of this, open a new file editor window, enter the following code, and save it as *passingReference.py*:

```
def eggs(someParameter):
    someParameter.append('Hello')

spam = [1, 2, 3]
eggs(spam)
print(spam)
```

Notice that when eggs() is called, a return value is not used to assign a new value to spam. Instead, it modifies the list in place, directly. When run, this program produces the following output:

```
[1, 2, 3, 'Hello']
```

Even though spam and someParameter contain separate references, they both refer to the same list. This is why the append('Hello') method call inside the function affects the list even after the function call has returned.

Keep this behavior in mind: forgetting that Python handles list and dictionary variables this way can lead to confusing bugs.

## The copy Module's copy() and deepcopy() Functions

Although passing around references is often the handiest way to deal with lists and dictionaries, if the function modifies the list or dictionary that is passed, you may not want these changes in the original list or dictionary value. For this, Python provides a module named copy that provides both the copy() and deepcopy() functions. The first of these, copy.copy(), can be used to make a duplicate copy of a mutable value like a list or dictionary, not just a copy of a reference. Enter the following into the interactive shell:

```
>>> import copy
>>> spam = ['A', 'B', 'C', 'D']
>>> id(spam)
44684232
>>> cheese = copy.copy(spam)
>>> id(cheese) # cheese is a different list with different identity.
44685832
>>> cheese[1] = 42
>>> spam
['A', 'B', 'C', 'D']
>>> cheese
['A', 42, 'C', 'D']
```

Now the spam and cheese variables refer to separate lists, which is why only the list in cheese is modified when you assign 42 at index 1. As you can see in Figure 4-7, the reference ID numbers are no longer the same for both variables because the variables refer to independent lists.

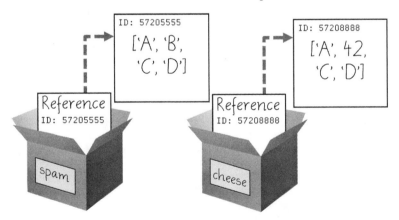

*Figure 4-7: cheese = copy.copy(spam) creates a second list that can be modified independently of the first.*

If the list you need to copy contains lists, then use the copy.deepcopy() function instead of copy.copy(). The deepcopy() function will copy these inner lists as well.

# A Short Program: Conway's Game of Life

Conway's Game of Life is an example of *cellular automata*: a set of rules governing the behavior of a field made up of discrete cells. In practice, it creates a pretty animation to look at. You can draw out each step on graph paper, using the squares as cells. A filled-in square will be "alive" and an empty square will be "dead." If a living square has two or three living neighbors, it continues to live on the next step. If a dead square has exactly three living neighbors, it comes alive on the next step. Every other square dies or remains dead on the next step. You can see an example of the progression of steps in Figure 4-8.

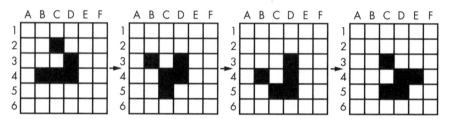

*Figure 4-8: Four steps in a Conway's Game of Life simulation*

Even though the rules are simple, there are many surprising behaviors that emerge. Patterns in Conway's Game of Life can move, self-replicate, or even mimic CPUs. But at the foundation of all of this complex, advanced behavior is a rather simple program.

We can use a list of lists to represent the two-dimensional field. The inner list represents each column of squares and stores a '#' hash string for living squares and a ' ' space string for dead squares. Type the following source code into the file editor, and save the file as *conway.py*. It's fine if you don't quite understand how all of the code works; just enter it and follow along with comments and explanations provided here as close as you can:

```
# Conway's Game of Life
import random, time, copy
WIDTH = 60
HEIGHT = 20

# Create a list of list for the cells:
nextCells = []
for x in range(WIDTH):
    column = [] # Create a new column.
    for y in range(HEIGHT):
        if random.randint(0, 1) == 0:
            column.append('#') # Add a living cell.
        else:
            column.append(' ') # Add a dead cell.
    nextCells.append(column) # nextCells is a list of column lists.

while True: # Main program loop.
    print('\n\n\n\n\n') # Separate each step with newlines.
    currentCells = copy.deepcopy(nextCells)
```

```
    # Print currentCells on the screen:
    for y in range(HEIGHT):
        for x in range(WIDTH):
            print(currentCells[x][y], end='') # Print the # or space.
        print() # Print a newline at the end of the row.

    # Calculate the next step's cells based on current step's cells:
    for x in range(WIDTH):
        for y in range(HEIGHT):
            # Get neighboring coordinates:
            # `% WIDTH` ensures leftCoord is always between 0 and WIDTH - 1
            leftCoord  = (x - 1) % WIDTH
            rightCoord = (x + 1) % WIDTH
            aboveCoord = (y - 1) % HEIGHT
            belowCoord = (y + 1) % HEIGHT

            # Count number of living neighbors:
            numNeighbors = 0
            if currentCells[leftCoord][aboveCoord] == '#':
                numNeighbors += 1 # Top-left neighbor is alive.
            if currentCells[x][aboveCoord] == '#':
                numNeighbors += 1 # Top neighbor is alive.
            if currentCells[rightCoord][aboveCoord] == '#':
                numNeighbors += 1 # Top-right neighbor is alive.
            if currentCells[leftCoord][y] == '#':
                numNeighbors += 1 # Left neighbor is alive.
            if currentCells[rightCoord][y] == '#':
                numNeighbors += 1 # Right neighbor is alive.
            if currentCells[leftCoord][belowCoord] == '#':
                numNeighbors += 1 # Bottom-left neighbor is alive.
            if currentCells[x][belowCoord] == '#':
                numNeighbors += 1 # Bottom neighbor is alive.
            if currentCells[rightCoord][belowCoord] == '#':
                numNeighbors += 1 # Bottom-right neighbor is alive.

            # Set cell based on Conway's Game of Life rules:
            if currentCells[x][y] == '#' and (numNeighbors == 2 or
numNeighbors == 3):
                # Living cells with 2 or 3 neighbors stay alive:
                nextCells[x][y] = '#'
            elif currentCells[x][y] == ' ' and numNeighbors == 3:
                # Dead cells with 3 neighbors become alive:
                nextCells[x][y] = '#'
            else:
                # Everything else dies or stays dead:
                nextCells[x][y] = ' '
    time.sleep(1) # Add a 1-second pause to reduce flickering.
```

Let's look at this code line by line, starting at the top.

```
# Conway's Game of Life
import random, time, copy
WIDTH = 60
HEIGHT = 20
```

First we import modules that contain functions we'll need, namely the `random.randint()`, `time.sleep()`, and `copy.deepcopy()` functions.

```
# Create a list of list for the cells:
nextCells = []
for x in range(WIDTH):
    column = [] # Create a new column.
    for y in range(HEIGHT):
        if random.randint(0, 1) == 0:
            column.append('#') # Add a living cell.
        else:
            column.append(' ') # Add a dead cell.
    nextCells.append(column) # nextCells is a list of column lists.
```

The very first step of our cellular automata will be completely random. We need to create a list of lists data structure to store the '#' and ' ' strings that represent a living or dead cell, and their place in the list of lists reflects their position on the screen. The inner lists each represent a column of cells. The `random.randint(0, 1)` call gives an even 50/50 chance between the cell starting off alive or dead.

We put this list of lists in a variable called `nextCells`, because the first step in our main program loop will be to copy `nextCells` into `currentCells`. For our list of lists data structure, the x-coordinates start at 0 on the left and increase going right, while the y-coordinates start at 0 at the top and increase going down. So `nextCells[0][0]` will represent the cell at the top left of the screen, while `nextCells[1][0]` represents the cell to the right of that cell and `nextCells[0][1]` represents the cell beneath it.

```
while True: # Main program loop.
    print('\n\n\n\n\n') # Separate each step with newlines.
    currentCells = copy.deepcopy(nextCells)
```

Each iteration of our main program loop will be a single step of our cellular automata. On each step, we'll copy `nextCells` to `currentCells`, print `currentCells` on the screen, and then use the cells in `currentCells` to calculate the cells in `nextCells`.

```
# Print currentCells on the screen:
for y in range(HEIGHT):
    for x in range(WIDTH):
        print(currentCells[x][y], end='') # Print the # or space.
    print() # Print a newline at the end of the row.
```

These nested for loops ensure that we print a full row of cells to the screen, followed by a newline character at the end of the row. We repeat this for each row in `nextCells`.

```
# Calculate the next step's cells based on current step's cells:
for x in range(WIDTH):
    for y in range(HEIGHT):
        # Get neighboring coordinates:
```

```
# `% WIDTH` ensures leftCoord is always between 0 and WIDTH - 1
leftCoord  = (x - 1) % WIDTH
rightCoord = (x + 1) % WIDTH
aboveCoord = (y - 1) % HEIGHT
belowCoord = (y + 1) % HEIGHT
```

Next, we need to use two nested for loops to calculate each cell for the next step. The living or dead state of the cell depends on the neighbors, so let's first calculate the index of the cells to the left, right, above, and below the current x- and y-coordinates.

The % mod operator performs a "wraparound." The left neighbor of a cell in the leftmost column 0 would be 0 - 1 or -1. To wrap this around to the rightmost column's index, 59, we calculate (0 - 1) % WIDTH. Since WIDTH is 60, this expression evaluates to 59. This mod-wraparound technique works for the right, above, and below neighbors as well.

```
# Count number of living neighbors:
numNeighbors = 0
if currentCells[leftCoord][aboveCoord] == '#':
    numNeighbors += 1 # Top-left neighbor is alive.
if currentCells[x][aboveCoord] == '#':
    numNeighbors += 1 # Top neighbor is alive.
if currentCells[rightCoord][aboveCoord] == '#':
    numNeighbors += 1 # Top-right neighbor is alive.
if currentCells[leftCoord][y] == '#':
    numNeighbors += 1 # Left neighbor is alive.
if currentCells[rightCoord][y] == '#':
    numNeighbors += 1 # Right neighbor is alive.
if currentCells[leftCoord][belowCoord] == '#':
    numNeighbors += 1 # Bottom-left neighbor is alive.
if currentCells[x][belowCoord] == '#':
    numNeighbors += 1 # Bottom neighbor is alive.
if currentCells[rightCoord][belowCoord] == '#':
    numNeighbors += 1 # Bottom-right neighbor is alive.
```

To decide if the cell at nextCells[x][y] should be living or dead, we need to count the number of living neighbors currentCells[x][y] has. This series of if statements checks each of the eight neighbors of this cell, and adds 1 to numNeighbors for each living one.

```
# Set cell based on Conway's Game of Life rules:
if currentCells[x][y] == '#' and (numNeighbors == 2 or
numNeighbors == 3):
        # Living cells with 2 or 3 neighbors stay alive:
        nextCells[x][y] = '#'
    elif currentCells[x][y] == ' ' and numNeighbors == 3:
        # Dead cells with 3 neighbors become alive:
        nextCells[x][y] = '#'
    else:
        # Everything else dies or stays dead:
        nextCells[x][y] = ' '
time.sleep(1) # Add a 1-second pause to reduce flickering.
```

Now that we know the number of living neighbors for the cell at currentCells[x][y], we can set nextCells[x][y] to either '#' or ' '. After we loop over every possible x- and y-coordinate, the program takes a 1-second pause by calling time.sleep(1). Then the program execution goes back to the start of the main program loop to continue with the next step.

Several patterns have been discovered with names such as "glider," "propeller," or "heavyweight spaceship." The glider pattern, pictured in Figure 4-8, results in a pattern that "moves" diagonally every four steps. You can create a single glider by replacing this line in our *conway.py* program:

```
if random.randint(0, 1) == 0:
```

with this line:

```
if (x, y) in ((1, 0), (2, 1), (0, 2), (1, 2), (2, 2)):
```

You can find out more about the intriguing devices made using Conway's Game of Life by searching the web. And you can find other short, text-based Python programs like this one at *https://github.com/asweigart/pythonstdiogames*.

## Summary

Lists are useful data types since they allow you to write code that works on a modifiable number of values in a single variable. Later in this book, you will see programs using lists to do things that would be difficult or impossible to do without them.

Lists are a sequence data type that is mutable, meaning that their contents can change. Tuples and strings, though also sequence data types, are immutable and cannot be changed. A variable that contains a tuple or string value can be overwritten with a new tuple or string value, but this is not the same thing as modifying the existing value in place—like, say, the append() or remove() methods do on lists.

Variables do not store list values directly; they store *references* to lists. This is an important distinction when you are copying variables or passing lists as arguments in function calls. Because the value that is being copied is the list reference, be aware that any changes you make to the list might impact another variable in your program. You can use copy() or deepcopy() if you want to make changes to a list in one variable without modifying the original list.

## Practice Questions

1. What is []?
2. How would you assign the value 'hello' as the third value in a list stored in a variable named spam? (Assume spam contains [2, 4, 6, 8, 10].)

For the following three questions, let's say spam contains the list ['a', 'b', 'c', 'd'].

3. What does spam[int(int('3' * 2) // 11)] evaluate to?

4. What does spam[-1] evaluate to?

5. What does spam[:2] evaluate to?

For the following three questions, let's say bacon contains the list [3.14, 'cat', 11, 'cat', True].

6. What does bacon.index('cat') evaluate to?

7. What does bacon.append(99) make the list value in bacon look like?

8. What does bacon.remove('cat') make the list value in bacon look like?

9. What are the operators for list concatenation and list replication?

10. What is the difference between the append() and insert() list methods?

11. What are two ways to remove values from a list?

12. Name a few ways that list values are similar to string values.

13. What is the difference between lists and tuples?

14. How do you type the tuple value that has just the integer value 42 in it?

15. How can you get the tuple form of a list value? How can you get the list form of a tuple value?

16. Variables that "contain" list values don't actually contain lists directly. What do they contain instead?

17. What is the difference between copy.copy() and copy.deepcopy()?

# Practice Projects

For practice, write programs to do the following tasks.

## Comma Code

Say you have a list value like this:

```
spam = ['apples', 'bananas', 'tofu', 'cats']
```

Write a function that takes a list value as an argument and returns a string with all the items separated by a comma and a space, with *and* inserted before the last item. For example, passing the previous spam list to the function would return 'apples, bananas, tofu, and cats'. But your function should be able to work with any list value passed to it. Be sure to test the case where an empty list [] is passed to your function.

## Coin Flip Streaks

For this exercise, we'll try doing an experiment. If you flip a coin 100 times and write down an "H" for each heads and "T" for each tails, you'll create a list that looks like "T T T T H H H H T T." If you ask a human to make up 100 random coin flips, you'll probably end up with alternating head-tail results like "H T H T H H T H T T," which looks random (to humans), but isn't mathematically random. A human will almost never write down a streak of six heads or six tails in a row, even though it is highly likely to happen in truly random coin flips. Humans are predictably bad at being random.

Write a program to find out how often a streak of six heads or a streak of six tails comes up in a randomly generated list of heads and tails. Your program breaks up the experiment into two parts: the first part generates a list of randomly selected 'heads' and 'tails' values, and the second part checks if there is a streak in it. Put all of this code in a loop that repeats the experiment 10,000 times so we can find out what percentage of the coin flips contains a streak of six heads or tails in a row. As a hint, the function call random.randint(0, 1) will return a 0 value 50% of the time and a 1 value the other 50% of the time.

You can start with the following template:

```
import random
numberOfStreaks = 0
for experimentNumber in range(10000):
    # Code that creates a list of 100 'heads' or 'tails' values.

    # Code that checks if there is a streak of 6 heads or tails in a row.
print('Chance of streak: %s%%' % (numberOfStreaks / 100))
```

Of course, this is only an estimate, but 10,000 is a decent sample size. Some knowledge of mathematics could give you the exact answer and save you the trouble of writing a program, but programmers are notoriously bad at math.

## Character Picture Grid

Say you have a list of lists where each value in the inner lists is a one-character string, like this:

```
grid = [['.', '.', '.', '.', '.', '.'],
        ['.', 'O', 'O', '.', '.', '.'],
        ['O', 'O', 'O', 'O', '.', '.'],
        ['O', 'O', 'O', 'O', 'O', '.'],
        ['.', 'O', 'O', 'O', 'O', 'O'],
        ['O', 'O', 'O', 'O', 'O', '.'],
        ['O', 'O', 'O', 'O', '.', '.'],
        ['.', 'O', 'O', '.', '.', '.'],
        ['.', '.', '.', '.', '.', '.']]
```

Think of grid[x][y] as being the character at the x- and y-coordinates of a "picture" drawn with text characters. The (0, 0) origin is in the upper-left corner, the x-coordinates increase going right, and the y-coordinates increase going down.

Copy the previous grid value, and write code that uses it to print the image.

```
..00.00..
.0000000.
.0000000.
..00000..
...000...
....0....
```

Hint: You will need to use a loop in a loop in order to print grid[0][0], then grid[1][0], then grid[2][0], and so on, up to grid[8][0]. This will finish the first row, so then print a newline. Then your program should print grid[0][1], then grid[1][1], then grid[2][1], and so on. The last thing your program will print is grid[8][5].

Also, remember to pass the end keyword argument to print() if you don't want a newline printed automatically after each print() call.

# 5

## DICTIONARIES AND STRUCTURING DATA

In this chapter, I will cover the dictionary data type, which provides a flexible way to access and organize data. Then, combining dictionaries with your knowledge of lists from the previous chapter, you'll learn how to create a data structure to model a tic-tac-toe board.

## The Dictionary Data Type

Like a list, a *dictionary* is a mutable collection of many values. But unlike indexes for lists, indexes for dictionaries can use many different data types, not just integers. Indexes for dictionaries are called *keys*, and a key with its associated value is called a *key-value pair*.

In code, a dictionary is typed with braces, {}. Enter the following into the interactive shell:

```
>>> myCat = {'size': 'fat', 'color': 'gray', 'disposition': 'loud'}
```

This assigns a dictionary to the myCat variable. This dictionary's keys are 'size', 'color', and 'disposition'. The values for these keys are 'fat', 'gray', and 'loud', respectively. You can access these values through their keys:

```
>>> myCat['size']
'fat'
>>> 'My cat has ' + myCat['color'] + ' fur.'
'My cat has gray fur.'
```

Dictionaries can still use integer values as keys, just like lists use integers for indexes, but they do not have to start at 0 and can be any number.

```
>>> spam = {12345: 'Luggage Combination', 42: 'The Answer'}
```

## Dictionaries vs. Lists

Unlike lists, items in dictionaries are unordered. The first item in a list named spam would be spam[0]. But there is no "first" item in a dictionary. While the order of items matters for determining whether two lists are the same, it does not matter in what order the key-value pairs are typed in a dictionary. Enter the following into the interactive shell:

```
>>> spam = ['cats', 'dogs', 'moose']
>>> bacon = ['dogs', 'moose', 'cats']
>>> spam == bacon
False
>>> eggs = {'name': 'Zophie', 'species': 'cat', 'age': '8'}
>>> ham = {'species': 'cat', 'age': '8', 'name': 'Zophie'}
>>> eggs == ham
True
```

Because dictionaries are not ordered, they can't be sliced like lists.

Trying to access a key that does not exist in a dictionary will result in a KeyError error message, much like a list's "out-of-range" IndexError error message. Enter the following into the interactive shell, and notice the error message that shows up because there is no 'color' key:

```
>>> spam = {'name': 'Zophie', 'age': 7}
>>> spam['color']
Traceback (most recent call last):
  File "<pyshell#1>", line 1, in <module>
    spam['color']
KeyError: 'color'
```

Though dictionaries are not ordered, the fact that you can have arbitrary values for the keys allows you to organize your data in powerful ways. Say you wanted your program to store data about your friends' birthdays. You can use a dictionary with the names as keys and the birthdays as values. Open a new file editor window and enter the following code. Save it as *birthdays.py*.

```
❶ birthdays = {'Alice': 'Apr 1', 'Bob': 'Dec 12', 'Carol': 'Mar 4'}

while True:
    print('Enter a name: (blank to quit)')
    name = input()
    if name == '':
        break

❷   if name in birthdays:
❸       print(birthdays[name] + ' is the birthday of ' + name)
    else:
        print('I do not have birthday information for ' + name)
        print('What is their birthday?')
        bday = input()
❹       birthdays[name] = bday
        print('Birthday database updated.')
```

You can view the execution of this program at *https://autbor.com/bdaydb*. You create an initial dictionary and store it in birthdays ❶. You can see if the entered name exists as a key in the dictionary with the in keyword ❷, just as you did for lists. If the name is in the dictionary, you access the associated value using square brackets ❸; if not, you can add it using the same square bracket syntax combined with the assignment operator ❹.

When you run this program, it will look like this:

```
Enter a name: (blank to quit)
Alice
Apr 1 is the birthday of Alice
Enter a name: (blank to quit)
Eve
I do not have birthday information for Eve
What is their birthday?
Dec 5
Birthday database updated.
Enter a name: (blank to quit)
Eve
Dec 5 is the birthday of Eve
Enter a name: (blank to quit)
```

Of course, all the data you enter in this program is forgotten when the program terminates. You'll learn how to save data to files on the hard drive in Chapter 9.

While they're still not ordered and have no "first" key-value pair, dictionaries in Python 3.7 and later will remember the insertion order of their key-value pairs if you create a sequence value from them. For example, notice the order of items in the lists made from the eggs and ham dictionaries matches the order in which they were entered:

```
>>> eggs = {'name': 'Zophie', 'species': 'cat', 'age': '8'}
>>> list(eggs)
['name', 'species', 'age']
>>> ham = {'species': 'cat', 'age': '8', 'name': 'Zophie'}
>>> list(ham)
['species', 'age', 'name']
```

The dictionaries are still unordered, as you can't access items in them using integer indexes like eggs[0] or ham[2]. You shouldn't rely on this behavior, as dictionaries in older versions of Python don't remember the insertion order of key-value pairs. For example, notice how the list doesn't match the insertion order of the dictionary's key-value pairs when I run this code in Python 3.5:

```
>>> spam = {}
>>> spam['first key'] = 'value'
>>> spam['second key'] = 'value'
>>> spam['third key'] = 'value'
>>> list(spam)
['first key', 'third key', 'second key']
```

## The keys(), values(), and items() Methods

There are three dictionary methods that will return list-like values of the dictionary's keys, values, or both keys and values: keys(), values(), and items(). The values returned by these methods are not true lists: they cannot be modified and do not have an append() method. But these data types (dict_keys, dict_values, and dict_items, respectively) *can* be used in for loops. To see how these methods work, enter the following into the interactive shell:

```
>>> spam = {'color': 'red', 'age': 42}
>>> for v in spam.values():
...     print(v)

red
42
```

Here, a for loop iterates over each of the values in the spam dictionary. A for loop can also iterate over the keys or both keys and values:

```
>>> for k in spam.keys():
...     print(k)

color
age
>>> for i in spam.items():
...     print(i)

('color', 'red')
('age', 42)
```

When you use the keys(), values(), and items() methods, a for loop can iterate over the keys, values, or key-value pairs in a dictionary, respectively. Notice that the values in the dict_items value returned by the items() method are tuples of the key and value.

If you want a true list from one of these methods, pass its list-like return value to the list() function. Enter the following into the interactive shell:

```
>>> spam = {'color': 'red', 'age': 42}
>>> spam.keys()
dict_keys(['color', 'age'])
>>> list(spam.keys())
['color', 'age']
```

The list(spam.keys()) line takes the dict_keys value returned from keys() and passes it to list(), which then returns a list value of ['color', 'age'].

You can also use the multiple assignment trick in a for loop to assign the key and value to separate variables. Enter the following into the interactive shell:

```
>>> spam = {'color': 'red', 'age': 42}
>>> for k, v in spam.items():
...     print('Key: ' + k + ' Value: ' + str(v))

Key: age Value: 42
Key: color Value: red
```

## Checking Whether a Key or Value Exists in a Dictionary

Recall from the previous chapter that the in and not in operators can check whether a value exists in a list. You can also use these operators to see whether a certain key or value exists in a dictionary. Enter the following into the interactive shell:

```
>>> spam = {'name': 'Zophie', 'age': 7}
>>> 'name' in spam.keys()
True
```

```
>>> 'Zophie' in spam.values()
True
>>> 'color' in spam.keys()
False
>>> 'color' not in spam.keys()
True
>>> 'color' in spam
False
```

In the previous example, notice that 'color' in spam is essentially a shorter version of writing 'color' in spam.keys(). This is always the case: if you ever want to check whether a value is (or isn't) a key in the dictionary, you can simply use the in (or not in) keyword with the dictionary value itself.

## The get() Method

It's tedious to check whether a key exists in a dictionary before accessing that key's value. Fortunately, dictionaries have a get() method that takes two arguments: the key of the value to retrieve and a fallback value to return if that key does not exist.

Enter the following into the interactive shell:

```
>>> picnicItems = {'apples': 5, 'cups': 2}
>>> 'I am bringing ' + str(picnicItems.get('cups', 0)) + ' cups.'
'I am bringing 2 cups.'
>>> 'I am bringing ' + str(picnicItems.get('eggs', 0)) + ' eggs.'
'I am bringing 0 eggs.'
```

Because there is no 'eggs' key in the picnicItems dictionary, the default value 0 is returned by the get() method. Without using get(), the code would have caused an error message, such as in the following example:

```
>>> picnicItems = {'apples': 5, 'cups': 2}
>>> 'I am bringing ' + str(picnicItems['eggs']) + ' eggs.'
Traceback (most recent call last):
  File "<pyshell#34>", line 1, in <module>
    'I am bringing ' + str(picnicItems['eggs']) + ' eggs.'
KeyError: 'eggs'
```

## The setdefault() Method

You'll often have to set a value in a dictionary for a certain key only if that key does not already have a value. The code looks something like this:

```
spam = {'name': 'Pooka', 'age': 5}
if 'color' not in spam:
    spam['color'] = 'black'
```

The setdefault() method offers a way to do this in one line of code. The first argument passed to the method is the key to check for, and the second argument is the value to set at that key if the key does not exist. If the key does exist, the setdefault() method returns the key's value. Enter the following into the interactive shell:

```
>>> spam = {'name': 'Pooka', 'age': 5}
>>> spam.setdefault('color', 'black')
'black'
>>> spam
{'color': 'black', 'age': 5, 'name': 'Pooka'}
>>> spam.setdefault('color', 'white')
'black'
>>> spam
{'color': 'black', 'age': 5, 'name': 'Pooka'}
```

The first time setdefault() is called, the dictionary in spam changes to {'color': 'black', 'age': 5, 'name': 'Pooka'}. The method returns the value 'black' because this is now the value set for the key 'color'. When spam.setdefault('color', 'white') is called next, the value for that key is *not* changed to 'white', because spam already has a key named 'color'.

The setdefault() method is a nice shortcut to ensure that a key exists. Here is a short program that counts the number of occurrences of each letter in a string. Open the file editor window and enter the following code, saving it as *characterCount.py*:

```
message = 'It was a bright cold day in April, and the clocks were striking
thirteen.'
count = {}

for character in message:
❶ count.setdefault(character, 0)
❷ count[character] = count[character] + 1

print(count)
```

You can view the execution of this program at *https://autbor.com/setdefault*. The program loops over each character in the message variable's string, counting how often each character appears. The setdefault() method call ❶ ensures that the key is in the count dictionary (with a default value of 0) so the program doesn't throw a KeyError error when count[character] = count[character] + 1 is executed ❷. When you run this program, the output will look like this:

```
{' ': 13, ',': 1, '.': 1, 'A': 1, 'I': 1, 'a': 4, 'c': 3, 'b': 1, 'e': 5, 'd': 3, 'g': 2,
'i': 6, 'h': 3, 'k': 2, 'l': 3, 'o': 2, 'n': 4, 'p': 1, 's': 3, 'r': 5, 't': 6, 'w': 2, 'y': 1}
```

From the output, you can see that the lowercase letter *c* appears 3 times, the space character appears 13 times, and the uppercase letter *A* appears 1 time. This program will work no matter what string is inside the message variable, even if the string is millions of characters long!

## Pretty Printing

If you import the pprint module into your programs, you'll have access to the pprint() and pformat() functions that will "pretty print" a dictionary's values. This is helpful when you want a cleaner display of the items in a dictionary than what print() provides. Modify the previous *characterCount.py* program and save it as *prettyCharacterCount.py*.

```
import pprint
message = 'It was a bright cold day in April, and the clocks were striking
thirteen.'
count = {}

for character in message:
    count.setdefault(character, 0)
    count[character] = count[character] + 1

pprint.pprint(count)
```

You can view the execution of this program at *https://autbor.com/pprint/*. This time, when the program is run, the output looks much cleaner, with the keys sorted.

```
{' ': 13,
 ',': 1,
 '.': 1,
 'A': 1,
 'I': 1,
 --snip--
 't': 6,
 'w': 2,
 'y': 1}
```

The pprint.pprint() function is especially helpful when the dictionary itself contains nested lists or dictionaries.

If you want to obtain the prettified text as a string value instead of displaying it on the screen, call pprint.pformat() instead. These two lines are equivalent to each other:

```
pprint.pprint(someDictionaryValue)
print(pprint.pformat(someDictionaryValue))
```

## Using Data Structures to Model Real-World Things

Even before the internet, it was possible to play a game of chess with someone on the other side of the world. Each player would set up a chessboard at their home and then take turns mailing a postcard to each other describing each move. To do this, the players needed a way to unambiguously describe the state of the board and their moves.

In *algebraic chess notation*, the spaces on the chessboard are identified by a number and letter coordinate, as in Figure 5-1.

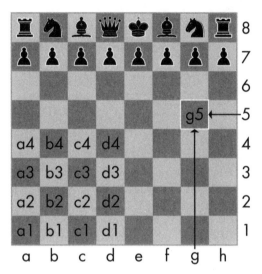

*Figure 5-1: The coordinates of a chessboard in algebraic chess notation*

The chess pieces are identified by letters: *K* for king, *Q* for queen, *R* for rook, *B* for bishop, and *N* for knight. Describing a move uses the letter of the piece and the coordinates of its destination. A pair of these moves describes what happens in a single turn (with white going first); for instance, the notation *2. Nf3 Nc6* indicates that white moved a knight to f3 and black moved a knight to c6 on the second turn of the game.

There's a bit more to algebraic notation than this, but the point is that you can unambiguously describe a game of chess without needing to be in front of a chessboard. Your opponent can even be on the other side of the world! In fact, you don't even need a physical chess set if you have a good memory: you can just read the mailed chess moves and update boards you have in your imagination.

Computers have good memories. A program on a modern computer can easily store billions of strings like `'2. Nf3 Nc6'`. This is how computers can play chess without having a physical chessboard. They model data to represent a chessboard, and you can write code to work with this model.

This is where lists and dictionaries can come in. For example, the dictionary `{'1h': 'bking', '6c': 'wqueen', '2g': 'bbishop', '5h': 'bqueen', '3e': 'wking'}` could represent the chess board in Figure 5-2.

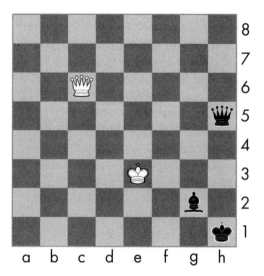

Figure 5-2: A chess board modeled by the dictionary
{'1h': 'bking', '6c': 'wqueen', '2g': 'bbishop', '5h': 'bqueen', '3e': 'wking'}

But for another example, you'll use a game that's a little simpler than chess: tic-tac-toe.

### A Tic-Tac-Toe Board

A tic-tac-toe board looks like a large hash symbol (#) with nine slots that can each contain an *X*, an *O*, or a blank. To represent the board with a dictionary, you can assign each slot a string-value key, as shown in Figure 5-3.

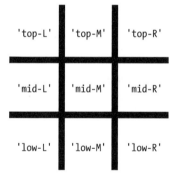

Figure 5-3: The slots of a tic-tac-toe board
with their corresponding keys

You can use string values to represent what's in each slot on the board: 'X', '0', or ' ' (a space). Thus, you'll need to store nine strings. You can use a dictionary of values for this. The string value with the key 'top-R' can represent the top-right corner, the string value with the key 'low-L' can represent the bottom-left corner, the string value with the key 'mid-M' can represent the middle, and so on.

This dictionary is a data structure that represents a tic-tac-toe board. Store this board-as-a-dictionary in a variable named theBoard. Open a new file editor window, and enter the following source code, saving it as *ticTacToe.py*:

```
theBoard = {'top-L': ' ', 'top-M': ' ', 'top-R': ' ',
            'mid-L': ' ', 'mid-M': ' ', 'mid-R': ' ',
            'low-L': ' ', 'low-M': ' ', 'low-R': ' '}
```

The data structure stored in the theBoard variable represents the tic-tac-toe board in Figure 5-4.

*Figure 5-4: An empty tic-tac-toe board*

Since the value for every key in theBoard is a single-space string, this dictionary represents a completely clear board. If player X went first and chose the middle space, you could represent that board with this dictionary:

```
theBoard = {'top-L': ' ', 'top-M': ' ', 'top-R': ' ',
            'mid-L': ' ', 'mid-M': 'X', 'mid-R': ' ',
            'low-L': ' ', 'low-M': ' ', 'low-R': ' '}
```

The data structure in theBoard now represents the tic-tac-toe board in Figure 5-5.

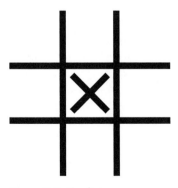

*Figure 5-5: The first move*

A board where player O has won by placing *O*s across the top might look like this:

```
theBoard = {'top-L': 'O', 'top-M': 'O', 'top-R': 'O',
            'mid-L': 'X', 'mid-M': 'X', 'mid-R': ' ',
            'low-L': ' ', 'low-M': ' ', 'low-R': 'X'}
```

The data structure in theBoard now represents the tic-tac-toe board in Figure 5-6.

Figure 5-6: Player O wins.

Of course, the player sees only what is printed to the screen, not the contents of variables. Let's create a function to print the board dictionary onto the screen. Make the following addition to *ticTacToe.py* (new code is in bold):

```
theBoard = {'top-L': ' ', 'top-M': ' ', 'top-R': ' ',
            'mid-L': ' ', 'mid-M': ' ', 'mid-R': ' ',
            'low-L': ' ', 'low-M': ' ', 'low-R': ' '}
def printBoard(board):
    print(board['top-L'] + '|' + board['top-M'] + '|' + board['top-R'])
    print('-+-+-')
    print(board['mid-L'] + '|' + board['mid-M'] + '|' + board['mid-R'])
    print('-+-+-')
    print(board['low-L'] + '|' + board['low-M'] + '|' + board['low-R'])
printBoard(theBoard)
```

You can view the execution of this program at *https://autbor.com/tictactoe1/*. When you run this program, printBoard() will print out a blank tic-tac-toe board.

```
 | |
-+-+-
 | |
-+-+-
 | |
```

The printBoard() function can handle any tic-tac-toe data structure you pass it. Try changing the code to the following:

```
theBoard = {'top-L': 'O', 'top-M': 'O', 'top-R': 'O', 'mid-L': 'X', 'mid-M':
'X', 'mid-R': ' ', 'low-L': ' ', 'low-M': ' ', 'low-R': 'X'}

def printBoard(board):
    print(board['top-L'] + '|' + board['top-M'] + '|' + board['top-R'])
    print('-+-+-')
    print(board['mid-L'] + '|' + board['mid-M'] + '|' + board['mid-R'])
    print('-+-+-')
    print(board['low-L'] + '|' + board['low-M'] + '|' + board['low-R'])
printBoard(theBoard)
```

You can view the execution of this program at *https://autbor.com/tictactoe2/*. Now when you run this program, the new board will be printed to the screen.

```
O|O|O
-+-+-
X|X|
-+-+-
 | |X
```

Because you created a data structure to represent a tic-tac-toe board and wrote code in printBoard() to interpret that data structure, you now have a program that "models" the tic-tac-toe board. You could have organized your data structure differently (for example, using keys like 'TOP-LEFT' instead of 'top-L'), but as long as the code works with your data structures, you will have a correctly working program.

For example, the printBoard() function expects the tic-tac-toe data structure to be a dictionary with keys for all nine slots. If the dictionary you passed was missing, say, the 'mid-L' key, your program would no longer work.

```
O|O|O
-+-+-
Traceback (most recent call last):
  File "ticTacToe.py", line 10, in <module>
    printBoard(theBoard)
  File "ticTacToe.py", line 6, in printBoard
    print(board['mid-L'] + '|' + board['mid-M'] + '|' + board['mid-R'])
KeyError: 'mid-L'
```

Now let's add code that allows the players to enter their moves. Modify the *ticTacToe.py* program to look like this:

```
theBoard = {'top-L': ' ', 'top-M': ' ', 'top-R': ' ', 'mid-L': ' ', 'mid-M': '
', 'mid-R': ' ', 'low-L': ' ', 'low-M': ' ', 'low-R': ' '}

def printBoard(board):
    print(board['top-L'] + '|' + board['top-M'] + '|' + board['top-R'])
    print('-+-+-')
```

```
        print(board['mid-L'] + '|' + board['mid-M'] + '|' + board['mid-R'])
        print('-+-+-')
        print(board['low-L'] + '|' + board['low-M'] + '|' + board['low-R'])
turn = 'X'
for i in range(9):
 ❶ printBoard(theBoard)
    print('Turn for ' + turn + '. Move on which space?')
 ❷ move = input()
 ❸ theBoard[move] = turn
 ❹ if turn == 'X':
        turn = 'O'
    else:
        turn = 'X'
printBoard(theBoard)
```

You can view the execution of this program at *https://autbor.com/tictactoe3/*. The new code prints out the board at the start of each new turn ❶, gets the active player's move ❷, updates the game board accordingly ❸, and then swaps the active player ❹ before moving on to the next turn.

When you run this program, it will look something like this:

```
 | |
-+-+-
 | |
-+-+-
 | |
Turn for X. Move on which space?
mid-M
 | |
-+-+-
 |X|
-+-+-
 | |

--snip--

O|O|X
-+-+-
X|X|O
-+-+-
O| |X
Turn for X. Move on which space?
low-M
O|O|X
-+-+-
X|X|O
-+-+-
O|X|X
```

This isn't a complete tic-tac-toe game—for instance, it doesn't ever check whether a player has won—but it's enough to see how data structures can be used in programs.

**NOTE** *If you are curious, the source code for a complete tic-tac-toe program is described in the resources available from* https://nostarch.com/automatestuff2/.

## Nested Dictionaries and Lists

Modeling a tic-tac-toe board was fairly simple: the board needed only a single dictionary value with nine key-value pairs. As you model more complicated things, you may find you need dictionaries and lists that contain other dictionaries and lists. Lists are useful to contain an ordered series of values, and dictionaries are useful for associating keys with values. For example, here's a program that uses a dictionary that contains other dictionaries of what items guests are bringing to a picnic. The totalBrought() function can read this data structure and calculate the total number of an item being brought by all the guests.

```python
allGuests = {'Alice': {'apples': 5, 'pretzels': 12},
             'Bob': {'ham sandwiches': 3, 'apples': 2},
             'Carol': {'cups': 3, 'apple pies': 1}}

def totalBrought(guests, item):
    numBrought = 0
❶   for k, v in guests.items():
❷       numBrought = numBrought + v.get(item, 0)
    return numBrought

print('Number of things being brought:')
print(' - Apples        ' + str(totalBrought(allGuests, 'apples')))
print(' - Cups          ' + str(totalBrought(allGuests, 'cups')))
print(' - Cakes         ' + str(totalBrought(allGuests, 'cakes')))
print(' - Ham Sandwiches ' + str(totalBrought(allGuests, 'ham sandwiches')))
print(' - Apple Pies    ' + str(totalBrought(allGuests, 'apple pies')))
```

You can view the execution of this program at *https://autbor.com /guestpicnic/*. Inside the totalBrought() function, the for loop iterates over the key-value pairs in guests ❶. Inside the loop, the string of the guest's name is assigned to k, and the dictionary of picnic items they're bringing is assigned to v. If the item parameter exists as a key in this dictionary, its value (the quantity) is added to numBrought ❷. If it does not exist as a key, the get() method returns 0 to be added to numBrought.

The output of this program looks like this:

```
Number of things being brought:
- Apples 7
- Cups 3
- Cakes 0
- Ham Sandwiches 3
- Apple Pies 1
```

This may seem like such a simple thing to model that you wouldn't need to bother with writing a program to do it. But realize that this same totalBrought() function could easily handle a dictionary that contains thousands of guests, each bringing *thousands* of different picnic items. Then having this information in a data structure along with the totalBrought() function would save you a lot of time!

You can model things with data structures in whatever way you like, as long as the rest of the code in your program can work with the data model correctly. When you first begin programming, don't worry so much about the "right" way to model data. As you gain more experience, you may come up with more efficient models, but the important thing is that the data model works for your program's needs.

## Summary

You learned all about dictionaries in this chapter. Lists and dictionaries are values that can contain multiple values, including other lists and dictionaries. Dictionaries are useful because you can map one item (the key) to another (the value), as opposed to lists, which simply contain a series of values in order. Values inside a dictionary are accessed using square brackets just as with lists. Instead of an integer index, dictionaries can have keys of a variety of data types: integers, floats, strings, or tuples. By organizing a program's values into data structures, you can create representations of real-world objects. You saw an example of this with a tic-tac-toe board.

## Practice Questions

1. What does the code for an empty dictionary look like?
2. What does a dictionary value with a key 'foo' and a value 42 look like?
3. What is the main difference between a dictionary and a list?
4. What happens if you try to access spam['foo'] if spam is {'bar': 100}?
5. If a dictionary is stored in spam, what is the difference between the expressions 'cat' in spam and 'cat' in spam.keys()?
6. If a dictionary is stored in spam, what is the difference between the expressions 'cat' in spam and 'cat' in spam.values()?
7. What is a shortcut for the following code?

```
if 'color' not in spam:
    spam['color'] = 'black'
```

8. What module and function can be used to "pretty print" dictionary values?

# Practice Projects

For practice, write programs to do the following tasks.

## Chess Dictionary Validator

In this chapter, we used the dictionary value {'1h': 'bking', '6c': 'wqueen', '2g': 'bbishop', '5h': 'bqueen', '3e': 'wking'} to represent a chess board. Write a function named isValidChessBoard() that takes a dictionary argument and returns True or False depending on if the board is valid.

A valid board will have exactly one black king and exactly one white king. Each player can only have at most 16 pieces, at most 8 pawns, and all pieces must be on a valid space from '1a' to '8h'; that is, a piece can't be on space '9z'. The piece names begin with either a 'w' or 'b' to represent white or black, followed by 'pawn', 'knight', 'bishop', 'rook', 'queen', or 'king'. This function should detect when a bug has resulted in an improper chess board.

## Fantasy Game Inventory

You are creating a fantasy video game. The data structure to model the player's inventory will be a dictionary where the keys are string values describing the item in the inventory and the value is an integer value detailing how many of that item the player has. For example, the dictionary value {'rope': 1, 'torch': 6, 'gold coin': 42, 'dagger': 1, 'arrow': 12} means the player has 1 rope, 6 torches, 42 gold coins, and so on.

Write a function named displayInventory() that would take any possible "inventory" and display it like the following:

```
Inventory:
12 arrow
42 gold coin
1 rope
6 torch
1 dagger
Total number of items: 62
```

Hint: You can use a for loop to loop through all the keys in a dictionary.

```
# inventory.py
stuff = {'rope': 1, 'torch': 6, 'gold coin': 42, 'dagger': 1, 'arrow': 12}

def displayInventory(inventory):
    print("Inventory:")
    item_total = 0
    for k, v in inventory.items():
        # FILL THIS PART IN
    print("Total number of items: " + str(item_total))

displayInventory(stuff)
```

## List to Dictionary Function for Fantasy Game Inventory

Imagine that a vanquished dragon's loot is represented as a list of strings like this:

```
dragonLoot = ['gold coin', 'dagger', 'gold coin', 'gold coin', 'ruby']
```

Write a function named addToInventory(inventory, addedItems), where the inventory parameter is a dictionary representing the player's inventory (like in the previous project) and the addedItems parameter is a list like dragonLoot. The addToInventory() function should return a dictionary that represents the updated inventory. Note that the addedItems list can contain multiples of the same item. Your code could look something like this:

```
def addToInventory(inventory, addedItems):
    # your code goes here

inv = {'gold coin': 42, 'rope': 1}
dragonLoot = ['gold coin', 'dagger', 'gold coin', 'gold coin', 'ruby']
inv = addToInventory(inv, dragonLoot)
displayInventory(inv)
```

The previous program (with your displayInventory() function from the previous project) would output the following:

```
Inventory:
45 gold coin
1 rope
1 ruby
1 dagger

Total number of items: 48
```

# 6

## MANIPULATING STRINGS

Text is one of the most common forms of data your programs will handle. You already know how to concatenate two string values together with the + operator, but you can do much more than that. You can extract partial strings from string values, add or remove spacing, convert letters to lowercase or uppercase, and check that strings are formatted correctly. You can even write Python code to access the clipboard for copying and pasting text.

In this chapter, you'll learn all this and more. Then you'll work through two different programming projects: a simple clipboard that stores multiple strings of text and a program to automate the boring chore of formatting pieces of text.

# Working with Strings

Let's look at some of the ways Python lets you write, print, and access strings in your code.

## String Literals

Typing string values in Python code is fairly straightforward: they begin and end with a single quote. But then how can you use a quote inside a string? Typing `'That is Alice's cat.'` won't work, because Python thinks the string ends after `Alice`, and the rest (`s cat.'`) is invalid Python code. Fortunately, there are multiple ways to type strings.

### Double Quotes

Strings can begin and end with double quotes, just as they do with single quotes. One benefit of using double quotes is that the string can have a single quote character in it. Enter the following into the interactive shell:

```
>>> spam = "That is Alice's cat."
```

Since the string begins with a double quote, Python knows that the single quote is part of the string and not marking the end of the string. However, if you need to use both single quotes and double quotes in the string, you'll need to use escape characters.

### Escape Characters

An *escape character* lets you use characters that are otherwise impossible to put into a string. An escape character consists of a backslash (\) followed by the character you want to add to the string. (Despite consisting of two characters, it is commonly referred to as a singular escape character.) For example, the escape character for a single quote is \'. You can use this inside a string that begins and ends with single quotes. To see how escape characters work, enter the following into the interactive shell:

```
>>> spam = 'Say hi to Bob\'s mother.'
```

Python knows that since the single quote in `Bob\'s` has a backslash, it is not a single quote meant to end the string value. The escape characters \' and \" let you put single quotes and double quotes inside your strings, respectively.

Table 6-1 lists the escape characters you can use.

**Table 6-1:** Escape Characters

| Escape character | Prints as |
|---|---|
| \' | Single quote |
| \" | Double quote |
| \t | Tab |
| \n | Newline (line break) |
| \\ | Backslash |

Enter the following into the interactive shell:

```
>>> print("Hello there!\nHow are you?\nI\'m doing fine.")
Hello there!
How are you?
I'm doing fine.
```

## Raw Strings

You can place an r before the beginning quotation mark of a string to make it a raw string. A *raw string* completely ignores all escape characters and prints any backslash that appears in the string. For example, enter the following into the interactive shell:

```
>>> print(r'That is Carol\'s cat.')
That is Carol\'s cat.
```

Because this is a raw string, Python considers the backslash as part of the string and not as the start of an escape character. Raw strings are helpful if you are typing string values that contain many backslashes, such as the strings used for Windows file paths like r'C:\Users\Al\Desktop' or regular expressions described in the next chapter.

## Multiline Strings with Triple Quotes

While you can use the \n escape character to put a newline into a string, it is often easier to use multiline strings. A multiline string in Python begins and ends with either three single quotes or three double quotes. Any quotes, tabs, or newlines in between the "triple quotes" are considered part of the string. Python's indentation rules for blocks do not apply to lines inside a multiline string.

Open the file editor and write the following:

```
print('''Dear Alice,

Eve's cat has been arrested for catnapping, cat burglary, and extortion.

Sincerely,
Bob''')
```

Save this program as *catnapping.py* and run it. The output will look like this:

```
Dear Alice,

Eve's cat has been arrested for catnapping, cat burglary, and extortion.

Sincerely,
Bob
```

Notice that the single quote character in `Eve's` does not need to be escaped. Escaping single and double quotes is optional in multiline strings. The following print() call would print identical text but doesn't use a multi-line string:

```
print('Dear Alice,\n\nEve\'s cat has been arrested for catnapping, cat
burglary, and extortion.\n\nSincerely,\nBob')
```

### Multiline Comments

While the hash character (#) marks the beginning of a comment for the rest of the line, a multiline string is often used for comments that span multiple lines. The following is perfectly valid Python code:

```
"""This is a test Python program.
Written by Al Sweigart al@inventwithpython.com

This program was designed for Python 3, not Python 2.
"""

def spam():
    """This is a multiline comment to help
    explain what the spam() function does."""
    print('Hello!')
```

## *Indexing and Slicing Strings*

Strings use indexes and slices the same way lists do. You can think of the string `'Hello, world!'` as a list and each character in the string as an item with a corresponding index.

```
'   H   e   l   l   o   ,       w   o   r   l   d   !   '
    0   1   2   3   4   5   6   7   8   9   10  11  12
```

The space and exclamation point are included in the character count, so `'Hello, world!'` is 13 characters long, from H at index 0 to ! at index 12.

Enter the following into the interactive shell:

```
>>> spam = 'Hello, world!'
>>> spam[0]
'H'
>>> spam[4]
'o'
>>> spam[-1]
'!'
>>> spam[0:5]
'Hello'
>>> spam[:5]
'Hello'
>>> spam[7:]
'world!'
```

If you specify an index, you'll get the character at that position in the string. If you specify a range from one index to another, the starting index is included and the ending index is not. That's why, if spam is 'Hello, world!', spam[0:5] is 'Hello'. The substring you get from spam[0:5] will include everything from spam[0] to spam[4], leaving out the comma at index 5 and the space at index 6. This is similar to how range(5) will cause a for loop to iterate up to, but not including, 5.

Note that slicing a string does not modify the original string. You can capture a slice from one variable in a separate variable. Try entering the following into the interactive shell:

```
>>> spam = 'Hello, world!'
>>> fizz = spam[0:5]
>>> fizz
'Hello'
```

By slicing and storing the resulting substring in another variable, you can have both the whole string and the substring handy for quick, easy access.

## The in and not in Operators with Strings

The in and not in operators can be used with strings just like with list values. An expression with two strings joined using in or not in will evaluate to a Boolean True or False. Enter the following into the interactive shell:

```
>>> 'Hello' in 'Hello, World'
True
>>> 'Hello' in 'Hello'
True
>>> 'HELLO' in 'Hello, World'
False
>>> '' in 'spam'
True
>>> 'cats' not in 'cats and dogs'
False
```

These expressions test whether the first string (the exact string, case-sensitive) can be found within the second string.

## Putting Strings Inside Other Strings

Putting strings inside other strings is a common operation in programming. So far, we've been using the + operator and string concatenation to do this:

```
>>> name = 'Al'
>>> age = 4000
>>> 'Hello, my name is ' + name + '. I am ' + str(age) + ' years old.'
'Hello, my name is Al. I am 4000 years old.'
```

However, this requires a lot of tedious typing. A simpler approach is to use *string interpolation*, in which the %s operator inside the string acts as a marker to be replaced by values following the string. One benefit of string interpolation is that str() doesn't have to be called to convert values to strings. Enter the following into the interactive shell:

```
>>> name = 'Al'
>>> age = 4000
>>> 'My name is %s. I am %s years old.' % (name, age)
'My name is Al. I am 4000 years old.'
```

Python 3.6 introduced *f-strings*, which is similar to string interpolation except that braces are used instead of %s, with the expressions placed directly inside the braces. Like raw strings, f-strings have an f prefix before the starting quotation mark. Enter the following into the interactive shell:

```
>>> name = 'Al'
>>> age = 4000
>>> f'My name is {name}. Next year I will be {age + 1}.'
'My name is Al. Next year I will be 4001.'
```

Remember to include the f prefix; otherwise, the braces and their contents will be a part of the string value:

```
>>> 'My name is {name}. Next year I will be {age + 1}.'
'My name is {name}. Next year I will be {age + 1}.'
```

## Useful String Methods

Several string methods analyze strings or create transformed string values. This section describes the methods you'll be using most often.

## The upper(), lower(), isupper(), and islower() Methods

The upper() and lower() string methods return a new string where all the letters in the original string have been converted to uppercase or lowercase, respectively. Nonletter characters in the string remain unchanged. Enter the following into the interactive shell:

```
>>> spam = 'Hello, world!'
>>> spam = spam.upper()
>>> spam
'HELLO, WORLD!'
>>> spam = spam.lower()
>>> spam
'hello, world!'
```

Note that these methods do not change the string itself but return new string values. If you want to change the original string, you have to call upper() or lower() on the string and then assign the new string to the variable where the original was stored. This is why you must use spam = spam.upper() to change the string in spam instead of simply spam.upper(). (This is just like if a variable eggs contains the value 10. Writing eggs + 3 does not change the value of eggs, but eggs = eggs + 3 does.)

The upper() and lower() methods are helpful if you need to make a case-insensitive comparison. For example, the strings 'great' and 'GREat' are not equal to each other. But in the following small program, it does not matter whether the user types Great, GREAT, or grEAT, because the string is first converted to lowercase.

```
print('How are you?')
feeling = input()
if feeling.lower() == 'great':
    print('I feel great too.')
else:
    print('I hope the rest of your day is good.')
```

When you run this program, the question is displayed, and entering a variation on great, such as GREat, will still give the output I feel great too. Adding code to your program to handle variations or mistakes in user input, such as inconsistent capitalization, will make your programs easier to use and less likely to fail.

```
How are you?
GREat
I feel great too.
```

You can view the execution of this program at *https://autbor.com /convertlowercase/*. The isupper() and islower() methods will return a Boolean True value if the string has at least one letter and all the letters are uppercase or lowercase, respectively. Otherwise, the method returns False. Enter the following into the interactive shell, and notice what each method call returns:

```
>>> spam = 'Hello, world!'
>>> spam.islower()
False
>>> spam.isupper()
False
>>> 'HELLO'.isupper()
True
>>> 'abc12345'.islower()
True
>>> '12345'.islower()
False
>>> '12345'.isupper()
False
```

Since the upper() and lower() string methods themselves return strings, you can call string methods on *those* returned string values as well. Expressions that do this will look like a chain of method calls. Enter the following into the interactive shell:

```
>>> 'Hello'.upper()
'HELLO'
>>> 'Hello'.upper().lower()
'hello'
>>> 'Hello'.upper().lower().upper()
'HELLO'
>>> 'HELLO'.lower()
'hello'
>>> 'HELLO'.lower().islower()
True
```

## The isX() Methods

Along with islower() and isupper(), there are several other string methods that have names beginning with the word *is*. These methods return a Boolean value that describes the nature of the string. Here are some common isX string methods:

**isalpha()**   Returns True if the string consists only of letters and isn't blank

**isalnum()**   Returns True if the string consists only of letters and numbers and is not blank

**isdecimal()**   Returns True if the string consists only of numeric characters and is not blank

**isspace()**   Returns True if the string consists only of spaces, tabs, and newlines and is not blank

**istitle()**   Returns True if the string consists only of words that begin with an uppercase letter followed by only lowercase letters

Enter the following into the interactive shell:

```
>>> 'hello'.isalpha()
True
>>> 'hello123'.isalpha()
False
>>> 'hello123'.isalnum()
True
>>> 'hello'.isalnum()
True
>>> '123'.isdecimal()
True
>>> '    '.isspace()
True
>>> 'This Is Title Case'.istitle()
True
>>> 'This Is Title Case 123'.istitle()
True
>>> 'This Is not Title Case'.istitle()
False
>>> 'This Is NOT Title Case Either'.istitle()
False
```

The isX() string methods are helpful when you need to validate user input. For example, the following program repeatedly asks users for their age and a password until they provide valid input. Open a new file editor window and enter this program, saving it as *validateInput.py*:

```
while True:
    print('Enter your age:')
    age = input()
    if age.isdecimal():
        break
    print('Please enter a number for your age.')

while True:
    print('Select a new password (letters and numbers only):')
    password = input()
    if password.isalnum():
        break
    print('Passwords can only have letters and numbers.')
```

In the first while loop, we ask the user for their age and store their input in age. If age is a valid (decimal) value, we break out of this first while loop and move on to the second, which asks for a password. Otherwise, we inform the user that they need to enter a number and again ask them to enter their age. In the second while loop, we ask for a password, store the user's input in password, and break out of the loop if the input was alphanumeric. If it wasn't, we're not satisfied, so we tell the user the password needs to be alphanumeric and again ask them to enter a password.

When run, the program's output looks like this:

```
Enter your age:
forty two
Please enter a number for your age.
Enter your age:
42
Select a new password (letters and numbers only):
secr3t!
Passwords can only have letters and numbers.
Select a new password (letters and numbers only):
secr3t
```

You can view the execution of this program at *https://autbor.com /validateinput/*. Calling isdecimal() and isalnum() on variables, we're able to test whether the values stored in those variables are decimal or not, alphanumeric or not. Here, these tests help us reject the input forty two but accept 42, and reject secr3t! but accept secr3t.

## The startswith() and endswith() Methods

The startswith() and endswith() methods return True if the string value they are called on begins or ends (respectively) with the string passed to the method; otherwise, they return False. Enter the following into the interactive shell:

```
>>> 'Hello, world!'.startswith('Hello')
True
>>> 'Hello, world!'.endswith('world!')
True
>>> 'abc123'.startswith('abcdef')
False
>>> 'abc123'.endswith('12')
False
>>> 'Hello, world!'.startswith('Hello, world!')
True
>>> 'Hello, world!'.endswith('Hello, world!')
True
```

These methods are useful alternatives to the == equals operator if you need to check only whether the first or last part of the string, rather than the whole thing, is equal to another string.

## The join() and split() Methods

The join() method is useful when you have a list of strings that need to be joined together into a single string value. The join() method is called on a string, gets passed a list of strings, and returns a string. The returned string is the concatenation of each string in the passed-in list. For example, enter the following into the interactive shell:

```
>>> ', '.join(['cats', 'rats', 'bats'])
'cats, rats, bats'
>>> ' '.join(['My', 'name', 'is', 'Simon'])
'My name is Simon'
>>> 'ABC'.join(['My', 'name', 'is', 'Simon'])
'MyABCnameABCisABCSimon'
```

Notice that the string join() calls on is inserted between each string of the list argument. For example, when join(['cats', 'rats', 'bats']) is called on the ', ' string, the returned string is 'cats, rats, bats'.

Remember that join() is called on a string value and is passed a list value. (It's easy to accidentally call it the other way around.) The split() method does the opposite: It's called on a string value and returns a list of strings. Enter the following into the interactive shell:

```
>>> 'My name is Simon'.split()
['My', 'name', 'is', 'Simon']
```

By default, the string 'My name is Simon' is split wherever whitespace characters such as the space, tab, or newline characters are found. These whitespace characters are not included in the strings in the returned list. You can pass a delimiter string to the split() method to specify a different string to split upon. For example, enter the following into the interactive shell:

```
>>> 'MyABCnameABCisABCSimon'.split('ABC')
['My', 'name', 'is', 'Simon']
>>> 'My name is Simon'.split('m')
['My na', 'e is Si', 'on']
```

A common use of split() is to split a multiline string along the newline characters. Enter the following into the interactive shell:

```
>>> spam = '''Dear Alice,
How have you been? I am fine.
There is a container in the fridge
that is labeled "Milk Experiment."

Please do not drink it.
Sincerely,
Bob'''
>>> spam.split('\n')
['Dear Alice,', 'How have you been? I am fine.', 'There is a container in the
fridge', 'that is labeled "Milk Experiment."', '', 'Please do not drink it.',
'Sincerely,', 'Bob']
```

Passing split() the argument '\n' lets us split the multiline string stored in spam along the newlines and return a list in which each item corresponds to one line of the string.

### Splitting Strings with the partition() Method

The partition() string method can split a string into the text before and after a separator string. This method searches the string it is called on for the separator string it is passed, and returns a tuple of three substrings for the "before," "separator," and "after" substrings. Enter the following into the interactive shell:

```
>>> 'Hello, world!'.partition('w')
('Hello, ', 'w', 'orld!')
>>> 'Hello, world!'.partition('world')
('Hello, ', 'world', '!')
```

If the separator string you pass to partition() occurs multiple times in the string that partition() calls on, the method splits the string only on the first occurrence:

```
>>> 'Hello, world!'.partition('o')
('Hell', 'o', ', world!')
```

If the separator string can't be found, the first string returned in the tuple will be the entire string, and the other two strings will be empty:

```
>>> 'Hello, world!'.partition('XYZ')
('Hello, world!', '', '')
```

You can use the multiple assignment trick to assign the three returned strings to three variables:

```
>>> before, sep, after = 'Hello, world!'.partition(' ')
>>> before
'Hello,'
>>> after
'world!'
```

The partition() method is useful for splitting a string whenever you need the parts before, including, and after a particular separator string.

### Justifying Text with the rjust(), ljust(), and center() Methods

The rjust() and ljust() string methods return a padded version of the string they are called on, with spaces inserted to justify the text. The first argument to both methods is an integer length for the justified string. Enter the following into the interactive shell:

```
>>> 'Hello'.rjust(10)
'     Hello'
>>> 'Hello'.rjust(20)
'               Hello'
>>> 'Hello, World'.rjust(20)
```

```
'        Hello, World'
>>> 'Hello'.ljust(10)
'Hello     '
```

'Hello'.rjust(10) says that we want to right-justify 'Hello' in a string of total length 10. 'Hello' is five characters, so five spaces will be added to its left, giving us a string of 10 characters with 'Hello' justified right.

An optional second argument to rjust() and ljust() will specify a fill character other than a space character. Enter the following into the interactive shell:

```
>>> 'Hello'.rjust(20, '*')
'***************Hello'
>>> 'Hello'.ljust(20, '-')
'Hello---------------'
```

The center() string method works like ljust() and rjust() but centers the text rather than justifying it to the left or right. Enter the following into the interactive shell:

```
>>> 'Hello'.center(20)
'       Hello        '
>>> 'Hello'.center(20, '=')
'=======Hello========'
```

These methods are especially useful when you need to print tabular data that has correct spacing. Open a new file editor window and enter the following code, saving it as *picnicTable.py*:

```
def printPicnic(itemsDict, leftWidth, rightWidth):
    print('PICNIC ITEMS'.center(leftWidth + rightWidth, '-'))
    for k, v in itemsDict.items():
        print(k.ljust(leftWidth, '.') + str(v).rjust(rightWidth))

picnicItems = {'sandwiches': 4, 'apples': 12, 'cups': 4, 'cookies': 8000}
printPicnic(picnicItems, 12, 5)
printPicnic(picnicItems, 20, 6)
```

You can view the execution of this program at *https://autbor.com /picnictable/*. In this program, we define a printPicnic() method that will take in a dictionary of information and use center(), ljust(), and rjust() to display that information in a neatly aligned table-like format.

The dictionary that we'll pass to printPicnic() is picnicItems. In picnicItems, we have 4 sandwiches, 12 apples, 4 cups, and 8,000 cookies. We want to organize this information into two columns, with the name of the item on the left and the quantity on the right.

To do this, we decide how wide we want the left and right columns to be. Along with our dictionary, we'll pass these values to printPicnic().

The `printPicnic()` function takes in a dictionary, a `leftWidth` for the left column of a table, and a `rightWidth` for the right column. It prints a title, PICNIC ITEMS, centered above the table. Then, it loops through the dictionary, printing each key-value pair on a line with the key justified left and padded by periods, and the value justified right and padded by spaces.

After defining `printPicnic()`, we define the dictionary `picnicItems` and call `printPicnic()` twice, passing it different widths for the left and right table columns.

When you run this program, the picnic items are displayed twice. The first time the left column is 12 characters wide, and the right column is 5 characters wide. The second time they are 20 and 6 characters wide, respectively.

```
---PICNIC ITEMS--
sandwiches..    4
apples......   12
cups........    4
cookies..... 8000
-------PICNIC ITEMS-------
sandwiches..........    4
apples..............   12
cups................    4
cookies.............  8000
```

Using `rjust()`, `ljust()`, and `center()` lets you ensure that strings are neatly aligned, even if you aren't sure how many characters long your strings are.

### Removing Whitespace with the strip(), rstrip(), and lstrip() Methods

Sometimes you may want to strip off whitespace characters (space, tab, and newline) from the left side, right side, or both sides of a string. The `strip()` string method will return a new string without any whitespace characters at the beginning or end. The `lstrip()` and `rstrip()` methods will remove whitespace characters from the left and right ends, respectively. Enter the following into the interactive shell:

```
>>> spam = '    Hello, World    '
>>> spam.strip()
'Hello, World'
>>> spam.lstrip()
'Hello, World    '
>>> spam.rstrip()
'    Hello, World'
```

Optionally, a string argument will specify which characters on the ends should be stripped. Enter the following into the interactive shell:

```
>>> spam = 'SpamSpamBaconSpamEggsSpamSpam'
>>> spam.strip('ampS')
'BaconSpamEggs'
```

Passing strip() the argument 'ampS' will tell it to strip occurrences of a, m, p, and capital S from the ends of the string stored in spam. The order of the characters in the string passed to strip() does not matter: strip('ampS') will do the same thing as strip('mapS') or strip('Spam').

## Numeric Values of Characters with the ord() and chr() Functions

Computers store information as bytes—strings of binary numbers, which means we need to be able to convert text to numbers. Because of this, every text character has a corresponding numeric value called a *Unicode code point*. For example, the numeric code point is 65 for 'A', 52 for '4', and 33 for '!'. You can use the ord() function to get the code point of a one-character string, and the chr() function to get the one-character string of an integer code point. Enter the following into the interactive shell:

```
>>> ord('A')
65
>>> ord('4')
52
>>> ord('!')
33
>>> chr(65)
'A'
```

These functions are useful when you need to do an ordering or mathematical operation on characters:

```
>>> ord('B')
66
>>> ord('A') < ord('B')
True
>>> chr(ord('A'))
'A'
>>> chr(ord('A') + 1)
'B'
```

There is more to Unicode and code points, but those details are beyond the scope of this book. If you'd like to know more, I recommend watching Ned Batchelder's 2012 PyCon talk, "Pragmatic Unicode, or, How Do I Stop the Pain?" at *https://youtu.be/sgHbC6udIqc*.

## Copying and Pasting Strings with the pyperclip Module

The pyperclip module has copy() and paste() functions that can send text to and receive text from your computer's clipboard. Sending the output of your program to the clipboard will make it easy to paste it into an email, word processor, or some other software.

The pyperclip module does not come with Python. To install it, follow the directions for installing third-party modules in Appendix A. After installing pyperclip, enter the following into the interactive shell:

```
>>> import pyperclip
>>> pyperclip.copy('Hello, world!')
>>> pyperclip.paste()
'Hello, world!'
```

Of course, if something outside of your program changes the clipboard contents, the paste() function will return it. For example, if I copied this sentence to the clipboard and then called paste(), it would look like this:

```
>>> pyperclip.paste()
'For example, if I copied this sentence to the clipboard and then called
paste(), it would look like this:'
```

## Project: Multi-Clipboard Automatic Messages

If you've responded to a large number of emails with similar phrasing, you've probably had to do a lot of repetitive typing. Maybe you keep a text document with these phrases so you can easily copy and paste them using the clipboard. But your clipboard can only store one message at a time, which isn't very convenient. Let's make this process a bit easier with a program that stores multiple phrases.

### Step 1: Program Design and Data Structures

You want to be able to run this program with a command line argument that is a short key phrase—for instance, *agree* or *busy*. The message associated with that key phrase will be copied to the clipboard so that the user can paste it into an email. This way, the user can have long, detailed messages without having to retype them.

Open a new file editor window and save the program as *mclip.py*. You need to start the program with a #! (*shebang*) line (see Appendix B) and should also write a comment that briefly describes the program. Since you want to associate each piece of text with its key phrase, you can store these as strings in a dictionary. The dictionary will be the data structure that organizes your key phrases and text. Make your program look like the following:

```
#! python3
# mclip.py - A multi-clipboard program.

TEXT = {'agree': """Yes, I agree. That sounds fine to me.""",
        'busy': """Sorry, can we do this later this week or next week?""",
        'upsell': """Would you consider making this a monthly donation?"""}
```

## Step 2: Handle Command Line Arguments

The command line arguments will be stored in the variable sys.argv. (See Appendix B for more information on how to use command line arguments in your programs.) The first item in the sys.argv list should always be a string containing the program's filename ('mclip.py'), and the second item should be the first command line argument. For this program, this argument is the key phrase of the message you want. Since the command line argument is mandatory, you display a usage message to the user if they forget to add it (that is, if the sys.argv list has fewer than two values in it). Make your program look like the following:

```
#! python3
# mclip.py - A multi-clipboard program.

TEXT = {'agree': """Yes, I agree. That sounds fine to me.""",
        'busy': """Sorry, can we do this later this week or next week?""",
        'upsell': """Would you consider making this a monthly donation?"""}

import sys
if len(sys.argv) < 2:
    print('Usage: python mclip.py [keyphrase] - copy phrase text')
    sys.exit()

keyphrase = sys.argv[1]    # first command line arg is the keyphrase
```

### Step 3: Copy the Right Phrase

Now that the key phrase is stored as a string in the variable keyphrase, you need to see whether it exists in the TEXT dictionary as a key. If so, you want to copy the key's value to the clipboard using pyperclip.copy(). (Since you're using the pyperclip module, you need to import it.) Note that you don't actually *need* the keyphrase variable; you could just use sys.argv[1] everywhere keyphrase is used in this program. But a variable named keyphrase is much more readable than something cryptic like sys.argv[1].

Make your program look like the following:

```python
#! python3
# mclip.py - A multi-clipboard program.

TEXT = {'agree': """Yes, I agree. That sounds fine to me.""",
        'busy': """Sorry, can we do this later this week or next week?""",
        'upsell': """Would you consider making this a monthly donation?"""}

import sys, pyperclip
if len(sys.argv) < 2:
    print('Usage: py mclip.py [keyphrase] - copy phrase text')
    sys.exit()

keyphrase = sys.argv[1]    # first command line arg is the keyphrase

if keyphrase in TEXT:
    pyperclip.copy(TEXT[keyphrase])
    print('Text for ' + keyphrase + ' copied to clipboard.')
else:
    print('There is no text for ' + keyphrase)
```

This new code looks in the TEXT dictionary for the key phrase. If the key phrase is a key in the dictionary, we get the value corresponding to that key, copy it to the clipboard, and print a message saying that we copied the value. Otherwise, we print a message saying there's no key phrase with that name.

That's the complete script. Using the instructions in Appendix B for launching command line programs easily, you now have a fast way to copy messages to the clipboard. You will have to modify the TEXT dictionary value in the source whenever you want to update the program with a new message.

On Windows, you can create a batch file to run this program with the WIN-R Run window. (For more about batch files, see Appendix B.) Enter the following into the file editor and save the file as *mclip.bat* in the *C:\Windows* folder:

```
@py.exe C:\path_to_file\mclip.py %*
@pause
```

With this batch file created, running the multi-clipboard program on Windows is just a matter of pressing WIN-R and typing mclip *key phrase*.

# Project: Adding Bullets to Wiki Markup

When editing a Wikipedia article, you can create a bulleted list by putting each list item on its own line and placing a star in front. But say you have a really large list that you want to add bullet points to. You could just type those stars at the beginning of each line, one by one. Or you could automate this task with a short Python script.

The *bulletPointAdder.py* script will get the text from the clipboard, add a star and space to the beginning of each line, and then paste this new text to the clipboard. For example, if I copied the following text (for the Wikipedia article "List of Lists of Lists") to the clipboard:

```
Lists of animals
Lists of aquarium life
Lists of biologists by author abbreviation
Lists of cultivars
```

and then ran the *bulletPointAdder.py* program, the clipboard would then contain the following:

```
* Lists of animals
* Lists of aquarium life
* Lists of biologists by author abbreviation
* Lists of cultivars
```

This star-prefixed text is ready to be pasted into a Wikipedia article as a bulleted list.

## Step 1: Copy and Paste from the Clipboard

You want the *bulletPointAdder.py* program to do the following:

1. Paste text from the clipboard.
2. Do something to it.
3. Copy the new text to the clipboard.

That second step is a little tricky, but steps 1 and 3 are pretty straightforward: they just involve the pyperclip.copy() and pyperclip.paste() functions. For now, let's just write the part of the program that covers steps 1 and 3. Enter the following, saving the program as *bulletPointAdder.py*:

```
#! python3
# bulletPointAdder.py - Adds Wikipedia bullet points to the start
# of each line of text on the clipboard.

import pyperclip
text = pyperclip.paste()

# TODO: Separate lines and add stars.

pyperclip.copy(text)
```

The TODO comment is a reminder that you should complete this part of the program eventually. The next step is to actually implement that piece of the program.

## Step 2: Separate the Lines of Text and Add the Star

The call to pyperclip.paste() returns all the text on the clipboard as one big string. If we used the "List of Lists of Lists" example, the string stored in text would look like this:

```
'Lists of animals\nLists of aquarium life\nLists of biologists by author
abbreviation\nLists of cultivars'
```

The \n newline characters in this string cause it to be displayed with multiple lines when it is printed or pasted from the clipboard. There are many "lines" in this one string value. You want to add a star to the start of each of these lines.

You could write code that searches for each \n newline character in the string and then adds the star just after that. But it would be easier to use the split() method to return a list of strings, one for each line in the original string, and then add the star to the front of each string in the list.

Make your program look like the following:

```python
#! python3
# bulletPointAdder.py - Adds Wikipedia bullet points to the start
# of each line of text on the clipboard.

import pyperclip
text = pyperclip.paste()

# Separate lines and add stars.
lines = text.split('\n')
for i in range(len(lines)):    # loop through all indexes in the "lines" list
    lines[i] = '* ' + lines[i] # add star to each string in "lines" list

pyperclip.copy(text)
```

We split the text along its newlines to get a list in which each item is one line of the text. We store the list in lines and then loop through the items in lines. For each line, we add a star and a space to the start of the line. Now each string in lines begins with a star.

## Step 3: Join the Modified Lines

The lines list now contains modified lines that start with stars. But pyperclip .copy() is expecting a single string value, however, not a list of string values. To make this single string value, pass lines into the join() method to get a single string joined from the list's strings. Make your program look like the following:

```python
#! python3
# bulletPointAdder.py - Adds Wikipedia bullet points to the start
```

```
# of each line of text on the clipboard.

import pyperclip
text = pyperclip.paste()

# Separate lines and add stars.
lines = text.split('\n')
for i in range(len(lines)):    # loop through all indexes for "lines" list
    lines[i] = '* ' + lines[i] # add star to each string in "lines" list
text = '\n'.join(lines)
pyperclip.copy(text)
```

When this program is run, it replaces the text on the clipboard with text that has stars at the start of each line. Now the program is complete, and you can try running it with text copied to the clipboard.

Even if you don't need to automate this specific task, you might want to automate some other kind of text manipulation, such as removing trailing spaces from the end of lines or converting text to uppercase or lowercase. Whatever your needs, you can use the clipboard for input and output.

## A Short Progam: Pig Latin

Pig Latin is a silly made-up language that alters English words. If a word begins with a vowel, the word *yay* is added to the end of it. If a word begins with a consonant or consonant cluster (like *ch* or *gr*), that consonant or cluster is moved to the end of the word followed by *ay*.

Let's write a Pig Latin program that will output something like this:

```
Enter the English message to translate into Pig Latin:
My name is AL SWEIGART and I am 4,000 years old.
Ymay amenay isyay ALYAY EIGARTSWAY andyay Iyay amyay 4,000 yearsyay oldyay.
```

This program works by altering a string using the methods introduced in this chapter. Type the following source code into the file editor, and save the file as *pigLat.py*:

```
# English to Pig Latin
print('Enter the English message to translate into Pig Latin:')
message = input()

VOWELS = ('a', 'e', 'i', 'o', 'u', 'y')

pigLatin = [] # A list of the words in Pig Latin.
for word in message.split():
    # Separate the non-letters at the start of this word:
    prefixNonLetters = ''
    while len(word) > 0 and not word[0].isalpha():
        prefixNonLetters += word[0]
        word = word[1:]
```

```
    if len(word) == 0:
        pigLatin.append(prefixNonLetters)
        continue

    # Separate the non-letters at the end of this word:
    suffixNonLetters = ''
    while not word[-1].isalpha():
        suffixNonLetters += word[-1]
        word = word[:-1]

    # Remember if the word was in uppercase or title case.
    wasUpper = word.isupper()
    wasTitle = word.istitle()

    word = word.lower() # Make the word lowercase for translation.

    # Separate the consonants at the start of this word:
    prefixConsonants = ''
    while len(word) > 0 and not word[0] in VOWELS:
        prefixConsonants += word[0]
        word = word[1:]

    # Add the Pig Latin ending to the word:
    if prefixConsonants != '':
        word += prefixConsonants + 'ay'
    else:
        word += 'yay'

    # Set the word back to uppercase or title case:
    if wasUpper:
        word = word.upper()
    if wasTitle:
        word = word.title()

    # Add the non-letters back to the start or end of the word.
    pigLatin.append(prefixNonLetters + word + suffixNonLetters)

# Join all the words back together into a single string:
print(' '.join(pigLatin))
```

Let's look at this code line by line, starting at the top:

```
# English to Pig Latin
print('Enter the English message to translate into Pig Latin:')
message = input()

VOWELS = ('a', 'e', 'i', 'o', 'u', 'y')
```

First, we ask the user to enter the English text to translate into Pig Latin. Also, we create a constant that holds every lowercase vowel letter (and *y*) as a tuple of strings. This will be used later in our program.

Next, we're going to create the pigLatin variable to store the words as we translate them into Pig Latin:

```
pigLatin = []  # A list of the words in Pig Latin.
for word in message.split():
    # Separate the non-letters at the start of this word:
    prefixNonLetters = ''
    while len(word) > 0 and not word[0].isalpha():
        prefixNonLetters += word[0]
        word = word[1:]
    if len(word) == 0:
        pigLatin.append(prefixNonLetters)
        continue
```

We need each word to be its own string, so we call message.split() to get a list of the words as separate strings. The string 'My name is AL SWEIGART and I am 4,000 years old.' would cause split() to return ['My', 'name', 'is', 'AL', 'SWEIGART', 'and', 'I', 'am', '4,000', 'years', 'old.'].

We need to remove any non-letters from the start and end of each word so that strings like 'old.' translate to 'oldyay.' instead of 'old.yay'. We'll save these non-letters to a variable named prefixNonLetters.

```
    # Separate the non-letters at the end of this word:
    suffixNonLetters = ''
    while not word[-1].isalpha():
        suffixNonLetters += word[-1]
        word = word[:-1]
```

A loop that calls isalpha() on the first character in the word will determine if we should remove a character from a word and concatenate it to the end of prefixNonLetters. If the entire word is made of non-letter characters, like '4,000', we can simply append it to the pigLatin list and continue to the next word to translate. We also need to save the non-letters at the end of the word string. This code is similar to the previous loop.

Next, we'll make sure the program remembers if the word was in uppercase or title case so we can restore it after translating the word to Pig Latin:

```
    # Remember if the word was in uppercase or title case.
    wasUpper = word.isupper()
    wasTitle = word.istitle()

    word = word.lower()  # Make the word lowercase for translation.
```

For the rest of the code in the for loop, we'll work on a lowercase version of word.

To convert a word like *sweigart* to *eigart-sway*, we need to remove all of the consonants from the beginning of word:

```
# Separate the consonants at the start of this word:
prefixConsonants = ''
while len(word) > 0 and not word[0] in VOWELS:
    prefixConsonants += word[0]
    word = word[1:]
```

We use a loop similar to the loop that removed the non-letters from the start of word, except now we are pulling off consonants and storing them to a variable named prefixConsonants.

If there were any consonants at the start of the word, they are now in prefixConsonants and we should concatenate that variable and the string 'ay' to the end of word. Otherwise, we can assume word begins with a vowel and we only need to concatenate 'yay':

```
# Add the Pig Latin ending to the word:
if prefixConsonants != '':
    word += prefixConsonants + 'ay'
else:
    word += 'yay'
```

Recall that we set word to its lowercase version with word = word.lower(). If word was originally in uppercase or title case, this code will convert word back to its original case:

```
# Set the word back to uppercase or title case:
if wasUpper:
    word = word.upper()
if wasTitle:
    word = word.title()
```

At the end of the for loop, we append the word, along with any non-letter prefix or suffix it originally had, to the pigLatin list:

```
# Add the non-letters back to the start or end of the word.
pigLatin.append(prefixNonLetters + word + suffixNonLetters)

# Join all the words back together into a single string:
print(' '.join(pigLatin))
```

After this loop finishes, we combine the list of strings into a single string by calling the join() method. This single string is passed to print() to display our Pig Latin on the screen.

You can find other short, text-based Python programs like this one at *https://github.com/asweigart/pythonstdiogames/*.

# Summary

Text is a common form of data, and Python comes with many helpful string methods to process the text stored in string values. You will make use of indexing, slicing, and string methods in almost every Python program you write.

The programs you are writing now don't seem too sophisticated— they don't have graphical user interfaces with images and colorful text. So far, you're displaying text with print() and letting the user enter text with input(). However, the user can quickly enter large amounts of text through the clipboard. This ability provides a useful avenue for writing programs that manipulate massive amounts of text. These text-based programs might not have flashy windows or graphics, but they can get a lot of useful work done quickly.

Another way to manipulate large amounts of text is reading and writing files directly off the hard drive. You'll learn how to do this with Python in Chapter 9.

That just about covers all the basic concepts of Python programming! You'll continue to learn new concepts throughout the rest of this book, but you now know enough to start writing some useful programs that can automate tasks. If you'd like to see a collection of short, simple Python programs built from the basic concepts you've learned so far, check out *https://github.com/asweigart/pythonstdiogames/*. Try copying the source code for each program by hand, and then make modifications to see how they affect the behavior of the program. Once you have an understanding of how the program works, try re-creating the program yourself from scratch. You don't need to re-create the source code exactly; just focus on what the program does rather than how it does it.

You might not think you have enough Python knowledge to do things such as download web pages, update spreadsheets, or send text messages, but that's where Python modules come in! These modules, written by other programmers, provide functions that make it easy for you to do all these things. So let's learn how to write real programs to do useful automated tasks.

## Practice Questions

1.  What are escape characters?

2.  What do the \n and \t escape characters represent?

3.  How can you put a \ backslash character in a string?

4.  The string value "Howl's Moving Castle" is a valid string. Why isn't it a problem that the single quote character in the word Howl's isn't escaped?

5.  If you don't want to put \n in your string, how can you write a string with newlines in it?

6. What do the following expressions evaluate to?
   - `'Hello, world!'[1]`
   - `'Hello, world!'[0:5]`
   - `'Hello, world!'[:5]`
   - `'Hello, world!'[3:]`

7. What do the following expressions evaluate to?
   - `'Hello'.upper()`
   - `'Hello'.upper().isupper()`
   - `'Hello'.upper().lower()`

8. What do the following expressions evaluate to?
   - `'Remember, remember, the fifth of November.'.split()`
   - `'-'.join('There can be only one.'.split())`

9. What string methods can you use to right-justify, left-justify, and center a string?

10. How can you trim whitespace characters from the beginning or end of a string?

## Practice Projects

For practice, write programs that do the following.

### *Table Printer*

Write a function named `printTable()` that takes a list of lists of strings and displays it in a well-organized table with each column right-justified. Assume that all the inner lists will contain the same number of strings. For example, the value could look like this:

```
tableData = [['apples', 'oranges', 'cherries', 'banana'],
             ['Alice', 'Bob', 'Carol', 'David'],
             ['dogs', 'cats', 'moose', 'goose']]
```

Your `printTable()` function would print the following:

```
  apples Alice  dogs
 oranges   Bob  cats
cherries Carol moose
  banana David goose
```

Hint: your code will first have to find the longest string in each of the inner lists so that the whole column can be wide enough to fit all the strings. You can store the maximum width of each column as a list of integers. The `printTable()` function can begin with `colWidths = [0] * len(tableData)`, which will create a list containing the same number of 0 values as the number of inner lists in `tableData`. That way, `colWidths[0]` can store the width of the longest string in `tableData[0]`, `colWidths[1]` can store the width of the longest string in `tableData[1]`, and so on. You can then find the largest value in the `colWidths` list to find out what integer width to pass to the `rjust()` string method.

### Zombie Dice Bots

*Programming games* are a game genre where instead of playing a game directly, players write bot programs to play the game autonomously. I've created a Zombie Dice simulator, which allows programmers to practice their skills while making game-playing AIs. Zombie Dice bots can be simple or incredibly complex, and are great for a class exercise or an individual programming challenge.

Zombie Dice is a quick, fun dice game from Steve Jackson Games. The players are zombies trying to eat as many human brains as possible without getting shot three times. There is a cup of 13 dice with brains, footsteps, and shotgun icons on their faces. The dice icons are colored, and each color has a different likelihood of each event occurring. Every die has two sides with footsteps, but dice with green icons have more sides with brains, red-icon dice have more shotguns, and yellow-icon dice have an even split of brains and shotguns. Do the following on each player's turn:

1. Place all 13 dice in the cup. The player randomly draws three dice from the cup and then rolls them. Players always roll exactly three dice.

2. They set aside and count up any brains (humans whose brains were eaten) and shotguns (humans who fought back). Accumulating three shotguns automatically ends a player's turn with zero points (regardless of how many brains they had). If they have between zero and two shotguns, they may continue rolling if they want. They may also choose to end their turn and collect one point per brain.

3. If the player decides to keep rolling, they must reroll all dice with footsteps. Remember that the player must always roll three dice; they must draw more dice out of the cup if they have fewer than three footsteps to roll. A player may keep rolling dice until either they get three shotguns—losing everything—or all 13 dice have been rolled. A player may not reroll only one or two dice, and may not stop mid-reroll.

4. When someone reaches 13 brains, the rest of the players finish out the round. The person with the most brains wins. If there's a tie, the tied players play one last tiebreaker round.

Zombie Dice has a push-your-luck game mechanic: the more you reroll the dice, the more brains you can get, but the more likely you'll eventually accrue three shotguns and lose everything. Once a player reaches 13 points, the rest of the players get one more turn (to potentially catch up) and the game ends. The player with the most points wins. You can find the complete rules at *https://github.com/asweigart/zombiedice/*.

Install the `zombiedice` module with pip by following the instructions in Appendix A. You can run a demo of the simulator with some pre-made bots by running the following in the interactive shell:

```
>>> import zombiedice
>>> zombiedice.demo()
Zombie Dice Visualization is running. Open your browser to http://
localhost:51810 to view it.
Press Ctrl-C to quit.
```

The program launches your web browser, which will look like Figure 6-1.

Figure 6-1: The web GUI for the Zombie Dice simulator

You'll create bots by writing a class with a turn() method, which is called by the simulator when it's your bot's turn to roll the dice. Classes are beyond the scope of this book, so the class code is already set up for you in the *myzombie.py* program, which is in the downloadable ZIP file for this book at *https://nostarch.com/automatestuff2/*. Writing a method is essentially the same as writing a function, and you can use the turn() code in the *myZombie.py* program as a template. Inside this turn() method, you'll call the zombiedice.roll() function as often as you want your bot to roll the dice.

```
import zombiedice

class MyZombie:
    def __init__(self, name):
        # All zombies must have a name:
        self.name = name

    def turn(self, gameState):
```

```
# gameState is a dict with info about the current state of the game.
# You can choose to ignore it in your code.

diceRollResults = zombiedice.roll() # first roll
# roll() returns a dictionary with keys 'brains', 'shotgun', and
# 'footsteps' with how many rolls of each type there were.
# The 'rolls' key is a list of (color, icon) tuples with the
# exact roll result information.
# Example of a roll() return value:
# {'brains': 1, 'footsteps': 1, 'shotgun': 1,
#  'rolls': [('yellow', 'brains'), ('red', 'footsteps'),
#            ('green', 'shotgun')]}

# REPLACE THIS ZOMBIE CODE WITH YOUR OWN:
brains = 0
while diceRollResults is not None:
    brains += diceRollResults['brains']

    if brains < 2:
        diceRollResults = zombiedice.roll() # roll again
    else:
        break

zombies = (
    zombiedice.examples.RandomCoinFlipZombie(name='Random'),
    zombiedice.examples.RollsUntilInTheLeadZombie(name='Until Leading'),
    zombiedice.examples.MinNumShotgunsThenStopsZombie(name='Stop at 2
Shotguns', minShotguns=2),
    zombiedice.examples.MinNumShotgunsThenStopsZombie(name='Stop at 1
Shotgun', minShotguns=1),
    MyZombie(name='My Zombie Bot'),
    # Add any other zombie players here.
)

# Uncomment one of the following lines to run in CLI or Web GUI mode:
#zombiedice.runTournament(zombies=zombies, numGames=1000)
zombiedice.runWebGui(zombies=zombies, numGames=1000)
```

The turn() method takes two parameters: self and gameState. You can ignore these in your first few zombie bots and consult the online documentation for details later if you want to learn more. The turn() method should call zombiedice.roll() at least once for the initial roll. Then, depending on the strategy the bot uses, it can call zombiedice.roll() again as many times as it wants. In *myZombie.py*, the turn() method calls zombiedice.roll() twice, which means the zombie bot will always roll its dice two times per turn regardless of the results of the roll.

The return value of zombiedice.roll() tells your code the results of the dice roll. It is a dictionary with four keys. Three of the keys, 'shotgun', 'brains', and 'footsteps', have integer values of how many dice came up with those icons. The fourth 'rolls' key has a value that is a list of tuples for each die roll. The tuples contain two strings: the color of the die at index 0 and the icon rolled at index 1. Look at the code comments in the turn()

method's definition for an example. If the bot has already rolled three shotguns, then `zombiedice.roll()` will return `None`.

Try writing some of your own bots to play Zombie Dice and see how they compare against the other bots. Specifically, try to create the following bots:

- A bot that, after the first roll, randomly decides if it will continue or stop
- A bot that stops rolling after it has rolled two brains
- A bot that stops rolling after it has rolled two shotguns
- A bot that initially decides it'll roll the dice one to four times, but will stop early if it rolls two shotguns
- A bot that stops rolling after it has rolled more shotguns than brains

Run these bots through the simulator and see how they compare to each other. You can also examine the code of some premade bots at *https://github.com/asweigart/zombiedice/*. If you find yourself playing this game in the real world, you'll have the benefit of thousands of simulated games telling you that one of the best strategies is to simply stop once you've rolled two shotguns. But you could always try pressing your luck . . .

# PART II

## AUTOMATING TASKS

# 7

## PATTERN MATCHING WITH REGULAR EXPRESSIONS

You may be familiar with searching for text by pressing CTRL-F and entering the words you're looking for. *Regular expressions* go one step further: they allow you to specify a *pattern* of text to search for. You may not know a business's exact phone number, but if you live in the United States or Canada, you know it will be three digits, followed by a hyphen, and then four more digits (and optionally, a three-digit area code at the start). This is how you, as a human, know a phone number when you see it: 415-555-1234 is a phone number, but 4,155,551,234 is not.

We also recognize all sorts of other text patterns every day: email addresses have @ symbols in the middle, US social security numbers have nine digits and two hyphens, website URLs often have periods and forward slashes, news headlines use title case, social media hashtags begin with # and contain no spaces, and more.

Regular expressions are helpful, but few non-programmers know about them even though most modern text editors and word processors, such as Microsoft Word or OpenOffice, have find and find-and-replace features that can search based on regular expressions. Regular expressions are huge time-savers, not just for software users but also for programmers. In fact, tech writer Cory Doctorow argues that we should be teaching regular expressions even before programming:

> Knowing [regular expressions] can mean the difference between solving a problem in 3 steps and solving it in 3,000 steps. When you're a nerd, you forget that the problems you solve with a couple keystrokes can take other people days of tedious, error-prone work to slog through.[1]

In this chapter, you'll start by writing a program to find text patterns *without* using regular expressions and then see how to use regular expressions to make the code much less bloated. I'll show you basic matching with regular expressions and then move on to some more powerful features, such as string substitution and creating your own character classes. Finally, at the end of the chapter, you'll write a program that can automatically extract phone numbers and email addresses from a block of text.

## Finding Patterns of Text Without Regular Expressions

Say you want to find an American phone number in a string. You know the pattern if you're American: three numbers, a hyphen, three numbers, a hyphen, and four numbers. Here's an example: 415-555-4242.

Let's use a function named isPhoneNumber() to check whether a string matches this pattern, returning either True or False. Open a new file editor tab and enter the following code; then save the file as *isPhoneNumber.py*:

```
def isPhoneNumber(text):
❶ if len(text) != 12:
        return False
    for i in range(0, 3):
      ❷ if not text[i].isdecimal():
            return False
❸ if text[3] != '-':
        return False
    for i in range(4, 7):
      ❹ if not text[i].isdecimal():
            return False
❺ if text[7] != '-':
        return False
```

---

1. Cory Doctorow, "Here's What ICT Should Really Teach Kids: How to Do Regular Expressions," *Guardian*, December 4, 2012, *http://www.theguardian.com/technology/2012 /dec/04/ict-teach-kids-regular-expressions/*.

```
    for i in range(8, 12):
  ❻ if not text[i].isdecimal():
            return False
❼ return True

print('Is 415-555-4242 a phone number?')
print(isPhoneNumber('415-555-4242'))
print('Is Moshi moshi a phone number?')
print(isPhoneNumber('Moshi moshi'))
```

When this program is run, the output looks like this:

```
Is 415-555-4242 a phone number?
True
Is Moshi moshi a phone number?
False
```

The isPhoneNumber() function has code that does several checks to see whether the string in text is a valid phone number. If any of these checks fail, the function returns False. First the code checks that the string is exactly 12 characters ❶. Then it checks that the area code (that is, the first three characters in text) consists of only numeric characters ❷. The rest of the function checks that the string follows the pattern of a phone number: the number must have the first hyphen after the area code ❸, three more numeric characters ❹, then another hyphen ❺, and finally four more numbers ❻. If the program execution manages to get past all the checks, it returns True ❼.

Calling isPhoneNumber() with the argument '415-555-4242' will return True. Calling isPhoneNumber() with 'Moshi moshi' will return False; the first test fails because 'Moshi moshi' is not 12 characters long.

If you wanted to find a phone number within a larger string, you would have to add even more code to find the phone number pattern. Replace the last four print() function calls in *isPhoneNumber.py* with the following:

```
message = 'Call me at 415-555-1011 tomorrow. 415-555-9999 is my office.'
for i in range(len(message)):
  ❶ chunk = message[i:i+12]
  ❷ if isPhoneNumber(chunk):
        print('Phone number found: ' + chunk)
print('Done')
```

When this program is run, the output will look like this:

```
Phone number found: 415-555-1011
Phone number found: 415-555-9999
Done
```

On each iteration of the for loop, a new chunk of 12 characters from message is assigned to the variable chunk ❶. For example, on the first iteration, i is 0, and chunk is assigned message[0:12] (that is, the string 'Call me at 4'). On the next iteration, i is 1, and chunk is assigned message[1:13] (the string 'all me at 41'). In other words, on each iteration of the for loop, chunk takes on the following values:

- 'Call me at 4'
- 'all me at 41'
- 'll me at 415'
- 'l me at 415-'
- . . . and so on.

You pass chunk to isPhoneNumber() to see whether it matches the phone number pattern ❷, and if so, you print the chunk.

Continue to loop through message, and eventually the 12 characters in chunk will be a phone number. The loop goes through the entire string, testing each 12-character piece and printing any chunk it finds that satisfies isPhoneNumber(). Once we're done going through message, we print Done.

While the string in message is short in this example, it could be millions of characters long and the program would still run in less than a second. A similar program that finds phone numbers using regular expressions would also run in less than a second, but regular expressions make it quicker to write these programs.

## Finding Patterns of Text with Regular Expressions

The previous phone number–finding program works, but it uses a lot of code to do something limited: the isPhoneNumber() function is 17 lines but can find only one pattern of phone numbers. What about a phone number formatted like 415.555.4242 or (415) 555-4242? What if the phone number had an extension, like 415-555-4242 x99? The isPhoneNumber() function would fail to validate them. You could add yet more code for these additional patterns, but there is an easier way.

Regular expressions, called *regexes* for short, are descriptions for a pattern of text. For example, a \d in a regex stands for a digit character—that is, any single numeral from 0 to 9. The regex \d\d\d-\d\d\d-\d\d\d\d is used by Python to match the same text pattern the previous isPhoneNumber() function did: a string of three numbers, a hyphen, three more numbers, another hyphen, and four numbers. Any other string would not match the \d\d\d-\d\d\d-\d\d\d\d regex.

But regular expressions can be much more sophisticated. For example, adding a 3 in braces ({3}) after a pattern is like saying, "Match this pattern three times." So the slightly shorter regex \d{3}-\d{3}-\d{4} also matches the correct phone number format.

### Creating Regex Objects

All the regex functions in Python are in the re module. Enter the following into the interactive shell to import this module:

```
>>> import re
```

*Most of the examples in this chapter will require the re module, so remember to import it at the beginning of any script you write or any time you restart Mu. Otherwise, you'll get a* NameError: name 're' is not defined *error message.*

Passing a string value representing your regular expression to re.compile() returns a Regex pattern object (or simply, a Regex object).

To create a Regex object that matches the phone number pattern, enter the following into the interactive shell. (Remember that \d means "a digit character" and \d\d\d-\d\d\d-\d\d\d\d is the regular expression for a phone number pattern.)

```
>>> phoneNumRegex = re.compile(r'\d\d\d-\d\d\d-\d\d\d\d')
```

Now the phoneNumRegex variable contains a Regex object.

### Matching Regex Objects

A Regex object's search() method searches the string it is passed for any matches to the regex. The search() method will return None if the regex pattern is not found in the string. If the pattern *is* found, the search() method returns a Match object, which have a group() method that will return the actual matched text from the searched string. (I'll explain groups shortly.) For example, enter the following into the interactive shell:

```
>>> phoneNumRegex = re.compile(r'\d\d\d-\d\d\d-\d\d\d\d')
>>> mo = phoneNumRegex.search('My number is 415-555-4242.')
>>> print('Phone number found: ' + mo.group())
Phone number found: 415-555-4242
```

The mo variable name is just a generic name to use for Match objects. This example might seem complicated at first, but it is much shorter than the earlier *isPhoneNumber.py* program and does the same thing.

Here, we pass our desired pattern to re.compile() and store the resulting Regex object in phoneNumRegex. Then we call search() on phoneNumRegex and pass search() the string we want to match for during the search. The result of the search gets stored in the variable mo. In this example, we know that our pattern will be found in the string, so we know that a Match object will be returned. Knowing that mo contains a Match object and not the null value None, we can call group() on mo to return the match. Writing mo.group() inside our print() function call displays the whole match, 415-555-4242.

### Review of Regular Expression Matching

While there are several steps to using regular expressions in Python, each step is fairly simple.

1.  Import the regex module with import re.
2.  Create a Regex object with the re.compile() function. (Remember to use a raw string.)
3.  Pass the string you want to search into the Regex object's search() method. This returns a Match object.
4.  Call the Match object's group() method to return a string of the actual matched text.

**NOTE**  *While I encourage you to enter the example code into the interactive shell, you should also make use of web-based regular expression testers, which can show you exactly how a regex matches a piece of text that you enter. I recommend the tester at* https://pythex.org/.

# More Pattern Matching with Regular Expressions

Now that you know the basic steps for creating and finding regular expression objects using Python, you're ready to try some of their more powerful pattern-matching capabilities.

### Grouping with Parentheses

Say you want to separate the area code from the rest of the phone number. Adding parentheses will create *groups* in the regex: (\d\d\d)-(\d\d\d-\d\d\d\d). Then you can use the group() match object method to grab the matching text from just one group.

The first set of parentheses in a regex string will be group 1. The second set will be group 2. By passing the *integer* 1 or 2 to the group() match object method, you can grab different parts of the matched text. Passing 0 or nothing to the group() method will return the entire matched text. Enter the following into the interactive shell:

```
>>> phoneNumRegex = re.compile(r'(\d\d\d)-(\d\d\d-\d\d\d\d)')
>>> mo = phoneNumRegex.search('My number is 415-555-4242.')
>>> mo.group(1)
'415'
>>> mo.group(2)
'555-4242'
>>> mo.group(0)
'415-555-4242'
>>> mo.group()
'415-555-4242'
```

If you would like to retrieve all the groups at once, use the groups() method—note the plural form for the name.

```
>>> mo.groups()
('415', '555-4242')
>>> areaCode, mainNumber = mo.groups()
>>> print(areaCode)
415
>>> print(mainNumber)
555-4242
```

Since mo.groups() returns a tuple of multiple values, you can use the multiple-assignment trick to assign each value to a separate variable, as in the previous areaCode, mainNumber = mo.groups() line.

Parentheses have a special meaning in regular expressions, but what do you do if you need to match a parenthesis in your text? For instance, maybe the phone numbers you are trying to match have the area code set in parentheses. In this case, you need to escape the ( and ) characters with a backslash. Enter the following into the interactive shell:

```
>>> phoneNumRegex = re.compile(r'(\(\d\d\d\)) (\d\d\d-\d\d\d\d)')
>>> mo = phoneNumRegex.search('My phone number is (415) 555-4242.')
>>> mo.group(1)
'(415)'
>>> mo.group(2)
'555-4242'
```

The \( and \) escape characters in the raw string passed to re.compile() will match actual parenthesis characters. In regular expressions, the following characters have special meanings:

```
. ^ $ * + ? { } [ ] \ | ( )
```

If you want to detect these characters as part of your text pattern, you need to escape them with a backslash:

```
\. \^ \$ \* \+ \? \{ \} \[ \] \\ \| \( \)
```

Make sure to double-check that you haven't mistaken escaped parentheses \( and \) for parentheses ( and ) in a regular expression. If you receive an error message about "missing )" or "unbalanced parenthesis," you may have forgotten to include the closing unescaped parenthesis for a group, like in this example:

```
>>> re.compile(r'(\(Parentheses\)')
Traceback (most recent call last):
    --snip--
re.error: missing ), unterminated subpattern at position 0
```

The error message tells you that there is an opening parenthesis at index 0 of the r'(\(Parentheses\)' string that is missing its corresponding closing parenthesis.

## Matching Multiple Groups with the Pipe

The | character is called a *pipe*. You can use it anywhere you want to match one of many expressions. For example, the regular expression r'Batman|Tina Fey' will match either 'Batman' or 'Tina Fey'.

When *both* Batman and Tina Fey occur in the searched string, the first occurrence of matching text will be returned as the Match object. Enter the following into the interactive shell:

```
>>> heroRegex = re.compile (r'Batman|Tina Fey')
>>> mo1 = heroRegex.search('Batman and Tina Fey')
>>> mo1.group()
'Batman'

>>> mo2 = heroRegex.search('Tina Fey and Batman')
>>> mo2.group()
'Tina Fey'
```

**NOTE** *You can find all matching occurrences with the findall() method that's discussed in "The findall() Method" on page 171.*

You can also use the pipe to match one of several patterns as part of your regex. For example, say you wanted to match any of the strings 'Batman', 'Batmobile', 'Batcopter', and 'Batbat'. Since all these strings start with Bat, it would be nice if you could specify that prefix only once. This can be done with parentheses. Enter the following into the interactive shell:

```
>>> batRegex = re.compile(r'Bat(man|mobile|copter|bat)')
>>> mo = batRegex.search('Batmobile lost a wheel')
>>> mo.group()
'Batmobile'
>>> mo.group(1)
'mobile'
```

The method call mo.group() returns the full matched text 'Batmobile', while mo.group(1) returns just the part of the matched text inside the first parentheses group, 'mobile'. By using the pipe character and grouping parentheses, you can specify several alternative patterns you would like your regex to match.

If you need to match an actual pipe character, escape it with a backslash, like \|.

## Optional Matching with the Question Mark

Sometimes there is a pattern that you want to match only optionally. That is, the regex should find a match regardless of whether that bit of text is there. The ? character flags the group that precedes it as an optional part of the pattern. For example, enter the following into the interactive shell:

```
>>> batRegex = re.compile(r'Bat(wo)?man')
>>> mo1 = batRegex.search('The Adventures of Batman')
```

```
>>> mo1.group()
'Batman'

>>> mo2 = batRegex.search('The Adventures of Batwoman')
>>> mo2.group()
'Batwoman'
```

The (wo)? part of the regular expression means that the pattern wo is an optional group. The regex will match text that has zero instances or one instance of *wo* in it. This is why the regex matches both 'Batwoman' and 'Batman'. This is why the regex matches both 'Batwoman' and 'Batman'.

Using the earlier phone number example, you can make the regex look for phone numbers that do or do not have an area code. Enter the following into the interactive shell:

```
>>> phoneRegex = re.compile(r'(\d\d\d-)?\d\d\d-\d\d\d\d')
>>> mo1 = phoneRegex.search('My number is 415-555-4242')
>>> mo1.group()
'415-555-4242'

>>> mo2 = phoneRegex.search('My number is 555-4242')
>>> mo2.group()
'555-4242'
```

You can think of the ? as saying, "Match zero or one of the group preceding this question mark."

If you need to match an actual question mark character, escape it with \?.

## Matching Zero or More with the Star

The * (called the *star* or *asterisk*) means "match zero or more"—the group that precedes the star can occur any number of times in the text. It can be completely absent or repeated over and over again. Let's look at the Batman example again.

```
>>> batRegex = re.compile(r'Bat(wo)*man')
>>> mo1 = batRegex.search('The Adventures of Batman')
>>> mo1.group()
'Batman'

>>> mo2 = batRegex.search('The Adventures of Batwoman')
>>> mo2.group()
'Batwoman'

>>> mo3 = batRegex.search('The Adventures of Batwowowowoman')
>>> mo3.group()
'Batwowowowoman'
```

For 'Batman', the (wo)* part of the regex matches zero instances of wo in the string; for 'Batwoman', the (wo)* matches one instance of wo; and for 'Batwowowowoman', (wo)* matches four instances of wo.

If you need to match an actual star character, prefix the star in the regular expression with a backslash, \*.

## Matching One or More with the Plus

While * means "match zero or more," the + (or *plus*) means "match one or more." Unlike the star, which does not require its group to appear in the matched string, the group preceding a plus must appear *at least once*. It is not optional. Enter the following into the interactive shell, and compare it with the star regexes in the previous section:

```
>>> batRegex = re.compile(r'Bat(wo)+man')
>>> mo1 = batRegex.search('The Adventures of Batwoman')
>>> mo1.group()
'Batwoman'

>>> mo2 = batRegex.search('The Adventures of Batwowowowoman')
>>> mo2.group()
'Batwowowowoman'

>>> mo3 = batRegex.search('The Adventures of Batman')
>>> mo3 == None
True
```

The regex Bat(wo)+man will not match the string 'The Adventures of Batman', because at least one wo is required by the plus sign.

If you need to match an actual plus sign character, prefix the plus sign with a backslash to escape it: \+.

## Matching Specific Repetitions with Braces

If you have a group that you want to repeat a specific number of times, follow the group in your regex with a number in braces. For example, the regex (Ha){3} will match the string 'HaHaHa', but it will not match 'HaHa', since the latter has only two repeats of the (Ha) group.

Instead of one number, you can specify a range by writing a minimum, a comma, and a maximum in between the braces. For example, the regex (Ha){3,5} will match 'HaHaHa', 'HaHaHaHa', and 'HaHaHaHaHa'.

You can also leave out the first or second number in the braces to leave the minimum or maximum unbounded. For example, (Ha){3,} will match three or more instances of the (Ha) group, while (Ha){,5} will match zero to five instances. Braces can help make your regular expressions shorter. These two regular expressions match identical patterns:

```
(Ha){3}
(Ha)(Ha)(Ha)
```

And these two regular expressions also match identical patterns:

```
(Ha){3,5}
((Ha)(Ha)(Ha))|((Ha)(Ha)(Ha)(Ha))|((Ha)(Ha)(Ha)(Ha)(Ha))
```

Enter the following into the interactive shell:

```
>>> haRegex = re.compile(r'(Ha){3}')
>>> mo1 = haRegex.search('HaHaHa')
>>> mo1.group()
'HaHaHa'

>>> mo2 = haRegex.search('Ha')
>>> mo2 == None
True
```

Here, (Ha){3} matches 'HaHaHa' but not 'Ha'. Since it doesn't match 'Ha', search() returns None.

## Greedy and Non-greedy Matching

Since (Ha){3,5} can match three, four, or five instances of Ha in the string 'HaHaHaHaHa', you may wonder why the Match object's call to group() in the previous brace example returns 'HaHaHaHaHa' instead of the shorter possibilities. After all, 'HaHaHa' and 'HaHaHaHa' are also valid matches of the regular expression (Ha){3,5}.

Python's regular expressions are *greedy* by default, which means that in ambiguous situations they will match the longest string possible. The *non-greedy* (also called *lazy*) version of the braces, which matches the shortest string possible, has the closing brace followed by a question mark.

Enter the following into the interactive shell, and notice the difference between the greedy and non-greedy forms of the braces searching the same string:

```
>>> greedyHaRegex = re.compile(r'(Ha){3,5}')
>>> mo1 = greedyHaRegex.search('HaHaHaHaHa')
>>> mo1.group()
'HaHaHaHaHa'

>>> nongreedyHaRegex = re.compile(r'(Ha){3,5}?')
>>> mo2 = nongreedyHaRegex.search('HaHaHaHaHa')
>>> mo2.group()
'HaHaHa'
```

Note that the question mark can have two meanings in regular expressions: declaring a non-greedy match or flagging an optional group. These meanings are entirely unrelated.

## The findall() Method

In addition to the search() method, Regex objects also have a findall() method. While search() will return a Match object of the *first* matched text in the searched string, the findall() method will return the strings of *every* match in the searched string. To see how search() returns a Match object only on the first instance of matching text, enter the following into the interactive shell:

```
>>> phoneNumRegex = re.compile(r'\d\d\d-\d\d\d-\d\d\d\d')
>>> mo = phoneNumRegex.search('Cell: 415-555-9999 Work: 212-555-0000')
>>> mo.group()
'415-555-9999'
```

On the other hand, findall() will not return a Match object but a list of strings—*as long as there are no groups in the regular expression.* Each string in the list is a piece of the searched text that matched the regular expression. Enter the following into the interactive shell:

```
>>> phoneNumRegex = re.compile(r'\d\d\d-\d\d\d-\d\d\d\d') # has no groups
>>> phoneNumRegex.findall('Cell: 415-555-9999 Work: 212-555-0000')
['415-555-9999', '212-555-0000']
```

If there *are* groups in the regular expression, then findall() will return a list of tuples. Each tuple represents a found match, and its items are the matched strings for each group in the regex. To see findall() in action, enter the following into the interactive shell (notice that the regular expression being compiled now has groups in parentheses):

```
>>> phoneNumRegex = re.compile(r'(\d\d\d)-(\d\d\d)-(\d\d\d\d)') # has groups
>>> phoneNumRegex.findall('Cell: 415-555-9999 Work: 212-555-0000')
[('415', '555', '9999'), ('212', '555', '0000')]
```

To summarize what the findall() method returns, remember the following:

- When called on a regex with no groups, such as \d\d\d-\d\d\d-\d\d\d\d, the method findall() returns a list of string matches, such as ['415-555-9999', '212-555-0000'].

- When called on a regex that has groups, such as (\d\d\d)-(\d\d\d)-(\d\d\d\d), the method findall() returns a list of tuples of strings (one string for each group), such as [('415', '555', '9999'), ('212', '555', '0000')].

## Character Classes

In the earlier phone number regex example, you learned that \d could stand for any numeric digit. That is, \d is shorthand for the regular expression (0|1|2|3|4|5|6|7|8|9). There are many such *shorthand character classes*, as shown in Table 7-1.

**Table 7-1:** Shorthand Codes for Common Character Classes

| Shorthand character class | Represents |
| --- | --- |
| \d | Any numeric digit from 0 to 9. |
| \D | Any character that is *not* a numeric digit from 0 to 9. |
| \w | Any letter, numeric digit, or the underscore character. (Think of this as matching "word" characters.) |
| \W | Any character that is *not* a letter, numeric digit, or the underscore character. |
| \s | Any space, tab, or newline character. (Think of this as matching "space" characters.) |
| \S | Any character that is *not* a space, tab, or newline. |

Character classes are nice for shortening regular expressions. The character class [0-5] will match only the numbers 0 to 5; this is much shorter than typing (0|1|2|3|4|5). Note that while \d matches digits and \w matches digits, letters, and the underscore, there is no shorthand character class that matches only letters. (Though you can use the [a-zA-Z] character class, as explained next.)

For example, enter the following into the interactive shell:

```
>>> xmasRegex = re.compile(r'\d+\s\w+')
>>> xmasRegex.findall('12 drummers, 11 pipers, 10 lords, 9 ladies, 8 maids, 7
swans, 6 geese, 5 rings, 4 birds, 3 hens, 2 doves, 1 partridge')
['12 drummers', '11 pipers', '10 lords', '9 ladies', '8 maids', '7 swans', '6
geese', '5 rings', '4 birds', '3 hens', '2 doves', '1 partridge']
```

The regular expression \d+\s\w+ will match text that has one or more numeric digits (\d+), followed by a whitespace character (\s), followed by one or more letter/digit/underscore characters (\w+). The findall() method returns all matching strings of the regex pattern in a list.

## Making Your Own Character Classes

There are times when you want to match a set of characters but the shorthand character classes (\d, \w, \s, and so on) are too broad. You can define your own character class using square brackets. For example, the character class [aeiouAEIOU] will match any vowel, both lowercase and uppercase. Enter the following into the interactive shell:

```
>>> vowelRegex = re.compile(r'[aeiouAEIOU]')
>>> vowelRegex.findall('RoboCop eats baby food. BABY FOOD.')
['o', 'o', 'o', 'e', 'a', 'a', 'o', 'o', 'A', 'O', 'O']
```

You can also include ranges of letters or numbers by using a hyphen. For example, the character class [a-zA-Z0-9] will match all lowercase letters, uppercase letters, and numbers.

Note that inside the square brackets, the normal regular expression symbols are not interpreted as such. This means you do not need to escape the ., *, ?, or () characters with a preceding backslash. For example, the character class [0-5.] will match digits 0 to 5 and a period. You do not need to write it as [0-5\.].

By placing a caret character (^) just after the character class's opening bracket, you can make a *negative character class*. A negative character class will match all the characters that are *not* in the character class. For example, enter the following into the interactive shell:

```
>>> consonantRegex = re.compile(r'[^aeiouAEIOU]')
>>> consonantRegex.findall('RoboCop eats baby food. BABY FOOD.')
['R', 'b', 'C', 'p', ' ', 't', 's', ' ', 'b', 'b', 'y', ' ', 'f', 'd', '.', ' ', ' ', 'B', 'B', 'Y', ' ', 'F', 'D', '.']
```

Now, instead of matching every vowel, we're matching every character that isn't a vowel.

## The Caret and Dollar Sign Characters

You can also use the caret symbol (^) at the start of a regex to indicate that a match must occur at the *beginning* of the searched text. Likewise, you can put a dollar sign ($) at the end of the regex to indicate the string must *end* with this regex pattern. And you can use the ^ and $ together to indicate that the entire string must match the regex—that is, it's not enough for a match to be made on some subset of the string.

For example, the r'^Hello' regular expression string matches strings that begin with 'Hello'. Enter the following into the interactive shell:

```
>>> beginsWithHello = re.compile(r'^Hello')
>>> beginsWithHello.search('Hello, world!')
<re.Match object; span=(0, 5), match='Hello'>
>>> beginsWithHello.search('He said hello.') == None
True
```

The r'\d$' regular expression string matches strings that end with a numeric character from 0 to 9. Enter the following into the interactive shell:

```
>>> endsWithNumber = re.compile(r'\d$')
>>> endsWithNumber.search('Your number is 42')
<re.Match object; span=(16, 17), match='2'>
>>> endsWithNumber.search('Your number is forty two.') == None
True
```

The r'^\d+$' regular expression string matches strings that both begin and end with one or more numeric characters. Enter the following into the interactive shell:

```
>>> wholeStringIsNum = re.compile(r'^\d+$')
>>> wholeStringIsNum.search('1234567890')
<re.Match object; span=(0, 10), match='1234567890'>
>>> wholeStringIsNum.search('12345xyz67890') == None
True
>>> wholeStringIsNum.search('12  34567890') == None
True
```

The last two search() calls in the previous interactive shell example demonstrate how the entire string must match the regex if ^ and $ are used.

I always confuse the meanings of these two symbols, so I use the mnemonic "Carrots cost dollars" to remind myself that the caret comes first and the dollar sign comes last.

# The Wildcard Character

The . (or *dot*) character in a regular expression is called a *wildcard* and will match any character except for a newline. For example, enter the following into the interactive shell:

```
>>> atRegex = re.compile(r'.at')
>>> atRegex.findall('The cat in the hat sat on the flat mat.')
['cat', 'hat', 'sat', 'lat', 'mat']
```

Remember that the dot character will match just one character, which is why the match for the text flat in the previous example matched only lat. To match an actual dot, escape the dot with a backslash: \..

## Matching Everything with Dot-Star

Sometimes you will want to match everything and anything. For example, say you want to match the string 'First Name:', followed by any and all text, followed by 'Last Name:', and then followed by anything again. You can use the dot-star (.*) to stand in for that "anything." Remember that the dot character means "any single character except the newline," and the star character means "zero or more of the preceding character."

Enter the following into the interactive shell:

```
>>> nameRegex = re.compile(r'First Name: (.*) Last Name: (.*)')
>>> mo = nameRegex.search('First Name: Al Last Name: Sweigart')
>>> mo.group(1)
'Al'
>>> mo.group(2)
'Sweigart'
```

The dot-star uses *greedy* mode: It will always try to match as much text as possible. To match any and all text in a *non-greedy* fashion, use the dot, star, and question mark (.*?). Like with braces, the question mark tells Python to match in a non-greedy way.

Enter the following into the interactive shell to see the difference between the greedy and non-greedy versions:

```
>>> nongreedyRegex = re.compile(r'<.*?>')
>>> mo = nongreedyRegex.search('<To serve man> for dinner.>')
>>> mo.group()
'<To serve man>'

>>> greedyRegex = re.compile(r'<.*>')
>>> mo = greedyRegex.search('<To serve man> for dinner.>')
>>> mo.group()
'<To serve man> for dinner.>'
```

Both regexes roughly translate to "Match an opening angle bracket, followed by anything, followed by a closing angle bracket." But the string '<To serve man> for dinner.>' has two possible matches for the closing angle bracket. In the non-greedy version of the regex, Python matches the shortest possible string: '<To serve man>'. In the greedy version, Python matches the longest possible string: '<To serve man> for dinner.>'.

## Matching Newlines with the Dot Character

The dot-star will match everything except a newline. By passing re.DOTALL as the second argument to re.compile(), you can make the dot character match *all* characters, including the newline character.

Enter the following into the interactive shell:

```
>>> noNewlineRegex = re.compile('.*')
>>> noNewlineRegex.search('Serve the public trust.\nProtect the innocent.
\nUphold the law.').group()
'Serve the public trust.'

>>> newlineRegex = re.compile('.*', re.DOTALL)
>>> newlineRegex.search('Serve the public trust.\nProtect the innocent.
\nUphold the law.').group()
'Serve the public trust.\nProtect the innocent.\nUphold the law.'
```

The regex noNewlineRegex, which did not have re.DOTALL passed to the re.compile() call that created it, will match everything only up to the first newline character, whereas newlineRegex, which *did* have re.DOTALL passed to re.compile(), matches everything. This is why the newlineRegex.search() call matches the full string, including its newline characters.

## Review of Regex Symbols

This chapter covered a lot of notation, so here's a quick review of what you learned about basic regular expression syntax:

- The ? matches zero or one of the preceding group.
- The * matches zero or more of the preceding group.
- The + matches one or more of the preceding group.
- The {n} matches exactly *n* of the preceding group.
- The {n,} matches *n* or more of the preceding group.
- The {,m} matches 0 to *m* of the preceding group.
- The {n,m} matches at least *n* and at most *m* of the preceding group.
- {n,m}? or *? or +? performs a non-greedy match of the preceding group.
- ^spam means the string must begin with *spam*.
- spam$ means the string must end with *spam*.
- The . matches any character, except newline characters.
- \d, \w, and \s match a digit, word, or space character, respectively.
- \D, \W, and \S match anything except a digit, word, or space character, respectively.
- [abc] matches any character between the brackets (such as *a*, *b*, or *c*).
- [^abc] matches any character that isn't between the brackets.

## Case-Insensitive Matching

Normally, regular expressions match text with the exact casing you specify. For example, the following regexes match completely different strings:

```
>>> regex1 = re.compile('RoboCop')
>>> regex2 = re.compile('ROBOCOP')
>>> regex3 = re.compile('robOcop')
>>> regex4 = re.compile('RobocOp')
```

But sometimes you care only about matching the letters without worrying whether they're uppercase or lowercase. To make your regex case-insensitive, you can pass re.IGNORECASE or re.I as a second argument to re.compile(). Enter the following into the interactive shell:

```
>>> robocop = re.compile(r'robocop', re.I)
>>> robocop.search('RoboCop is part man, part machine, all cop.').group()
'RoboCop'

>>> robocop.search('ROBOCOP protects the innocent.').group()
'ROBOCOP'

>>> robocop.search('Al, why does your programming book talk about robocop so much?').group()
'robocop'
```

## Substituting Strings with the sub() Method

Regular expressions can not only find text patterns but can also substitute new text in place of those patterns. The sub() method for Regex objects is passed two arguments. The first argument is a string to replace any matches. The second is the string for the regular expression. The sub() method returns a string with the substitutions applied.

For example, enter the following into the interactive shell:

```
>>> namesRegex = re.compile(r'Agent \w+')
>>> namesRegex.sub('CENSORED', 'Agent Alice gave the secret documents to Agent Bob.')
'CENSORED gave the secret documents to CENSORED.'
```

Sometimes you may need to use the matched text itself as part of the substitution. In the first argument to sub(), you can type \1, \2, \3, and so on, to mean "Enter the text of group 1, 2, 3, and so on, in the substitution."

For example, say you want to censor the names of the secret agents by showing just the first letters of their names. To do this, you could use the regex Agent (\w)\w* and pass r'\1****' as the first argument to sub(). The \1 in that string will be replaced by whatever text was matched by group 1— that is, the (\w) group of the regular expression.

```
>>> agentNamesRegex = re.compile(r'Agent (\w)\w*')
>>> agentNamesRegex.sub(r'\1****', 'Agent Alice told Agent Carol that Agent
Eve knew Agent Bob was a double agent.')
A**** told C**** that E**** knew B**** was a double agent.'
```

## Managing Complex Regexes

Regular expressions are fine if the text pattern you need to match is simple. But matching complicated text patterns might require long, convoluted regular expressions. You can mitigate this by telling the re.compile() function to ignore whitespace and comments inside the regular expression string. This "verbose mode" can be enabled by passing the variable re.VERBOSE as the second argument to re.compile().

Now instead of a hard-to-read regular expression like this:

```
phoneRegex = re.compile(r'((\d{3}|\(\d{3}\))?(\s|-|\.)?\d{3}(\s|-|\.)\d{4}
(\s*(ext|x|ext.)\s*\d{2,5})?)')
```

you can spread the regular expression over multiple lines with comments like this:

```
phoneRegex = re.compile(r'''(
    (\d{3}|\(\d{3}\))?            # area code
    (\s|-|\.)?                   # separator
    \d{3}                        # first 3 digits
    (\s|-|\.)                    # separator
    \d{4}                        # last 4 digits
```

```
(\s*(ext|x|ext.)\s*\d{2,5})?  # extension
)''', re.VERBOSE)
```

Note how the previous example uses the triple-quote syntax (''') to create a multiline string so that you can spread the regular expression definition over many lines, making it much more legible.

The comment rules inside the regular expression string are the same as regular Python code: the # symbol and everything after it to the end of the line are ignored. Also, the extra spaces inside the multiline string for the regular expression are not considered part of the text pattern to be matched. This lets you organize the regular expression so it's easier to read.

## Combining re.IGNORECASE, re.DOTALL, and re.VERBOSE

What if you want to use re.VERBOSE to write comments in your regular expression but also want to use re.IGNORECASE to ignore capitalization? Unfortunately, the re.compile() function takes only a single value as its second argument. You can get around this limitation by combining the re.IGNORECASE, re.DOTALL, and re.VERBOSE variables using the pipe character (|), which in this context is known as the *bitwise or* operator.

So if you want a regular expression that's case-insensitive *and* includes newlines to match the dot character, you would form your re.compile() call like this:

```
>>> someRegexValue = re.compile('foo', re.IGNORECASE | re.DOTALL)
```

Including all three options in the second argument will look like this:

```
>>> someRegexValue = re.compile('foo', re.IGNORECASE | re.DOTALL | re.VERBOSE)
```

This syntax is a little old-fashioned and originates from early versions of Python. The details of the bitwise operators are beyond the scope of this book, but check out the resources at *https://nostarch.com/automatestuff2/* for more information. You can also pass other options for the second argument; they're uncommon, but you can read more about them in the resources, too.

## Project: Phone Number and Email Address Extractor

Say you have the boring task of finding every phone number and email address in a long web page or document. If you manually scroll through the page, you might end up searching for a long time. But if you had a program that could search the text in your clipboard for phone numbers and email addresses, you could simply press CTRL-A to select all the text, press CTRL-C to copy it to the clipboard, and then run your program. It could replace the text on the clipboard with just the phone numbers and email addresses it finds.

Whenever you're tackling a new project, it can be tempting to dive right into writing code. But more often than not, it's best to take a step back and consider the bigger picture. I recommend first drawing up a high-level plan for what your program needs to do. Don't think about the actual code yet—you can worry about that later. Right now, stick to broad strokes.

For example, your phone and email address extractor will need to do the following:

1. Get the text off the clipboard.
2. Find all phone numbers and email addresses in the text.
3. Paste them onto the clipboard.

Now you can start thinking about how this might work in code. The code will need to do the following:

1. Use the pyperclip module to copy and paste strings.
2. Create two regexes, one for matching phone numbers and the other for matching email addresses.
3. Find all matches, not just the first match, of both regexes.
4. Neatly format the matched strings into a single string to paste.
5. Display some kind of message if no matches were found in the text.

This list is like a road map for the project. As you write the code, you can focus on each of these steps separately. Each step is fairly manageable and expressed in terms of things you already know how to do in Python.

### Step 1: Create a Regex for Phone Numbers

First, you have to create a regular expression to search for phone numbers. Create a new file, enter the following, and save it as *phoneAndEmail.py*:

```python
#! python3
# phoneAndEmail.py - Finds phone numbers and email addresses on the clipboard.

import pyperclip, re

phoneRegex = re.compile(r'''(
    (\d{3}|\(\d{3}\))?            # area code
    (\s|-|\.)?                   # separator
    (\d{3})                     # first 3 digits
    (\s|-|\.)                   # separator
    (\d{4})                     # last 4 digits
    (\s*(ext|x|ext.)\s*(\d{2,5}))?   # extension
    )''', re.VERBOSE)

# TODO: Create email regex.

# TODO: Find matches in clipboard text.

# TODO: Copy results to the clipboard.
```

The TODO comments are just a skeleton for the program. They'll be replaced as you write the actual code.

The phone number begins with an *optional* area code, so the area code group is followed with a question mark. Since the area code can be just three digits (that is, \d{3}) *or* three digits within parentheses (that is, \(\d{3}\)), you should have a pipe joining those parts. You can add the regex comment # Area code to this part of the multiline string to help you remember what (\d{3}|\(\d{3}\))? is supposed to match.

The phone number separator character can be a space (\s), hyphen (-), or period (.), so these parts should also be joined by pipes. The next few parts of the regular expression are straightforward: three digits, followed by another separator, followed by four digits. The last part is an optional extension made up of any number of spaces followed by ext, x, or ext., followed by two to five digits.

**NOTE**    *It's easy to get mixed up with regular expressions that contain groups with parentheses ( ) and escaped parentheses \( \). Remember to double-check that you're using the correct one if you get a "missing ), unterminated subpattern" error message.*

## Step 2: Create a Regex for Email Addresses

You will also need a regular expression that can match email addresses. Make your program look like the following:

```
#! python3
# phoneAndEmail.py - Finds phone numbers and email addresses on the clipboard.

import pyperclip, re

phoneRegex = re.compile(r'''(
--snip--

# Create email regex.
emailRegex = re.compile(r'''(
❶ [a-zA-Z0-9._%+-]+        # username
❷ @                       # @ symbol
❸ [a-zA-Z0-9.-]+          # domain name
  (\.[a-zA-Z]{2,4})       # dot-something
  )''', re.VERBOSE)

# TODO: Find matches in clipboard text.

# TODO: Copy results to the clipboard.
```

The username part of the email address ❶ is one or more characters that can be any of the following: lowercase and uppercase letters, numbers, a dot, an underscore, a percent sign, a plus sign, or a hyphen. You can put all of these into a character class: [a-zA-Z0-9._%+-].

The domain and username are separated by an @ symbol ❷. The domain name ❸ has a slightly less permissive character class with only letters, numbers, periods, and hyphens: [a-zA-Z0-9.-]. And last will be the "dot-com" part (technically known as the *top-level domain*), which can really be dot-anything. This is between two and four characters.

The format for email addresses has a lot of weird rules. This regular expression won't match every possible valid email address, but it'll match almost any typical email address you'll encounter.

### Step 3: Find All Matches in the Clipboard Text

Now that you have specified the regular expressions for phone numbers and email addresses, you can let Python's re module do the hard work of finding all the matches on the clipboard. The pyperclip.paste() function will get a string value of the text on the clipboard, and the findall() regex method will return a list of tuples.

Make your program look like the following:

```
#! python3
# phoneAndEmail.py - Finds phone numbers and email addresses on the clipboard.

import pyperclip, re

phoneRegex = re.compile(r'''(
--snip--

# Find matches in clipboard text.
text = str(pyperclip.paste())

❶ matches = []
❷ for groups in phoneRegex.findall(text):
      phoneNum = '-'.join([groups[1], groups[3], groups[5]])
      if groups[8] != '':
          phoneNum += ' x' + groups[8]
      matches.append(phoneNum)
❸ for groups in emailRegex.findall(text):
      matches.append(groups[0])

# TODO: Copy results to the clipboard.
```

There is one tuple for each match, and each tuple contains strings for each group in the regular expression. Remember that group 0 matches the entire regular expression, so the group at index 0 of the tuple is the one you are interested in.

As you can see at ❶, you'll store the matches in a list variable named matches. It starts off as an empty list, and a couple for loops. For the email addresses, you append group 0 of each match ❸. For the matched phone numbers, you don't want to just append group 0. While the program *detects* phone numbers in several formats, you want the phone number appended to be in a single, standard format. The phoneNum variable contains a string built from groups 1, 3, 5, and 8 of the matched text ❷. (These groups are the area code, first three digits, last four digits, and extension.)

### Step 4: Join the Matches into a String for the Clipboard

Now that you have the email addresses and phone numbers as a list of strings in matches, you want to put them on the clipboard. The pyperclip.copy() function takes only a single string value, not a list of strings, so you call the join() method on matches.

To make it easier to see that the program is working, let's print any matches you find to the terminal. If no phone numbers or email addresses were found, the program should tell the user this.

Make your program look like the following:

```
#! python3
# phoneAndEmail.py - Finds phone numbers and email addresses on the clipboard.

--snip--
for groups in emailRegex.findall(text):
    matches.append(groups[0])

# Copy results to the clipboard.
if len(matches) > 0:
    pyperclip.copy('\n'.join(matches))
    print('Copied to clipboard:')
    print('\n'.join(matches))
else:
    print('No phone numbers or email addresses found.')
```

### Running the Program

For an example, open your web browser to the No Starch Press contact page at *https://nostarch.com/contactus/*, press CTRL-A to select all the text on the page, and press CTRL-C to copy it to the clipboard. When you run this program, the output will look something like this:

```
Copied to clipboard:
800-420-7240
415-863-9900
415-863-9950
info@nostarch.com
media@nostarch.com
academic@nostarch.com
info@nostarch.com
```

### Ideas for Similar Programs

Identifying patterns of text (and possibly substituting them with the sub() method) has many different potential applications. For example, you could:

- Find website URLs that begin with *http://* or *https://*.
- Clean up dates in different date formats (such as 3/14/2019, 03-14-2019, and 2015/3/19) by replacing them with dates in a single, standard format.

- Remove sensitive information such as Social Security or credit card numbers.
- Find common typos such as multiple   spaces  between    words, acciden- tally accidentally repeated words, or multiple exclamation marks at the end of sentences. Those are annoying!!

## Summary

While a computer can search for text quickly, it must be told precisely what to look for. Regular expressions allow you to specify the pattern of charac- ters you are looking for, rather than the exact text itself. In fact, some word processing and spreadsheet applications provide find-and-replace features that allow you to search using regular expressions.

The re module that comes with Python lets you compile Regex objects. These objects have several methods: search() to find a single match, findall() to find all matching instances, and sub() to do a find-and-replace substitution of text.

You can find out more in the official Python documentation at *https:// docs.python.org/3/library/re.html*. Another useful resource is the tutorial website *https://www.regular-expressions.info/*.

## Practice Questions

1.  What is the function that creates Regex objects?
2.  Why are raw strings often used when creating Regex objects?
3.  What does the search() method return?
4.  How do you get the actual strings that match the pattern from a Match object?
5.  In the regex created from r'(\d\d\d)-(\d\d\d-\d\d\d\d)', what does group 0 cover? Group 1? Group 2?
6.  Parentheses and periods have specific meanings in regular expression syntax. How would you specify that you want a regex to match actual parentheses and period characters?
7.  The findall() method returns a list of strings or a list of tuples of strings. What makes it return one or the other?
8.  What does the | character signify in regular expressions?
9.  What two things does the ? character signify in regular expressions?
10. What is the difference between the + and * characters in regular expressions?
11. What is the difference between {3} and {3,5} in regular expressions?
12. What do the \d, \w, and \s shorthand character classes signify in regular expressions?

13. What do the \D, \W, and \S shorthand character classes signify in regular expressions?

14. What is the difference between .* and .*??

15. What is the character class syntax to match all numbers and lowercase letters?

16. How do you make a regular expression case-insensitive?

17. What does the . character normally match? What does it match if re.DOTALL is passed as the second argument to re.compile()?

18. If numRegex = re.compile(r'\d+'), what will numRegex.sub('X', '12 drummers, 11 pipers, five rings, 3 hens') return?

19. What does passing re.VERBOSE as the second argument to re.compile() allow you to do?

20. How would you write a regex that matches a number with commas for every three digits? It must match the following:

    - '42'
    - '1,234'
    - '6,368,745'

    but not the following:

    - '12,34,567' (which has only two digits between the commas)
    - '1234' (which lacks commas)

21. How would you write a regex that matches the full name of someone whose last name is Watanabe? You can assume that the first name that comes before it will always be one word that begins with a capital letter. The regex must match the following:

    - 'Haruto Watanabe'
    - 'Alice Watanabe'
    - 'RoboCop Watanabe'

    but not the following:

    - 'haruto Watanabe' (where the first name is not capitalized)
    - 'Mr. Watanabe' (where the preceding word has a nonletter character)
    - 'Watanabe' (which has no first name)
    - 'Haruto watanabe' (where Watanabe is not capitalized)

22. How would you write a regex that matches a sentence where the first word is either *Alice*, *Bob*, or *Carol*; the second word is either *eats*, *pets*, or *throws*; the third word is *apples*, *cats*, or *baseballs*; and the sentence ends with a period? This regex should be case-insensitive. It must match the following:

    - 'Alice eats apples.'
    - 'Bob pets cats.'
    - 'Carol throws baseballs.'
    - 'Alice throws Apples.'
    - 'BOB EATS CATS.'

but not the following:

- `'RoboCop eats apples.'`
- `'ALICE THROWS FOOTBALLS.'`
- `'Carol eats 7 cats.'`

## Practice Projects

For practice, write programs to do the following tasks.

### Date Detection

Write a regular expression that can detect dates in the *DD/MM/YYYY* format. Assume that the days range from 01 to 31, the months range from 01 to 12, and the years range from 1000 to 2999. Note that if the day or month is a single digit, it'll have a leading zero.

The regular expression doesn't have to detect correct days for each month or for leap years; it will accept nonexistent dates like 31/02/2020 or 31/04/2021. Then store these strings into variables named month, day, and year, and write additional code that can detect if it is a valid date. April, June, September, and November have 30 days, February has 28 days, and the rest of the months have 31 days. February has 29 days in leap years. Leap years are every year evenly divisible by 4, except for years evenly divisible by 100, unless the year is also evenly divisible by 400. Note how this calculation makes it impossible to make a reasonably sized regular expression that can detect a valid date.

### Strong Password Detection

Write a function that uses regular expressions to make sure the password string it is passed is strong. A strong password is defined as one that is at least eight characters long, contains both uppercase and lowercase characters, and has at least one digit. You may need to test the string against multiple regex patterns to validate its strength.

### Regex Version of the strip() Method

Write a function that takes a string and does the same thing as the strip() string method. If no other arguments are passed other than the string to strip, then whitespace characters will be removed from the beginning and end of the string. Otherwise, the characters specified in the second argument to the function will be removed from the string.

# 8

## INPUT VALIDATION

*Input validation* code checks that values entered by the user, such as text from the `input()` function, are formatted correctly. For example, if you want users to enter their ages, your code shouldn't accept nonsensical answers such as negative numbers (which are outside the range of acceptable integers) or words (which are the wrong data type). Input validation can also prevent bugs or security vulnerabilities. If you implement a `withdrawFromAccount()` function that takes an argument for the amount to subtract from an account, you need to ensure the amount is a positive number. If the `withdrawFromAccount()` function subtracts a negative number from the account, the "withdrawal" will end up adding money!

Typically, we perform input validation by repeatedly asking the user for input until they enter valid text, as in the following example:

```
while True:
    print('Enter your age:')
    age = input()
    try:
        age = int(age)
    except:
        print('Please use numeric digits.')
        continue
    if age < 1:
        print('Please enter a positive number.')
        continue
    break

print(f'Your age is {age}.')
```

When you run this program, the output could look like this:

```
Enter your age:
five
Please use numeric digits.
Enter your age:
-2
Please enter a positive number.
Enter your age:
30
Your age is 30.
```

When you run this code, you'll be prompted for your age until you enter a valid one. This ensures that by the time the execution leaves the while loop, the age variable will contain a valid value that won't crash the program later on.

However, writing input validation code for every input() call in your program quickly becomes tedious. Also, you may miss certain cases and allow invalid input to pass through your checks. In this chapter, you'll learn how to use the third-party PyInputPlus module for input validation.

# The PyInputPlus Module

PyInputPlus contains functions similar to input() for several kinds of data: numbers, dates, email addresses, and more. If the user ever enters invalid input, such as a badly formatted date or a number that is outside of an intended range, PyInputPlus will reprompt them for input just like our code in the previous section did. PyInputPlus also has other useful features like a limit for the number of times it reprompts users and a timeout if users are required to respond within a time limit.

PyInputPlus is not a part of the Python Standard Library, so you must install it separately using Pip. To install PyInputPlus, run `pip install --user pyinputplus` from the command line. Appendix A has complete instructions for installing third-party modules. To check if PyInputPlus installed correctly, import it in the interactive shell:

```
>>> import pyinputplus
```

If no errors appear when you import the module, it has been successfully installed.

PyInputPlus has several functions for different kinds of input:

**inputStr()**   Is like the built-in `input()` function but has the general PyInputPlus features. You can also pass a custom validation function to it

**inputNum()**   Ensures the user enters a number and returns an int or float, depending on if the number has a decimal point in it

**inputChoice()**   Ensures the user enters one of the provided choices

**inputMenu()**   Is similar to `inputChoice()`, but provides a menu with numbered or lettered options

**inputDatetime()**   Ensures the user enters a date and time

**inputYesNo()**   Ensures the user enters a "yes" or "no" response

**inputBool()**   Is similar to `inputYesNo()`, but takes a "True" or "False" response and returns a Boolean value

**inputEmail()**   Ensures the user enters a valid email address

**inputFilepath()**   Ensures the user enters a valid file path and filename, and can optionally check that a file with that name exists

**inputPassword()**   Is like the built-in `input()`, but displays * characters as the user types so that passwords, or other sensitive information, aren't displayed on the screen

These functions will automatically reprompt the user for as long as they enter invalid input:

```
>>> import pyinputplus as pyip
>>> response = pyip.inputNum()
five
'five' is not a number.
42
>>> response
42
```

The `as pyip` code in the `import` statement saves us from typing `pyinputplus` each time we want to call a PyInputPlus function. Instead we can use the shorter `pyip` name. If you take a look at the example, you see that unlike `input()`, these functions return an int or float value: 42 and 3.14 instead of the strings '42' and '3.14'.

Just as you can pass a string to input() to provide a prompt, you can pass a string to a PyInputPlus function's prompt keyword argument to display a prompt:

```
>>> response = input('Enter a number: ')
Enter a number: 42
>>> response
'42'
>>> import pyinputplus as pyip
>>> response = pyip.inputInt(prompt='Enter a number: ')
Enter a number: cat
'cat' is not an integer.
Enter a number: 42
>>> response
42
```

Use Python's help() function to find out more about each of these functions. For example, help(pyip.inputChoice) displays help information for the inputChoice() function. Complete documentation can be found at *https://pyinputplus.readthedocs.io/*.

Unlike Python's built-in input(), PyInputPlus functions have several additional features for input validation, as shown in the next section.

### The min, max, greaterThan, and lessThan Keyword Arguments

The inputNum(), inputInt(), and inputFloat() functions, which accept int and float numbers, also have min, max, greaterThan, and lessThan keyword arguments for specifying a range of valid values. For example, enter the following into the interactive shell:

```
>>> import pyinputplus as pyip
>>> response = pyip.inputNum('Enter num: ', min=4)
Enter num:3
Input must be at minimum 4.
Enter num:4
>>> response
4
>>> response = pyip.inputNum('Enter num: ', greaterThan=4)
Enter num: 4
Input must be greater than 4.
Enter num: 5
>>> response
5
>>> response = pyip.inputNum('>', min=4, lessThan=6)
Enter num: 6
Input must be less than 6.
Enter num: 3
Input must be at minimum 4.
Enter num: 4
>>> response
4
```

These keyword arguments are optional, but if supplied, the input cannot be less than the `min` argument or greater than the `max` argument (though the input can be equal to them). Also, the input must be greater than the `greaterThan` and less than the `lessThan` arguments (that is, the input cannot be equal to them).

## The blank Keyword Argument

By default, blank input isn't allowed unless the `blank` keyword argument is set to True:

```
>>> import pyinputplus as pyip
>>> response = pyip.inputNum('Enter num: ')
Enter num:(blank input entered here)
Blank values are not allowed.
Enter num: 42
>>> response
42
>>> response = pyip.inputNum(blank=True)
(blank input entered here)
>>> response
''
```

Use `blank=True` if you'd like to make input optional so that the user doesn't need to enter anything.

## The limit, timeout, and default Keyword Arguments

By default, the PyInputPlus functions will continue to ask the user for valid input forever (or for as long as the program runs). If you'd like a function to stop asking the user for input after a certain number of tries or a certain amount of time, you can use the `limit` and `timeout` keyword arguments. Pass an integer for the `limit` keyword argument to determine how many attempts a PyInputPlus function will make to receive valid input before giving up, and pass an integer for the `timeout` keyword argument to determine how many seconds the user has to enter valid input before the PyInputPlus function gives up.

If the user fails to enter valid input, these keyword arguments will cause the function to raise a `RetryLimitException` or `TimeoutException`, respectively. For example, enter the following into the interactive shell:

```
>>> import pyinputplus as pyip
>>> response = pyip.inputNum(limit=2)
blah
'blah' is not a number.
Enter num: number
'number' is not a number.
Traceback (most recent call last):
    --snip--
pyinputplus.RetryLimitException
>>> response = pyip.inputNum(timeout=10)
42 (entered after 10 seconds of waiting)
```

```
Traceback (most recent call last):
    --snip--
pyinputplus.TimeoutException
```

When you use these keyword arguments and also pass a default keyword argument, the function returns the default value instead of raising an exception. Enter the following into the interactive shell:

```
>>> response = pyip.inputNum(limit=2, default='N/A')
hello
'hello' is not a number.
world
'world' is not a number.
>>> response
'N/A'
```

Instead of raising RetryLimitException, the inputNum() function simply returns the string 'N/A'.

## The allowRegexes and blockRegexes Keyword Arguments

You can also use regular expressions to specify whether an input is allowed or not. The allowRegexes and blockRegexes keyword arguments take a list of regular expression strings to determine what the PyInputPlus function will accept or reject as valid input. For example, enter the following code into the interactive shell so that inputNum() will accept Roman numerals in addition to the usual numbers:

```
>>> import pyinputplus as pyip
>>> response = pyip.inputNum(allowRegexes=[r'(I|V|X|L|C|D|M)+', r'zero'])
XLII
>>> response
'XLII'
>>> response = pyip.inputNum(allowRegexes=[r'(i|v|x|l|c|d|m)+', r'zero'])
xlii
>>> response
'xlii'
```

Of course, this regex affects only what letters the inputNum() function will accept from the user; the function will still accept Roman numerals with invalid ordering such as 'XVX' or 'MILLI' because the r'(I|V|X|L|C|D|M)+' regular expression accepts those strings.

You can also specify a list of regular expression strings that a PyInputPlus function won't accept by using the blockRegexes keyword argument. Enter the following into the interactive shell so that inputNum() won't accept even numbers:

```
>>> import pyinputplus as pyip
>>> response = pyip.inputNum(blockRegexes=[r'[02468]$'])
42
This response is invalid.
```

```
44
This response is invalid.
43
>>> response
43
```

If you specify both an allowRegexes and blockRegexes argument, the allow list overrides the block list. For example, enter the following into the interactive shell, which allows 'caterpillar' and 'category' but blocks anything else that has the word 'cat' in it:

```
>>> import pyinputplus as pyip
>>> response = pyip.inputStr(allowRegexes=[r'caterpillar', 'category'],
blockRegexes=[r'cat'])
cat
This response is invalid.
catastrophe
This response is invalid.
category
>>> response
'category'
```

The PyInputPlus module's functions can save you from writing tedious input validation code yourself. But there's more to the PyInputPlus module than what has been detailed here. You can examine its full documentation online at *https://pyinputplus.readthedocs.io/*.

### Passing a Custom Validation Function to inputCustom()

You can write a function to perform your own custom validation logic by passing the function to inputCustom(). For example, say you want the user to enter a series of digits that adds up to 10. There is no pyinputplus .inputAddsUpToTen() function, but you can create your own function that:

- Accepts a single string argument of what the user entered
- Raises an exception if the string fails validation
- Returns None (or has no return statement) if inputCustom() should return the string unchanged
- Returns a non-None value if inputCustom() should return a different string from the one the user entered
- Is passed as the first argument to inputCustom()

For example, we can create our own addsUpToTen() function, and then pass it to inputCustom(). Note that the function call looks like inputCustom(addsUpToTen) and not inputCustom(addsUpToTen()) because we are passing the addsUpToTen() function itself to inputCustom(), not calling addsUpToTen() and passing its return value.

```
>>> import pyinputplus as pyip
>>> def addsUpToTen(numbers):
```

```
...     numbersList = list(numbers)
...     for i, digit in enumerate(numbersList):
...         numbersList[i] = int(digit)
...     if sum(numbersList) != 10:
...         raise Exception('The digits must add up to 10, not %s.' %
(sum(numbersList)))
...     return int(numbers) # Return an int form of numbers.
...
>>> response = pyip.inputCustom(addsUpToTen) # No parentheses after
addsUpToTen here.
123
The digits must add up to 10, not 6.
1235
The digits must add up to 10, not 11.
1234
>>> response # inputStr() returned an int, not a string.
1234
>>> response = pyip.inputCustom(addsUpToTen)
hello
invalid literal for int() with base 10: 'h'
55
>>> response
```

The inputCustom() function also supports the general PyInputPlus features, such as the blank, limit, timeout, default, allowRegexes, and blockRegexes keyword arguments. Writing your own custom validation function is useful when it's otherwise difficult or impossible to write a regular expression for valid input, as in the "adds up to 10" example.

## Project: How to Keep an Idiot Busy for Hours

Let's use PyInputPlus to create a simple program that does the following:

1.  Ask the user if they'd like to know how to keep an idiot busy for hours.
2.  If the user answers no, quit.
3.  If the user answers yes, go to Step 1.

Of course, we don't know if the user will enter something besides "yes" or "no," so we need to perform input validation. It would also be convenient for the user to be able to enter "y" or "n" instead of the full words. PyInputPlus's inputYesNo() function will handle this for us and, no matter what case the user enters, return a lowercase 'yes' or 'no' string value.

When you run this program, it should look like the following:

```
Want to know how to keep an idiot busy for hours?
sure
'sure' is not a valid yes/no response.
Want to know how to keep an idiot busy for hours?
yes
Want to know how to keep an idiot busy for hours?
y
```

```
Want to know how to keep an idiot busy for hours?
Yes
Want to know how to keep an idiot busy for hours?
YES
Want to know how to keep an idiot busy for hours?
YES!!!!!!
'YES!!!!!!' is not a valid yes/no response.
Want to know how to keep an idiot busy for hours?
TELL ME HOW TO KEEP AN IDIOT BUSY FOR HOURS.
'TELL ME HOW TO KEEP AN IDIOT BUSY FOR HOURS.' is not a valid yes/no response.
Want to know how to keep an idiot busy for hours?
no
Thank you. Have a nice day.
```

Open a new file editor tab and save it as *idiot.py*. Then enter the following code:

```
import pyinputplus as pyip
```

This imports the PyInputPlus module. Since `pyinputplus` is a bit much to type, we'll use the name `pyip` for short.

```
while True:
    prompt = 'Want to know how to keep an idiot busy for hours?\n'
    response = pyip.inputYesNo(prompt)
```

Next, `while True:` creates an infinite loop that continues to run until it encounters a break statement. In this loop, we call `pyip.inputYesNo()` to ensure that this function call won't return until the user enters a valid answer.

```
    if response == 'no':
        break
```

The `pyip.inputYesNo()` call is guaranteed to only return either the string yes or the string no. If it returned no, then our program breaks out of the infinite loop and continues to the last line, which thanks the user:

```
print('Thank you. Have a nice day.')
```

Otherwise, the loop iterates once again.

You can also make use of the `inputYesNo()` function in non-English languages by passing `yesVal` and `noVal` keyword arguments. For example, the Spanish version of this program would have these two lines:

```
    prompt = '¿Quieres saber cómo mantener ocupado a un idiota durante horas?\n'
    response = pyip.inputYesNo(prompt, yesVal='sí', noVal='no')
    if response == 'sí':
```

Now the user can enter either sí or s (in lower- or uppercase) instead of yes or y for an affirmative answer.

# Project: Multiplication Quiz

PyInputPlus's features can be useful for creating a timed multiplication quiz. By setting the allowRegexes, blockRegexes, timeout, and limit keyword argument to pyip.inputStr(), you can leave most of the implementation to PyInputPlus. The less code you need to write, the faster you can write your programs. Let's create a program that poses 10 multiplication problems to the user, where the valid input is the problem's correct answer. Open a new file editor tab and save the file as *multiplicationQuiz.py*.

First, we'll import pyinputplus, random, and time. We'll keep track of how many questions the program asks and how many correct answers the user gives with the variables numberOfQuestions and correctAnswers. A for loop will repeatedly pose a random multiplication problem 10 times:

```
import pyinputplus as pyip
import random, time

numberOfQuestions = 10
correctAnswers = 0
for questionNumber in range(numberOfQuestions):
```

Inside the for loop, the program will pick two single-digit numbers to multiply. We'll use these numbers to create a #Q: N × N = prompt for the user, where Q is the question number (1 to 10) and N are the two numbers to multiply.

```
# Pick two random numbers:
num1 = random.randint(0, 9)
num2 = random.randint(0, 9)

prompt = '#%s: %s x %s = ' % (questionNumber, num1, num2)
```

The pyip.inputStr() function will handle most of the features of this quiz program. The argument we pass for allowRegexes is a list with the regex string '^%s$', where %s is replaced with the correct answer. The ^ and % characters ensure that the answer begins and ends with the correct number, though PyInputPlus trims any whitespace from the start and end of the user's response first just in case they inadvertently pressed the spacebar before or after their answer. The argument we pass for blocklistRegexes is a list with ('.*', 'Incorrect!'). The first string in the tuple is a regex that matches every possible string. Therefore, if the user response doesn't match the correct answer, the program will reject any other answer they provide. In that case, the 'Incorrect!' string is displayed and the user is prompted to answer again. Additionally, passing 8 for timeout and 3 for limit will ensure that the user only has 8 seconds and 3 tries to provide a correct answer:

```
try:
    # Right answers are handled by allowRegexes.
    # Wrong answers are handled by blockRegexes, with a custom message.
    pyip.inputStr(prompt, allowRegexes=['^%s$' % (num1 * num2)],
```

```
blockRegexes=[('.*', 'Incorrect!')],
timeout=8, limit=3)
```

If the user answers after the 8-second timeout has expired, even if they answer correctly, pyip.inputStr() raises a TimeoutException exception. If the user answers incorrectly more than 3 times, it raises a RetryLimitException exception. Both of these exception types are in the PyInputPlus module, so pyip. needs to prepend them:

```
except pyip.TimeoutException:
    print('Out of time!')
except pyip.RetryLimitException:
    print('Out of tries!')
```

Remember that, just like how else blocks can follow an if or elif block, they can optionally follow the last except block. The code inside the following else block will run if no exception was raised in the try block. In our case, that means the code runs if the user entered the correct answer:

```
else:
    # This block runs if no exceptions were raised in the try block.
    print('Correct!')
    correctAnswers += 1
```

No matter which of the three messages, "Out of time!", "Out of tries!", or "Correct!", displays, let's place a 1-second pause at the end of the for loop to give the user time to read it. After the program has asked 10 questions and the for loop continues, let's show the user how many correct answers they made:

```
    time.sleep(1) # Brief pause to let user see the result.
print('Score: %s / %s' % (correctAnswers, numberOfQuestions))
```

PyInputPlus is flexible enough that you can use it in a wide variety of programs that take keyboard input from the user, as demonstrated by the programs in this chapter.

## Summary

It's easy to forget to write input validation code, but without it, your programs will almost certainly have bugs. The values you expect users to enter and the values they actually enter can be completely different, and your programs need to be robust enough to handle these exceptional cases. You can use regular expressions to create your own input validation code, but for common cases, it's easier to use an existing module, such as PyInputPlus. You can import the module with import pyinputplus as pyip so that you can enter a shorter name when calling the module's functions.

PyInputPlus has functions for entering a variety of input, including strings, numbers, dates, yes/no, True/False, emails, and files. While input()

always returns a string, these functions return the value in an appropriate data type. The inputChoice() function allow you to select one of several pre-selected options, while inputMenu() also adds numbers or letters for quick selection.

All of these functions have the following standard features: stripping whitespace from the sides, setting timeout and retry limits with the timeout and limit keyword arguments, and passing lists of regular expression strings to allowRegexes or blockRegexes to include or exclude particular responses. You'll no longer need to write your own tedious while loops that check for valid input and reprompt the user.

If none of the PyInputPlus module's, functions fit your needs, but you'd still like the other features that PyInputPlus provides, you can call inputCustom() and pass your own custom validation function for PyInputPlus to use. The documentation at *https://pyinputplus.readthedocs.io/en/latest/* has a complete listing of PyInputPlus's functions and additional features. There's far more in the PyInputPlus online documentation than what was described in this chapter. There's no use in reinventing the wheel, and learning to use this module will save you from having to write and debug code for yourself.

Now that you have expertise manipulating and validating text, it's time to learn how to read from and write to files on your computer's hard drive.

## Practice Questions

1. Does PyInputPlus come with the Python Standard Library?
2. Why is PyInputPlus commonly imported with import pyinputplus as pyip?
3. What is the difference between inputInt() and inputFloat()?
4. How can you ensure that the user enters a whole number between 0 and 99 using PyInputPlus?
5. What is passed to the allowRegexes and blockRegexes keyword arguments?
6. What does inputStr(limit=3) do if blank input is entered three times?
7. What does inputStr(limit=3, default='hello') do if blank input is entered three times?

## Practice Projects

For practice, write programs to do the following tasks.

### Sandwich Maker

Write a program that asks users for their sandwich preferences. The program should use PyInputPlus to ensure that they enter valid input, such as:

- Using inputMenu() for a bread type: wheat, white, or sourdough.
- Using inputMenu() for a protein type: chicken, turkey, ham, or tofu.

- Using `inputYesNo()` to ask if they want cheese.
- If so, using `inputMenu()` to ask for a cheese type: cheddar, Swiss, or mozzarella.
- Using `inputYesNo()` to ask if they want mayo, mustard, lettuce, or tomato.
- Using `inputInt()` to ask how many sandwiches they want. Make sure this number is 1 or more.

Come up with prices for each of these options, and have your program display a total cost after the user enters their selection.

### Write Your Own Multiplication Quiz

To see how much PyInputPlus is doing for you, try re-creating the multiplication quiz project on your own without importing it. This program will prompt the user with 10 multiplication questions, ranging from $0 \times 0$ to $9 \times 9$. You'll need to implement the following features:

- If the user enters the correct answer, the program displays "Correct!" for 1 second and moves on to the next question.
- The user gets three tries to enter the correct answer before the program moves on to the next question.
- Eight seconds after first displaying the question, the question is marked as incorrect even if the user enters the correct answer after the 8-second limit.

Compare your code to the code using PyInputPlus in "Project: Multiplication Quiz" on page 196.

# 9

## READING AND WRITING FILES

Variables are a fine way to store data while your program is running, but if you want your data to persist even after your program has finished, you need to save it to a file. You can think of a file's contents as a single string value, potentially gigabytes in size. In this chapter, you will learn how to use Python to create, read, and save files on the hard drive.

### Files and File Paths

A file has two key properties: a *filename* (usually written as one word) and a *path*. The path specifies the location of a file on the computer. For example, there is a file on my Windows laptop with the filename *project.docx* in the path *C:\Users\Al\Documents*. The part of the filename after the last period is called the file's *extension* and tells you a file's type. The filename *project.docx* is a Word document, and *Users, Al,* and *Documents* all refer to *folders* (also called *directories*). Folders can contain files and other folders. For example, *project.docx* is in the *Documents* folder, which is inside the *Al* folder, which is inside the *Users* folder. Figure 9-1 shows this folder organization.

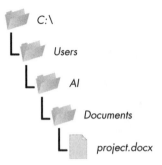

Figure 9-1: A file in a hierarchy of folders

The *C:\* part of the path is the *root folder*, which contains all other folders. On Windows, the root folder is named *C:\* and is also called the *C: drive*. On macOS and Linux, the root folder is */*. In this book, I'll use the Windows-style root folder, *C:\*. If you are entering the interactive shell examples on macOS or Linux, enter / instead.

Additional *volumes*, such as a DVD drive or USB flash drive, will appear differently on different operating systems. On Windows, they appear as new, lettered root drives, such as *D:\* or *E:\*. On macOS, they appear as new folders under the */Volumes* folder. On Linux, they appear as new folders under the */mnt* ("mount") folder. Also note that while folder names and filenames are not case-sensitive on Windows and macOS, they are case-sensitive on Linux.

**NOTE**     *Since your system probably has different files and folders on it than mine, you won't be able to follow every example in this chapter exactly. Still, try to follow along using folders that exist on your computer.*

## Backslash on Windows and Forward Slash on macOS and Linux

On Windows, paths are written using backslashes (\) as the separator between folder names. The macOS and Linux operating systems, however, use the forward slash (/) as their path separator. If you want your programs to work on all operating systems, you will have to write your Python scripts to handle both cases.

Fortunately, this is simple to do with the Path() function in the pathlib module. If you pass it the string values of individual file and folder names in your path, Path() will return a string with a file path using the correct path separators. Enter the following into the interactive shell:

```
>>> from pathlib import Path
>>> Path('spam', 'bacon', 'eggs')
```

```
WindowsPath('spam/bacon/eggs')
```

```
>>> str(Path('spam', 'bacon', 'eggs'))
'spam\\bacon\\eggs'
```

Note that the convention for importing pathlib is to run from pathlib import Path, since otherwise we'd have to enter pathlib.Path everywhere Path shows up in our code. Not only is this extra typing redundant, but it's also redundant.

I'm running this chapter's interactive shell examples on Windows, so Path('spam', 'bacon', 'eggs') returned a WindowsPath object for the joined path, represented as WindowsPath('spam/bacon/eggs'). Even though Windows uses backslashes, the WindowsPath representation in the interactive shell displays them using forward slashes, since open source software developers have historically favored the Linux operating system.

If you want to get a simple text string of this path, you can pass it to the str() function, which in our example returns 'spam\\bacon\\eggs'. (Notice that the backslashes are doubled because each backslash needs to be escaped by another backslash character.) If I had called this function on, say, Linux, Path() would have returned a PosixPath object that, when passed to str(), would have returned 'spam/bacon/eggs'. (*POSIX* is a set of standards for Unix-like operating systems such as Linux.)

These Path objects (really, WindowsPath or PosixPath objects, depending on your operating system) will be passed to several of the file-related functions introduced in this chapter. For example, the following code joins names from a list of filenames to the end of a folder's name:

```
>>> from pathlib import Path
>>> myFiles = ['accounts.txt', 'details.csv', 'invite.docx']
>>> for filename in myFiles:
        print(Path(r'C:\Users\Al', filename))
C:\Users\Al\accounts.txt
C:\Users\Al\details.csv
C:\Users\Al\invite.docx
```

On Windows, the backslash separates directories, so you can't use it in filenames. However, you can use backslashes in filenames on macOS and Linux. So while Path(r'spam\eggs') refers to two separate folders (or a file *eggs* in a folder *spam*) on Windows, the same command would refer to a single folder (or file) named *spam\eggs* on macOS and Linux. For this reason, it's usually a good idea to always use forward slashes in your Python code (and I'll be doing so for the rest of this chapter). The pathlib module will ensure that it always works on all operating systems.

Note that pathlib was introduced in Python 3.4 to replace older os.path functions. The Python Standard Library modules support it as of Python 3.6, but if you are working with legacy Python 2 versions, I recommend using pathlib2, which gives you pathlib's features on Python 2.7. Appendix A has instructions for installing pathlib2 using pip. Whenever I've replaced an older os.path function with pathlib, I've made a short note. You can look up the older functions at *https://docs.python.org/3/library/os.path.html*.

## Using the / Operator to Join Paths

We normally use the + operator to add two integer or floating-point numbers, such as in the expression 2 + 2, which evaluates to the integer value 4. But we can also use the + operator to concatenate two string values, like the expression 'Hello' + 'World', which evaluates to the string value 'HelloWorld'. Similarly, the / operator that we normally use for division can also combine Path objects and strings. This is helpful for modifying a Path object after you've already created it with the Path() function.

For example, enter the following into the interactive shell:

```
>>> from pathlib import Path
>>> Path('spam') / 'bacon' / 'eggs'
WindowsPath('spam/bacon/eggs')
>>> Path('spam') / Path('bacon/eggs')
WindowsPath('spam/bacon/eggs')
>>> Path('spam') / Path('bacon', 'eggs')
WindowsPath('spam/bacon/eggs')
```

Using the / operator with Path objects makes joining paths just as easy as string concatenation. It's also safer than using string concatenation or the join() method, like we do in this example:

```
>>> homeFolder = r'C:\Users\Al'
>>> subFolder = 'spam'
>>> homeFolder + '\\' + subFolder
'C:\\Users\\Al\\spam'
>>> '\\'.join([homeFolder, subFolder])
'C:\\Users\\Al\\spam'
```

A script that uses this code isn't safe, because its backslashes would only work on Windows. You could add an if statement that checks sys.platform (which contains a string describing the computer's operating system) to decide what kind of slash to use, but applying this custom code everywhere it's needed can be inconsistent and bug-prone.

The pathlib module solves these problems by reusing the / math division operator to join paths correctly, no matter what operating system your code is running on. The following example uses this strategy to join the same paths as in the previous example:

```
>>> homeFolder = Path('C:/Users/Al')
>>> subFolder = Path('spam')
>>> homeFolder / subFolder
WindowsPath('C:/Users/Al/spam')
>>> str(homeFolder / subFolder)
'C:\\Users\\Al\\spam'
```

The only thing you need to keep in mind when using the / operator for joining paths is that one of the first two values must be a Path object.

Python will give you an error if you try entering the following into the interactive shell:

```
>>> 'spam' / 'bacon' / 'eggs'
Traceback (most recent call last):
  File "<stdin>", line 1, in <module>
TypeError: unsupported operand type(s) for /: 'str' and 'str'
```

Python evaluates the / operator from left to right and evaluates to a Path object, so either the first or second leftmost value must be a Path object for the entire expression to evaluate to a Path object. Here's how the / operator and a Path object evaluate to the final Path object.

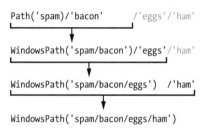

If you see the TypeError: unsupported operand type(s) for /: 'str' and 'str' error message shown previously, you need to put a Path object on the left side of the expression.

The / operator replaces the older os.path.join() function, which you can learn more about from *https://docs.python.org/3/library/os.path.html#os .path.join*.

### The Current Working Directory

Every program that runs on your computer has a *current working directory*, or *cwd*. Any filenames or paths that do not begin with the root folder are assumed to be under the current working directory.

**NOTE**      *While* folder *is the more modern name for directory, note that* current working directory *(or just* working directory*) is the standard term, not "current working folder."*

You can get the current working directory as a string value with the Path.cwd() function and change it using os.chdir(). Enter the following into the interactive shell:

```
>>> from pathlib import Path
>>> import os
>>> Path.cwd()
WindowsPath('C:/Users/Al/AppData/Local/Programs/Python/Python37')'
>>> os.chdir('C:\\Windows\\System32')
>>> Path.cwd()
WindowsPath('C:/Windows/System32')
```

Here, the current working directory is set to *C:\Users\Al\AppData\Local \Programs\Python\Python37*, so the filename *project.docx* refers to *C:\Users\Al \AppData\Local\Programs\Python\Python37\project.docx*. When we change the current working directory to *C:\Windows\System32*, the filename *project.docx* is interpreted as *C:\Windows\System32\project.docx*.

Python will display an error if you try to change to a directory that does not exist.

```
>>> os.chdir('C:/ThisFolderDoesNotExist')
Traceback (most recent call last):
  File "<stdin>", line 1, in <module>
FileNotFoundError: [WinError 2] The system cannot find the file specified:
'C:/ThisFolderDoesNotExist'
```

There is no pathlib function for changing the working directory, because changing the current working directory while a program is running can often lead to subtle bugs.

The os.getcwd() function is the older way of getting the current working directory as a string.

### The Home Directory

All users have a folder for their own files on the computer called the *home directory* or *home folder*. You can get a Path object of the home folder by calling Path.home():

```
>>> Path.home()
WindowsPath('C:/Users/Al')
```

The home directories are located in a set place depending on your operating system:

- On Windows, home directories are under *C:\Users*.
- On Mac, home directories are under */Users*.
- On Linux, home directories are often under */home*.

Your scripts will almost certainly have permissions to read and write the files under your home directory, so it's an ideal place to put the files that your Python programs will work with.

### Absolute vs. Relative Paths

There are two ways to specify a file path:

- An *absolute path*, which always begins with the root folder
- A *relative path*, which is relative to the program's current working directory

There are also the *dot* (.) and *dot-dot* (..) folders. These are not real folders but special names that can be used in a path. A single period ("dot") for a folder name is shorthand for "this directory." Two periods ("dot-dot") means "the parent folder."

Figure 9-2 is an example of some folders and files. When the current working directory is set to *C:\bacon*, the relative paths for the other folders and files are set as they are in the figure.

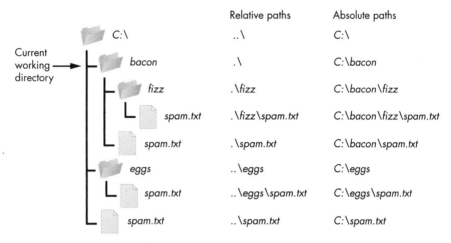

Figure 9-2: The relative paths for folders and files in the working directory C:\bacon

The .\ at the start of a relative path is optional. For example, *.\spam.txt* and *spam.txt* refer to the same file.

## Creating New Folders Using the os.makedirs() Function

Your programs can create new folders (directories) with the os.makedirs() function. Enter the following into the interactive shell:

```
>>> import os
>>> os.makedirs('C:\\delicious\\walnut\\waffles')
```

This will create not just the *C:\delicious* folder but also a *walnut* folder inside *C:\delicious* and a *waffles* folder inside *C:\delicious\walnut*. That is, os.makedirs() will create any necessary intermediate folders in order to ensure that the full path exists. Figure 9-3 shows this hierarchy of folders.

Figure 9-3: The result of os.makedirs('C:\\delicious\\walnut\\waffles')

To make a directory from a Path object, call the mkdir() method. For example, this code will create a *spam* folder under the home folder on my computer:

```
>>> from pathlib import Path
>>> Path(r'C:\Users\Al\spam').mkdir()
```

Note that mkdir() can only make one directory at a time; it won't make several subdirectories at once like os.makedirs().

## Handling Absolute and Relative Paths

The pathlib module provides methods for checking whether a given path is an absolute path and returning the absolute path of a relative path.

Calling the is_absolute() method on a Path object will return True if it represents an absolute path or False if it represents a relative path. For example, enter the following into the interactive shell, using your own files and folders instead of the exact ones listed here:

```
>>> Path.cwd()
WindowsPath('C:/Users/Al/AppData/Local/Programs/Python/Python37')
>>> Path.cwd().is_absolute()
True
>>> Path('spam/bacon/eggs').is_absolute()
False
```

To get an absolute path from a relative path, you can put Path.cwd() / in front of the relative Path object. After all, when we say "relative path," we almost always mean a path that is relative to the current working directory. Enter the following into the interactive shell:

```
>>> Path('my/relative/path')
WindowsPath('my/relative/path')
>>> Path.cwd() / Path('my/relative/path')
WindowsPath('C:/Users/Al/AppData/Local/Programs/Python/Python37/my/relative/
path')
```

If your relative path is relative to another path besides the current working directory, just replace Path.cwd() with that other path instead. The following example gets an absolute path using the home directory instead of the current working directory:

```
>>> Path('my/relative/path')
WindowsPath('my/relative/path')
>>> Path.home() / Path('my/relative/path')
WindowsPath('C:/Users/Al/my/relative/path')
```

The os.path module also has some useful functions related to absolute and relative paths:

- Calling os.path.abspath(*path*) will return a string of the absolute path of the argument. This is an easy way to convert a relative path into an absolute one.

- Calling os.path.isabs(*path*) will return True if the argument is an absolute path and False if it is a relative path.

- Calling os.path.relpath(*path*, *start*) will return a string of a relative path from the *start* path to *path*. If *start* is not provided, the current working directory is used as the start path.

Try these functions in the interactive shell:

```
>>> os.path.abspath('.')
'C:\\Users\\Al\\AppData\\Local\\Programs\\Python\\Python37'
>>> os.path.abspath('.\\Scripts')
'C:\\Users\\Al\\AppData\\Local\\Programs\\Python\\Python37\\Scripts'
>>> os.path.isabs('.')
False
>>> os.path.isabs(os.path.abspath('.'))
True
```

Since *C:\Users\Al\AppData\Local\Programs\Python\Python37* was the working directory when os.path.abspath() was called, the "single-dot" folder represents the absolute path 'C:\\Users\\Al\\AppData\\Local\\Programs\\Python\\Python37'.

Enter the following calls to os.path.relpath() into the interactive shell:

```
>>> os.path.relpath('C:\\Windows', 'C:\\')
'Windows'
>>> os.path.relpath('C:\\Windows', 'C:\\spam\\eggs')
'..\\..\\Windows'
```

When the relative path is within the same parent folder as the path, but is within subfolders of a different path, such as 'C:\\Windows' and 'C:\\spam\\eggs', you can use the "dot-dot" notation to return to the parent folder.

## Getting the Parts of a File Path

Given a Path object, you can extract the file path's different parts as strings using several Path object attributes. These can be useful for constructing new file paths based on existing ones. The attributes are diagrammed in Figure 9-4.

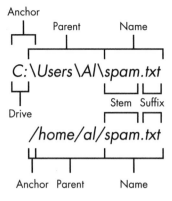

Figure 9-4: The parts of a Windows (top) and macOS/Linux (bottom) file path

The parts of a file path include the following:

- The *anchor*, which is the root folder of the filesystem
- On Windows, the *drive*, which is the single letter that often denotes a physical hard drive or other storage device
- The *parent*, which is the folder that contains the file
- The *name* of the file, made up of the *stem* (or *base name*) and the *suffix* (or *extension*)

Note that Windows Path objects have a drive attribute, but macOS and Linux Path objects don't. The drive attribute doesn't include the first backslash.

To extract each attribute from the file path, enter the following into the interactive shell:

```
>>> p = Path('C:/Users/Al/spam.txt')
>>> p.anchor
'C:\\'
>>> p.parent # This is a Path object, not a string.
WindowsPath('C:/Users/Al')
>>> p.name
'spam.txt'
>>> p.stem
'spam'
>>> p.suffix
'.txt'
>>> p.drive
'C:'
```

These attributes evaluate to simple string values, except for parent, which evaluates to another Path object.

The parents attribute (which is different from the parent attribute) evaluates to the ancestor folders of a Path object with an integer index:

```
>>> Path.cwd()
WindowsPath('C:/Users/Al/AppData/Local/Programs/Python/Python37')
>>> Path.cwd().parents[0]
WindowsPath('C:/Users/Al/AppData/Local/Programs/Python')
>>> Path.cwd().parents[1]
WindowsPath('C:/Users/Al/AppData/Local/Programs')
>>> Path.cwd().parents[2]
WindowsPath('C:/Users/Al/AppData/Local')
>>> Path.cwd().parents[3]
WindowsPath('C:/Users/Al/AppData')
>>> Path.cwd().parents[4]
WindowsPath('C:/Users/Al')
>>> Path.cwd().parents[5]
WindowsPath('C:/Users')
>>> Path.cwd().parents[6]
WindowsPath('C:/')
```

The older os.path module also has similar functions for getting the different parts of a path written in a string value. Calling os.path.dirname(*path*) will return a string of everything that comes before the last slash in the path argument. Calling os.path.basename(*path*) will return a string of everything that comes after the last slash in the path argument. The directory (or dir) name and base name of a path are outlined in Figure 9-5.

*C:\Windows\System32\calc.exe*

Dir name          Base name

*Figure 9-5: The base name follows the last slash in a path and is the same as the filename. The dir name is everything before the last slash.*

For example, enter the following into the interactive shell:

```
>>> calcFilePath = 'C:\\Windows\\System32\\calc.exe'
>>> os.path.basename(calcFilePath)
'calc.exe'
>>> os.path.dirname(calcFilePath)
'C:\\Windows\\System32'
```

If you need a path's dir name and base name together, you can just call os.path.split() to get a tuple value with these two strings, like so:

```
>>> calcFilePath = 'C:\\Windows\\System32\\calc.exe'
>>> os.path.split(calcFilePath)
('C:\\Windows\\System32', 'calc.exe')
```

Notice that you could create the same tuple by calling os.path.dirname()
and os.path.basename() and placing their return values in a tuple:

```
>>> (os.path.dirname(calcFilePath), os.path.basename(calcFilePath))
('C:\\Windows\\System32', 'calc.exe')
```

But os.path.split() is a nice shortcut if you need both values.

Also, note that os.path.split() does *not* take a file path and return a list
of strings of each folder. For that, use the split() string method and split on
the string in os.sep. (Note that sep is in os, not os.path.) The os.sep variable
is set to the correct folder-separating slash for the computer running the
program, '\\' on Windows and '/' on macOS and Linux, and splitting on
it will return a list of the individual folders.

For example, enter the following into the interactive shell:

```
>>> calcFilePath.split(os.sep)
['C:', 'Windows', 'System32', 'calc.exe']
```

This returns all the parts of the path as strings.

On macOS and Linux systems, the returned list of folders will begin
with a blank string, like this:

```
>>> '/usr/bin'.split(os. sep)
['', 'usr', 'bin']
```

The split() string method will work to return a list of each part of
the path.

## Finding File Sizes and Folder Contents

Once you have ways of handling file paths, you can then start gathering
information about specific files and folders. The os.path module provides
functions for finding the size of a file in bytes and the files and folders
inside a given folder.

- Calling os.path.getsize(*path*) will return the size in bytes of the file in
  the *path* argument.
- Calling os.listdir(*path*) will return a list of filename strings for each file
  in the *path* argument. (Note that this function is in the os module, not
  os.path.)

Here's what I get when I try these functions in the interactive shell:

```
>>> os.path.getsize('C:\\Windows\\System32\\calc.exe')
27648
>>> os.listdir('C:\\Windows\\System32')
['0409', '12520437.cpx', '12520850.cpx', '5U877.ax', 'aaclient.dll',
--snip--
'xwtpdui.dll', 'xwtpw32.dll', 'zh-CN', 'zh-HK', 'zh-TW', 'zipfldr.dll']
```

As you can see, the *calc.exe* program on my computer is 27,648 bytes in size, and I have a lot of files in *C:\Windows\system32*. If I want to find the total size of all the files in this directory, I can use os.path.getsize() and os.listdir() together.

```
>>> totalSize = 0
>>> for filename in os.listdir('C:\\Windows\\System32'):
        totalSize = totalSize + os.path.getsize(os.path.join('C:\\Windows\\System32', filename))
>>> print(totalSize)
2559970473
```

As I loop over each filename in the *C:\Windows\System32* folder, the totalSize variable is incremented by the size of each file. Notice how when I call os.path.getsize(), I use os.path.join() to join the folder name with the current filename. The integer that os.path.getsize() returns is added to the value of totalSize. After looping through all the files, I print totalSize to see the total size of the *C:\Windows\System32* folder.

## Modifying a List of Files Using Glob Patterns

If you want to work on specific files, the glob() method is simpler to use than listdir(). Path objects have a glob() method for listing the contents of a folder according to a *glob pattern*. Glob patterns are like a simplified form of regular expressions often used in command line commands. The glob() method returns a generator object (which are beyond the scope of this book) that you'll need to pass to list() to easily view in the interactive shell:

```
>>> p = Path('C:/Users/Al/Desktop')
>>> p.glob('*')
<generator object Path.glob at 0x000002A6E389DED0>
>>> list(p.glob('*')) # Make a list from the generator.
[WindowsPath('C:/Users/Al/Desktop/1.png'), WindowsPath('C:/Users/Al/
Desktop/22-ap.pdf'), WindowsPath('C:/Users/Al/Desktop/cat.jpg'),
  --snip--
WindowsPath('C:/Users/Al/Desktop/zzz.txt')]
```

The asterisk (*) stands for "multiple of any characters," so p.glob('*') returns a generator of all files in the path stored in p.

Like with regexes, you can create complex expressions:

```
>>> list(p.glob('*.txt') # Lists all text files.
[WindowsPath('C:/Users/Al/Desktop/foo.txt'),
  --snip--
WindowsPath('C:/Users/Al/Desktop/zzz.txt')]
```

The glob pattern '*.txt' will return files that start with any combination of characters as long as it ends with the string '.txt', which is the text file extension.

In contrast with the asterisk, the question mark (?) stands for any single character:

```
>>> list(p.glob('project?.docx'))
[WindowsPath('C:/Users/Al/Desktop/project1.docx'), WindowsPath('C:/Users/Al/
Desktop/project2.docx'),
  --snip--
WindowsPath('C:/Users/Al/Desktop/project9.docx')]
```

The glob expression 'project?.docx' will return 'project1.docx' or 'project5.docx', but it will not return 'project10.docx', because ? only matches to one character—so it will not match to the two-character string '10'.

Finally, you can also combine the asterisk and question mark to create even more complex glob expressions, like this:

```
>>> list(p.glob('*.?x?'))
[WindowsPath('C:/Users/Al/Desktop/calc.exe'), WindowsPath('C:/Users/Al/
Desktop/foo.txt'),
  --snip--
WindowsPath('C:/Users/Al/Desktop/zzz.txt')]
```

The glob expression '*.?x?' will return files with any name and any three-character extension where the middle character is an 'x'.

By picking out files with specific attributes, the glob() method lets you easily specify the files in a directory you want to perform some operation on. You can use a for loop to iterate over the generator that glob() returns:

```
>>> p = Path('C:/Users/Al/Desktop')
>>> for textFilePathObj in p.glob('*.txt'):
...     print(textFilePathObj) # Prints the Path object as a string.
...     # Do something with the text file.
...
C:\Users\Al\Desktop\foo.txt
C:\Users\Al\Desktop\spam.txt
C:\Users\Al\Desktop\zzz.txt
```

If you want to perform some operation on every file in a directory, you can use either os.listdir(p) or p.glob('*').

## Checking Path Validity

Many Python functions will crash with an error if you supply them with a path that does not exist. Luckily, Path objects have methods to check whether a given path exists and whether it is a file or folder. Assuming that a variable p holds a Path object, you could expect the following:

- Calling p.exists() returns True if the path exists or returns False if it doesn't exist.

- Calling p.is_file() returns True if the path exists and is a file, or returns False otherwise.

- Calling `p.is_dir()` returns True if the path exists and is a directory, or returns False otherwise.

On my computer, here's what I get when I try these methods in the interactive shell:

```
>>> winDir = Path('C:/Windows')
>>> notExistsDir = Path('C:/This/Folder/Does/Not/Exist')
>>> calcFile = Path('C:/Windows
/System32/calc.exe')
>>> winDir.exists()
True
>>> winDir.is_dir()
True
>>> notExistsDir.exists()
False
>>> calcFile.is_file()
True
>>> calcFile.is_dir()
False
```

You can determine whether there is a DVD or flash drive currently attached to the computer by checking for it with the exists() method. For instance, if I wanted to check for a flash drive with the volume named *D:\* on my Windows computer, I could do that with the following:

```
>>> dDrive = Path('D:/')
>>> dDrive.exists()
False
```

Oops! It looks like I forgot to plug in my flash drive.

The older os.path module can accomplish the same task with the os.path.exists(*path*), os.path.isfile(*path*), and os.path.isdir(*path*) functions, which act just like their Path function counterparts. As of Python 3.6, these functions can accept Path objects as well as strings of the file paths.

## The File Reading/Writing Process

Once you are comfortable working with folders and relative paths, you'll be able to specify the location of files to read and write. The functions covered in the next few sections will apply to plaintext files. *Plaintext files* contain only basic text characters and do not include font, size, or color information. Text files with the *.txt* extension or Python script files with the *.py* extension are examples of plaintext files. These can be opened with Windows's Notepad or macOS's TextEdit application. Your programs can easily read the contents of plaintext files and treat them as an ordinary string value.

*Binary files* are all other file types, such as word processing documents, PDFs, images, spreadsheets, and executable programs. If you open a binary file in Notepad or TextEdit, it will look like scrambled nonsense, like in Figure 9-6.

Figure 9-6: The Windows `calc.exe` program opened in Notepad

Since every different type of binary file must be handled in its own way, this book will not go into reading and writing raw binary files directly. Fortunately, many modules make working with binary files easier—you will explore one of them, the shelve module, later in this chapter. The pathlib module's read_text() method returns a string of the full contents of a text file. Its write_text() method creates a new text file (or overwrites an existing one) with the string passed to it. Enter the following into the interactive shell:

```
>>> from pathlib import Path
>>> p = Path('spam.txt')
>>> p.write_text('Hello, world!')
13
>>> p.read_text()
'Hello, world!'
```

These method calls create a *spam.txt* file with the content 'Hello, world!'. The 13 that write_text() returns indicates that 13 characters were written to the file. (You can often disregard this information.) The read_text() call reads and returns the contents of our new file as a string: 'Hello, world!'.

Keep in mind that these Path object methods only provide basic interactions with files. The more common way of writing to a file involves using the open() function and file objects. There are three steps to reading or writing files in Python:

1. Call the open() function to return a File object.

2. Call the read() or write() method on the File object.

3. Close the file by calling the close() method on the File object.

We'll go over these steps in the following sections.

### Opening Files with the open() Function

To open a file with the open() function, you pass it a string path indicating the file you want to open; it can be either an absolute or relative path. The open() function returns a File object.

Try it by creating a text file named *hello.txt* using Notepad or TextEdit. Type **Hello, world!** as the content of this text file and save it in your user home folder. Then enter the following into the interactive shell:

```
>>> helloFile = open(Path.home() / 'hello.txt')
```

The open() function can also accept strings. If you're using Windows, enter the following into the interactive shell:

```
>>> helloFile = open('C:\\Users\\your_home_folder\\hello.txt')
```

If you're using macOS, enter the following into the interactive shell instead:

```
>>> helloFile = open('/Users/your_home_folder/hello.txt')
```

Make sure to replace *your_home_folder* with your computer username. For example, my username is *Al*, so I'd enter 'C:\\Users\\Al\\hello.txt' on Windows. Note that the open() function only accepts Path objects as of Python 3.6. In previous versions, you always need to pass a string to open().

Both these commands will open the file in "reading plaintext" mode, or *read mode* for short. When a file is opened in read mode, Python lets you only read data from the file; you can't write or modify it in any way. Read mode is the default mode for files you open in Python. But if you don't want to rely on Python's defaults, you can explicitly specify the mode by passing the string value 'r' as a second argument to open(). So open('/Users/Al/hello .txt', 'r') and open('/Users/Al/hello.txt') do the same thing.

The call to open() returns a File object. A File object represents a file on your computer; it is simply another type of value in Python, much like the lists and dictionaries you're already familiar with. In the previous example, you stored the File object in the variable helloFile. Now, whenever you want to read from or write to the file, you can do so by calling methods on the File object in helloFile.

### Reading the Contents of Files

Now that you have a File object, you can start reading from it. If you want to read the entire contents of a file as a string value, use the File object's read() method. Let's continue with the *hello.txt* File object you stored in helloFile. Enter the following into the interactive shell:

```
>>> helloContent = helloFile.read()
>>> helloContent
'Hello, world!'
```

If you think of the contents of a file as a single large string value, the read() method returns the string that is stored in the file.

Alternatively, you can use the readlines() method to get a *list* of string values from the file, one string for each line of text. For example, create a file named *sonnet29.txt* in the same directory as *hello.txt* and write the following text in it:

```
When, in disgrace with fortune and men's eyes,
I all alone beweep my outcast state,
And trouble deaf heaven with my bootless cries,
And look upon myself and curse my fate,
```

Make sure to separate the four lines with line breaks. Then enter the following into the interactive shell:

```
>>> sonnetFile = open(Path.home() / 'sonnet29.txt')
>>> sonnetFile.readlines()
[When, in disgrace with fortune and men's eyes,\n', ' I all alone beweep my
outcast state,\n', And trouble deaf heaven with my bootless cries,\n', And
look upon myself and curse my fate,']
```

Note that, except for the last line of the file, each of the string values ends with a newline character \n. A list of strings is often easier to work with than a single large string value.

## Writing to Files

Python allows you to write content to a file in a way similar to how the print() function "writes" strings to the screen. You can't write to a file you've opened in read mode, though. Instead, you need to open it in "write plaintext" mode or "append plaintext" mode, or *write mode* and *append mode* for short.

Write mode will overwrite the existing file and start from scratch, just like when you overwrite a variable's value with a new value. Pass 'w' as the second argument to open() to open the file in write mode. Append mode, on the other hand, will append text to the end of the existing file. You can think of this as appending to a list in a variable, rather than overwriting the variable altogether. Pass 'a' as the second argument to open() to open the file in append mode.

If the filename passed to open() does not exist, both write and append mode will create a new, blank file. After reading or writing a file, call the close() method before opening the file again.

Let's put these concepts together. Enter the following into the interactive shell:

```
>>> baconFile = open('bacon.txt', 'w')
>>> baconFile.write('Hello, world!\n')
13
>>> baconFile.close()
>>> baconFile = open('bacon.txt', 'a')
>>> baconFile.write('Bacon is not a vegetable.')
```

```
25
>>> baconFile.close()
>>> baconFile = open('bacon.txt')
>>> content = baconFile.read()
>>> baconFile.close()
>>> print(content)
Hello, world!
Bacon is not a vegetable.
```

First, we open *bacon.txt* in write mode. Since there isn't a *bacon.txt* yet, Python creates one. Calling write() on the opened file and passing write() the string argument 'Hello, world! /n' writes the string to the file and returns the number of characters written, including the newline. Then we close the file.

To add text to the existing contents of the file instead of replacing the string we just wrote, we open the file in append mode. We write 'Bacon is not a vegetable.' to the file and close it. Finally, to print the file contents to the screen, we open the file in its default read mode, call read(), store the resulting File object in content, close the file, and print content.

Note that the write() method does not automatically add a newline character to the end of the string like the print() function does. You will have to add this character yourself.

As of Python 3.6, you can also pass a Path object to the open() function instead of a string for the filename.

## Saving Variables with the shelve Module

You can save variables in your Python programs to binary shelf files using the shelve module. This way, your program can restore data to variables from the hard drive. The shelve module will let you add Save and Open features to your program. For example, if you ran a program and entered some configuration settings, you could save those settings to a shelf file and then have the program load them the next time it is run.

Enter the following into the interactive shell:

```
>>> import shelve
>>> shelfFile = shelve.open('mydata')
>>> cats = ['Zophie', 'Pooka', 'Simon']
>>> shelfFile['cats'] = cats
>>> shelfFile.close()
```

To read and write data using the shelve module, you first import shelve. Call shelve.open() and pass it a filename, and then store the returned shelf value in a variable. You can make changes to the shelf value as if it were a dictionary. When you're done, call close() on the shelf value. Here, our shelf value is stored in shelfFile. We create a list cats and write shelfFile['cats'] = cats to store the list in shelfFile as a value associated with the key 'cats' (like in a dictionary). Then we call close() on shelfFile. Note that as of Python 3.7, you have to pass the open() shelf method filenames as strings. You can't pass it Path object.

After running the previous code on Windows, you will see three new files in the current working directory: *mydata.bak*, *mydata.dat*, and *mydata.dir*. On macOS, only a single *mydata.db* file will be created.

These binary files contain the data you stored in your shelf. The format of these binary files is not important; you only need to know what the shelve module does, not how it does it. The module frees you from worrying about how to store your program's data to a file.

Your programs can use the shelve module to later reopen and retrieve the data from these shelf files. Shelf values don't have to be opened in read or write mode—they can do both once opened. Enter the following into the interactive shell:

```
>>> shelfFile = shelve.open('mydata')
>>> type(shelfFile)
<class 'shelve.DbfilenameShelf'>
>>> shelfFile['cats']
['Zophie', 'Pooka', 'Simon']
>>> shelfFile.close()
```

Here, we open the shelf files to check that our data was stored correctly. Entering shelfFile['cats'] returns the same list that we stored earlier, so we know that the list is correctly stored, and we call close().

Just like dictionaries, shelf values have keys() and values() methods that will return list-like values of the keys and values in the shelf. Since these methods return list-like values instead of true lists, you should pass them to the list() function to get them in list form. Enter the following into the interactive shell:

```
>>> shelfFile = shelve.open('mydata')
>>> list(shelfFile.keys())
['cats']
>>> list(shelfFile.values())
[['Zophie', 'Pooka', 'Simon']]
>>> shelfFile.close()
```

Plaintext is useful for creating files that you'll read in a text editor such as Notepad or TextEdit, but if you want to save data from your Python programs, use the shelve module.

## Saving Variables with the pprint.pformat() Function

Recall from "Pretty Printing" on page 118 that the pprint.pprint() function will "pretty print" the contents of a list or dictionary to the screen, while the pprint.pformat() function will return this same text as a string instead of printing it. Not only is this string formatted to be easy to read, but it is also syntactically correct Python code. Say you have a dictionary stored in a variable and you want to save this variable and its contents for future use. Using `pprint.pformat()` will give you a string that you can write to a *.py* file. This file will be your very own module that you can import whenever you want to use the variable stored in it.

For example, enter the following into the interactive shell:

```
>>> import pprint
>>> cats = [{'name': 'Zophie', 'desc': 'chubby'}, {'name': 'Pooka', 'desc': 'fluffy'}]
>>> pprint.pformat(cats)
"[{'desc': 'chubby', 'name': 'Zophie'}, {'desc': 'fluffy', 'name': 'Pooka'}]"
>>> fileObj = open('myCats.py', 'w')
>>> fileObj.write('cats = ' + pprint.pformat(cats) + '\n')
83
>>> fileObj.close()
```

Here, we import `pprint` to let us use `pprint.pformat()`. We have a list of dictionaries, stored in a variable cats. To keep the list in cats available even after we close the shell, we use `pprint.pformat()` to return it as a string. Once we have the data in cats as a string, it's easy to write the string to a file, which we'll call *myCats.py*.

The modules that an `import` statement imports are themselves just Python scripts. When the string from `pprint.pformat()` is saved to a *.py* file, the file is a module that can be imported just like any other.

And since Python scripts are themselves just text files with the *.py* file extension, your Python programs can even generate other Python programs. You can then import these files into scripts.

```
>>> import myCats
>>> myCats.cats
[{'name': 'Zophie', 'desc': 'chubby'}, {'name': 'Pooka', 'desc': 'fluffy'}]
>>> myCats.cats[0]
{'name': 'Zophie', 'desc': 'chubby'}
>>> myCats.cats[0]['name']
'Zophie'
```

The benefit of creating a *.py* file (as opposed to saving variables with the `shelve` module) is that because it is a text file, the contents of the file can be read and modified by anyone with a simple text editor. For most applications, however, saving data using the `shelve` module is the preferred way to save variables to a file. Only basic data types such as integers, floats, strings, lists, and dictionaries can be written to a file as simple text. File objects, for example, cannot be encoded as text.

## Project: Generating Random Quiz Files

Say you're a geography teacher with 35 students in your class and you want to give a pop quiz on US state capitals. Alas, your class has a few bad eggs in it, and you can't trust the students not to cheat. You'd like to randomize the order of questions so that each quiz is unique, making it impossible for anyone to crib answers from anyone else. Of course, doing this by hand would be a lengthy and boring affair. Fortunately, you know some Python.

Here is what the program does:

1.  Creates 35 different quizzes
2.  Creates 50 multiple-choice questions for each quiz, in random order
3.  Provides the correct answer and three random wrong answers for each question, in random order
4.  Writes the quizzes to 35 text files
5.  Writes the answer keys to 35 text files

This means the code will need to do the following:

1.  Store the states and their capitals in a dictionary
2.  Call open(), write(), and close() for the quiz and answer key text files
3.  Use random.shuffle() to randomize the order of the questions and multiple-choice options

## Step 1: Store the Quiz Data in a Dictionary

The first step is to create a skeleton script and fill it with your quiz data. Create a file named *randomQuizGenerator.py*, and make it look like the following:

```python
#! python3
# randomQuizGenerator.py - Creates quizzes with questions and answers in
# random order, along with the answer key.

❶ import random

# The quiz data. Keys are states and values are their capitals.
❷ capitals = {'Alabama': 'Montgomery', 'Alaska': 'Juneau', 'Arizona': 'Phoenix',
'Arkansas': 'Little Rock', 'California': 'Sacramento', 'Colorado': 'Denver',
'Connecticut': 'Hartford', 'Delaware': 'Dover', 'Florida': 'Tallahassee',
'Georgia': 'Atlanta', 'Hawaii': 'Honolulu', 'Idaho': 'Boise', 'Illinois':
'Springfield', 'Indiana': 'Indianapolis', 'Iowa': 'Des Moines', 'Kansas':
'Topeka', 'Kentucky': 'Frankfort', 'Louisiana': 'Baton Rouge', 'Maine':
'Augusta', 'Maryland': 'Annapolis', 'Massachusetts': 'Boston', 'Michigan':
'Lansing', 'Minnesota': 'Saint Paul', 'Mississippi': 'Jackson', 'Missouri':
'Jefferson City', 'Montana': 'Helena', 'Nebraska': 'Lincoln', 'Nevada':
'Carson City', 'New Hampshire': 'Concord', 'New Jersey': 'Trenton', 'New
Mexico': 'Santa Fe', 'New York': 'Albany', 'North Carolina': 'Raleigh',
'North Dakota': 'Bismarck', 'Ohio': 'Columbus', 'Oklahoma': 'Oklahoma City',
'Oregon': 'Salem', 'Pennsylvania': 'Harrisburg', 'Rhode Island': 'Providence',
'South Carolina': 'Columbia', 'South Dakota': 'Pierre', 'Tennessee':
'Nashville', 'Texas': 'Austin', 'Utah': 'Salt Lake City', 'Vermont':
'Montpelier', 'Virginia': 'Richmond', 'Washington': 'Olympia', 'West
Virginia': 'Charleston', 'Wisconsin': 'Madison', 'Wyoming': 'Cheyenne'}

# Generate 35 quiz files.
```

```
❸ for quizNum in range(35):
        # TODO: Create the quiz and answer key files.

        # TODO: Write out the header for the quiz.

        # TODO: Shuffle the order of the states.

        # TODO: Loop through all 50 states, making a question for each.
```

Since this program will be randomly ordering the questions and answers, you'll need to import the random module ❶ to make use of its functions. The capitals variable ❷ contains a dictionary with US states as keys and their capitals as values. And since you want to create 35 quizzes, the code that actually generates the quiz and answer key files (marked with TODO comments for now) will go inside a for loop that loops 35 times ❸. (This number can be changed to generate any number of quiz files.)

### Step 2: Create the Quiz File and Shuffle the Question Order

Now it's time to start filling in those TODOs.

The code in the loop will be repeated 35 times—once for each quiz— so you have to worry about only one quiz at a time within the loop. First you'll create the actual quiz file. It needs to have a unique filename and should also have some kind of standard header in it, with places for the student to fill in a name, date, and class period. Then you'll need to get a list of states in randomized order, which can be used later to create the questions and answers for the quiz.

Add the following lines of code to *randomQuizGenerator.py*:

```
#! python3
# randomQuizGenerator.py - Creates quizzes with questions and answers in
# random order, along with the answer key.

--snip--

# Generate 35 quiz files.
for quizNum in range(35):
    # Create the quiz and answer key files.
  ❶ quizFile = open(f'capitalsquiz{quizNum + 1}.txt', 'w')
  ❷ answerKeyFile = open(f'capitalsquiz_answers{quizNum + 1}.txt', 'w')

    # Write out the header for the quiz.
  ❸ quizFile.write('Name:\n\nDate:\n\nPeriod:\n\n')
    quizFile.write((' ' * 20) + f'State Capitals Quiz (Form{quizNum + 1})')
    quizFile.write('\n\n')

    # Shuffle the order of the states.
    states = list(capitals.keys())
  ❹ random.shuffle(states)

    # TODO: Loop through all 50 states, making a question for each.
```

The filenames for the quizzes will be *capitalsquiz<N>.txt*, where *<N>* is a unique number for the quiz that comes from quizNum, the for loop's counter. The answer key for *capitalsquiz<N>.txt* will be stored in a text file named *capitalsquiz_answers<N>.txt*. Each time through the loop, the {quizNum + 1} placeholder in f'capitalsquiz{quizNum + 1}.txt' and f'capitalsquiz_answers{quizNum + 1}.txt' will be replaced by the unique number, so the first quiz and answer key created will be *capitalsquiz1.txt* and *capitalsquiz_answers1.txt*. These files will be created with calls to the open() function at ❶ and ❷, with 'w' as the second argument to open them in write mode.

The write() statements at ❸ create a quiz header for the student to fill out. Finally, a randomized list of US states is created with the help of the random.shuffle() function ❹, which randomly reorders the values in any list that is passed to it.

## Step 3: Create the Answer Options

Now you need to generate the answer options for each question, which will be multiple choice from A to D. You'll need to create another for loop—this one to generate the content for each of the 50 questions on the quiz. Then there will be a third for loop nested inside to generate the multiple-choice options for each question. Make your code look like the following:

```
#! python3
# randomQuizGenerator.py - Creates quizzes with questions and answers in
# random order, along with the answer key.

--snip--

    # Loop through all 50 states, making a question for each.
    for questionNum in range(50):

        # Get right and wrong answers.
❶       correctAnswer = capitals[states[questionNum]]
❷       wrongAnswers = list(capitals.values())
❸       del wrongAnswers[wrongAnswers.index(correctAnswer)]
❹       wrongAnswers = random.sample(wrongAnswers, 3)
❺       answerOptions = wrongAnswers + [correctAnswer]
❻       random.shuffle(answerOptions)

        # TODO: Write the question and answer options to the quiz file.

        # TODO: Write the answer key to a file.
```

The correct answer is easy to get—it's stored as a value in the capitals dictionary ❶. This loop will loop through the states in the shuffled states list, from states[0] to states[49], find each state in capitals, and store that state's corresponding capital in correctAnswer.

The list of possible wrong answers is trickier. You can get it by duplicating *all* the values in the capitals dictionary ❷, deleting the correct answer ❸, and selecting three random values from this list ❹. The random.sample() function makes it easy to do this selection. Its first argument is the list you want to select from; the second argument is the number of values you want to select. The full list of answer options is the combination of these three wrong answers with the correct answers ❺. Finally, the answers need to be randomized ❻ so that the correct response isn't always choice D.

## Step 4: Write Content to the Quiz and Answer Key Files

All that is left is to write the question to the quiz file and the answer to the answer key file. Make your code look like the following:

```
#! python3
# randomQuizGenerator.py - Creates quizzes with questions and answers in
# random order, along with the answer key.

--snip--

    # Loop through all 50 states, making a question for each.
    for questionNum in range(50):
        --snip--

        # Write the question and the answer options to the quiz file.
        quizFile.write(f'{questionNum + 1}. What is the capital of
{states[questionNum]}?\n')
      ❶ for i in range(4):
          ❷ quizFile.write(f"    {'ABCD'[i]}. { answerOptions[i]}\n")
        quizFile.write('\n')

        # Write the answer key to a file.
      ❸ answerKeyFile.write(f"{questionNum + 1}.
{'ABCD'[answerOptions.index(correctAnswer)]}")
    quizFile.close()
    answerKeyFile.close()
```

A for loop that goes through integers 0 to 3 will write the answer options in the answerOptions list ❶. The expression 'ABCD'[i] at ❷ treats the string 'ABCD' as an array and will evaluate to 'A','B', 'C', and then 'D' on each respective iteration through the loop.

In the final line ❸, the expression answerOptions.index(correctAnswer) will find the integer index of the correct answer in the randomly ordered answer options, and 'ABCD'[answerOptions.index(correctAnswer)] will evaluate to the correct answer's letter to be written to the answer key file.

After you run the program, this is how your *capitalsquiz1.txt* file will look, though of course your questions and answer options may be different from those shown here, depending on the outcome of your random.shuffle() calls:

```
Name:

Date:

Period:

                    State Capitals Quiz (Form 1)
```

```
1. What is the capital of West Virginia?
   A. Hartford
   B. Santa Fe
   C. Harrisburg
   D. Charleston

2. What is the capital of Colorado?
   A. Raleigh
   B. Harrisburg
   C. Denver
   D. Lincoln

--snip--
```

The corresponding *capitalsquiz_answers1.txt* text file will look like this:

```
1. D
2. C
3. A
4. C
--snip--
```

# Project: Updatable Multi-Clipboard

Let's rewrite the "multi-clipboard" program from Chapter 6 so that it uses the shelve module. The user will now be able to save new strings to load to the clipboard without having to modify the source code. We'll name this new program *mcb.pyw* (since "mcb" is shorter to type than "multi-clipboard"). The *.pyw* extension means that Python won't show a Terminal window when it runs this program. (See Appendix B for more details.)

The program will save each piece of clipboard text under a keyword. For example, when you run py mcb.pyw save spam, the current contents of the clipboard will be saved with the keyword *spam*. This text can later be loaded to the clipboard again by running py mcb.pyw spam. And if the user forgets what keywords they have, they can run py mcb.pyw list to copy a list of all keywords to the clipboard.

Here's what the program does:

1. The command line argument for the keyword is checked.
2. If the argument is save, then the clipboard contents are saved to the keyword.
3. If the argument is list, then all the keywords are copied to the clipboard.
4. Otherwise, the text for the keyword is copied to the clipboard.

This means the code will need to do the following:

1. Read the command line arguments from sys.argv.
2. Read and write to the clipboard.
3. Save and load to a shelf file.

If you use Windows, you can easily run this script from the Run...
window by creating a batch file named *mcb.bat* with the following content:

```
@pyw.exe C:\Python34\mcb.pyw %*
```

### Step 1: Comments and Shelf Setup

Let's start by making a skeleton script with some comments and basic setup.
Make your code look like the following:

```
#! python3
# mcb.pyw - Saves and loads pieces of text to the clipboard.
❶ # Usage: py.exe mcb.pyw save <keyword> - Saves clipboard to keyword.
#         py.exe mcb.pyw <keyword> - Loads keyword to clipboard.
#         py.exe mcb.pyw list - Loads all keywords to clipboard.

❷ import shelve, pyperclip, sys

❸ mcbShelf = shelve.open('mcb')

# TODO: Save clipboard content.

# TODO: List keywords and load content.

mcbShelf.close()
```

It's common practice to put general usage information in comments
at the top of the file ❶. If you ever forget how to run your script, you can
always look at these comments for a reminder. Then you import your modules ❷. Copying and pasting will require the pyperclip module, and reading
the command line arguments will require the sys module. The shelve module will also come in handy: Whenever the user wants to save a new piece
of clipboard text, you'll save it to a shelf file. Then, when the user wants to
paste the text back to their clipboard, you'll open the shelf file and load it
back into your program. The shelf file will be named with the prefix *mcb* ❸.

### Step 2: Save Clipboard Content with a Keyword

The program does different things depending on whether the user wants to
save text to a keyword, load text into the clipboard, or list all the existing keywords. Let's deal with that first case. Make your code look like the following:

```
#! python3
# mcb.pyw - Saves and loads pieces of text to the clipboard.
--snip--

# Save clipboard content.
❶ if len(sys.argv) == 3 and sys.argv[1].lower() == 'save':
        ❷ mcbShelf[sys.argv[2]] = pyperclip.paste()
    elif len(sys.argv) == 2:
```

```
❸ # TODO: List keywords and load content.

mcbShelf.close()
```

If the first command line argument (which will always be at index 1 of the sys.argv list) is 'save' ❶, the second command line argument is the keyword for the current content of the clipboard. The keyword will be used as the key for mcbShelf, and the value will be the text currently on the clipboard ❷.

If there is only one command line argument, you will assume it is either 'list' or a keyword to load content onto the clipboard. You will implement that code later. For now, just put a TODO comment there ❸.

### Step 3: List Keywords and Load a Keyword's Content

Finally, let's implement the two remaining cases: the user wants to load clipboard text in from a keyword, or they want a list of all available keywords. Make your code look like the following:

```
#! python3
# mcb.pyw - Saves and loads pieces of text to the clipboard.
--snip--

# Save clipboard content.
if len(sys.argv) == 3 and sys.argv[1].lower() == 'save':
        mcbShelf[sys.argv[2]] = pyperclip.paste()
elif len(sys.argv) == 2:
    # List keywords and load content.
  ❶ if sys.argv[1].lower() == 'list':
      ❷ pyperclip.copy(str(list(mcbShelf.keys())))
    elif sys.argv[1] in mcbShelf:
      ❸ pyperclip.copy(mcbShelf[sys.argv[1]])

mcbShelf.close()
```

If there is only one command line argument, first let's check whether it's 'list' ❶. If so, a string representation of the list of shelf keys will be copied to the clipboard ❷. The user can paste this list into an open text editor to read it.

Otherwise, you can assume the command line argument is a keyword. If this keyword exists in the mcbShelf shelf as a key, you can load the value onto the clipboard ❸.

And that's it! Launching this program has different steps depending on what operating system your computer uses. See Appendix B for details.

Recall the password locker program you created in Chapter 6 that stored the passwords in a dictionary. Updating the passwords required changing the source code of the program. This isn't ideal, because average users don't feel comfortable changing source code to update their software. Also, every time you modify the source code to a program, you run the risk of accidentally introducing new bugs. By storing the data for a program in a different place than the code, you can make your programs easier for others to use and more resistant to bugs.

# Summary

Files are organized into folders (also called directories), and a path describes the location of a file. Every program running on your computer has a current working directory, which allows you to specify file paths relative to the current location instead of always typing the full (or absolute) path. The pathlib and os.path modules have many functions for manipulating file paths.

Your programs can also directly interact with the contents of text files. The open() function can open these files to read in their contents as one large string (with the read() method) or as a list of strings (with the readlines() method). The open() function can open files in write or append mode to create new text files or add to existing text files, respectively.

In previous chapters, you used the clipboard as a way of getting large amounts of text into a program, rather than typing it all in. Now you can have your programs read files directly from the hard drive, which is a big improvement, since files are much less volatile than the clipboard.

In the next chapter, you will learn how to handle the files themselves, by copying them, deleting them, renaming them, moving them, and more.

# Practice Questions

1. What is a relative path relative to?
2. What does an absolute path start with?
3. What does Path('C:/Users') / 'Al' evaluate to on Windows?
4. What does 'C:/Users' / 'Al' evaluate to on Windows?
5. What do the os.getcwd() and os.chdir() functions do?
6. What are the . and .. folders?
7. In *C:\bacon\eggs\spam.txt*, which part is the dir name, and which part is the base name?
8. What are the three "mode" arguments that can be passed to the open() function?
9. What happens if an existing file is opened in write mode?
10. What is the difference between the read() and readlines() methods?
11. What data structure does a shelf value resemble?

# Practice Projects

For practice, design and write the following programs.

## Extending the Multi-Clipboard

Extend the multi-clipboard program in this chapter so that it has a delete <keyword> command line argument that will delete a keyword from the shelf. Then add a delete command line argument that will delete *all* keywords.

## Mad Libs

Create a Mad Libs program that reads in text files and lets the user add their own text anywhere the word *ADJECTIVE, NOUN, ADVERB,* or *VERB* appears in the text file. For example, a text file may look like this:

---

The ADJECTIVE panda walked to the NOUN and then VERB. A nearby NOUN was unaffected by these events.

---

The program would find these occurrences and prompt the user to replace them.

---

Enter an adjective:
**silly**
Enter a noun:
**chandelier**
Enter a verb:
**screamed**
Enter a noun:
**pickup truck**

---

The following text file would then be created:

---

The silly panda walked to the chandelier and then screamed. A nearby pickup truck was unaffected by these events.

---

The results should be printed to the screen and saved to a new text file.

## Regex Search

Write a program that opens all *.txt* files in a folder and searches for any line that matches a user-supplied regular expression. The results should be printed to the screen.

# 10

## ORGANIZING FILES

In the previous chapter, you learned how to create and write to new files in Python. Your programs can also organize preexisting files on the hard drive. Maybe you've had the experience of going through a folder full of dozens, hundreds, or even thousands of files and copying, renaming, moving, or compressing them all by hand. Or consider tasks such as these:

- Making copies of all PDF files (and *only* the PDF files) in every subfolder of a folder
- Removing the leading zeros in the filenames for every file in a folder of hundreds of files named *spam001.txt*, *spam002.txt*, *spam003.txt*, and so on
- Compressing the contents of several folders into one ZIP file (which could be a simple backup system)

All this boring stuff is just begging to be automated in Python. By programming your computer to do these tasks, you can transform it into a quick-working file clerk who never makes mistakes.

As you begin working with files, you may find it helpful to be able to quickly see what the extension (*.txt*, *.pdf*, *.jpg*, and so on) of a file is. With macOS and Linux, your file browser most likely shows extensions automatically. With Windows, file extensions may be hidden by default. To show extensions, go to **Start ▶ Control Panel ▶ Appearance and Personalization ▶ Folder Options**. On the View tab, under Advanced Settings, uncheck the **Hide extensions for known file types** checkbox.

# The shutil Module

The shutil (or shell utilities) module has functions to let you copy, move, rename, and delete files in your Python programs. To use the shutil functions, you will first need to use import shutil.

## Copying Files and Folders

The shutil module provides functions for copying files, as well as entire folders.

Calling shutil.copy(*source, destination*) will copy the file at the path *source* to the folder at the path *destination*. (Both *source* and *destination* can be strings or Path objects.) If *destination* is a filename, it will be used as the new name of the copied file. This function returns a string or Path object of the copied file.

Enter the following into the interactive shell to see how shutil.copy() works:

```
>>> import shutil, os
>>> from pathlib import Path
>>> p = Path.home()
❶ >>> shutil.copy(p / 'spam.txt', p / 'some_folder')
'C:\\Users\\Al\\some_folder\\spam.txt'
❷ >>> shutil.copy(p / 'eggs.txt', p / 'some_folder/eggs2.txt')
WindowsPath('C:/Users/Al/some_folder/eggs2.txt')
```

The first shutil.copy() call copies the file at *C:\Users\Al\spam.txt* to the folder *C:\Users\Al\some_folder*. The return value is the path of the newly copied file. Note that since a folder was specified as the destination ❶, the original *spam.txt* filename is used for the new, copied file's filename. The second shutil.copy() call ❷ also copies the file at *C:\Users\Al\eggs.txt* to the folder *C:\Users\Al\some_folder* but gives the copied file the name *eggs2.txt*.

While shutil.copy() will copy a single file, shutil.copytree() will copy an entire folder and every folder and file contained in it. Calling shutil .copytree(*source, destination*) will copy the folder at the path *source*, along with all of its files and subfolders, to the folder at the path *destination*. The *source* and *destination* parameters are both strings. The function returns a string of the path of the copied folder.

Enter the following into the interactive shell:

```
>>> import shutil, os
>>> from pathlib import Path
>>> p = Path.home()
>>> shutil.copytree(p / 'spam', p / 'spam_backup')
WindowsPath('C:/Users/Al/spam_backup')
```

The shutil.copytree() call creates a new folder named *spam_backup* with the same content as the original *spam* folder. You have now safely backed up your precious, precious spam.

## Moving and Renaming Files and Folders

Calling shutil.move(*source, destination*) will move the file or folder at the path *source* to the path *destination* and will return a string of the absolute path of the new location.

If *destination* points to a folder, the *source* file gets moved into *destination* and keeps its current filename. For example, enter the following into the interactive shell:

```
>>> import shutil
>>> shutil.move('C:\\bacon.txt', 'C:\\eggs')
'C:\\eggs\\bacon.txt'
```

Assuming a folder named *eggs* already exists in the *C:\* directory, this shutil.move() call says, "Move *C:\bacon.txt* into the folder *C:\eggs*."

If there had been a *bacon.txt* file already in *C:\eggs*, it would have been overwritten. Since it's easy to accidentally overwrite files in this way, you should take some care when using move().

The *destination* path can also specify a filename. In the following example, the *source* file is moved *and* renamed.

```
>>> shutil.move('C:\\bacon.txt', 'C:\\eggs\\new_bacon.txt')
'C:\\eggs\\new_bacon.txt'
```

This line says, "Move *C:\bacon.txt* into the folder *C:\eggs*, and while you're at it, rename that *bacon.txt* file to *new_bacon.txt*."

Both of the previous examples worked under the assumption that there was a folder *eggs* in the *C:\* directory. But if there is no *eggs* folder, then move() will rename *bacon.txt* to a file named *eggs*.

```
>>> shutil.move('C:\\bacon.txt', 'C:\\eggs')
'C:\\eggs'
```

Here, move() can't find a folder named *eggs* in the *C:\* directory and so assumes that *destination* must be specifying a filename, not a folder. So the *bacon.txt* text file is renamed to *eggs* (a text file without the *.txt* file extension)—probably not what you wanted! This can be a tough-to-spot bug in your programs since the move() call can happily do something that might be quite different from what you were expecting. This is yet another reason to be careful when using move().

Finally, the folders that make up the destination must already exist, or else Python will throw an exception. Enter the following into the interactive shell:

```
>>> shutil.move('spam.txt', 'c:\\does_not_exist\\eggs\\ham')
Traceback (most recent call last):
  --snip--
FileNotFoundError: [Errno 2] No such file or directory: 'c:\\does_not_exist\\
eggs\\ham'
```

Python looks for *eggs* and *ham* inside the directory *does_not_exist*. It doesn't find the nonexistent directory, so it can't move *spam.txt* to the path you specified.

## Permanently Deleting Files and Folders

You can delete a single file or a single empty folder with functions in the os module, whereas to delete a folder and all of its contents, you use the shutil module.

- Calling os.unlink(*path*) will delete the file at *path*.
- Calling os.rmdir(*path*) will delete the folder at *path*. This folder must be empty of any files or folders.
- Calling shutil.rmtree(*path*) will remove the folder at *path*, and all files and folders it contains will also be deleted.

Be careful when using these functions in your programs! It's often a good idea to first run your program with these calls commented out and with print() calls added to show the files that would be deleted. Here is a Python program that was intended to delete files that have the *.txt* file extension but has a typo (highlighted in bold) that causes it to delete *.rxt* files instead:

```
import os
from pathlib import Path
for filename in Path.home().glob('*.rxt'):
    os.unlink(filename)
```

If you had any important files ending with *.rxt*, they would have been accidentally, permanently deleted. Instead, you should have first run the program like this:

```
import os
from pathlib import Path
for filename in Path.home().glob('*.rxt'):
    #os.unlink(filename)
    print(filename)
```

Now the os.unlink() call is commented, so Python ignores it. Instead, you will print the filename of the file that would have been deleted. Running this version of the program first will show you that you've accidentally told the program to delete *.rxt* files instead of *.txt* files.

Once you are certain the program works as intended, delete the print(filename) line and uncomment the os.unlink(filename) line. Then run the program again to actually delete the files.

### Safe Deletes with the send2trash Module

Since Python's built-in shutil.rmtree() function irreversibly deletes files and folders, it can be dangerous to use. A much better way to delete files and folders is with the third-party send2trash module. You can install this module by running pip install --user send2trash from a Terminal window. (See Appendix A for a more in-depth explanation of how to install third-party modules.)

Using send2trash is much safer than Python's regular delete functions, because it will send folders and files to your computer's trash or recycle bin instead of permanently deleting them. If a bug in your program deletes something with send2trash you didn't intend to delete, you can later restore it from the recycle bin.

After you have installed send2trash, enter the following into the interactive shell:

```
>>> import send2trash
>>> baconFile = open('bacon.txt', 'a')   # creates the file
>>> baconFile.write('Bacon is not a vegetable.')
25
>>> baconFile.close()
>>> send2trash.send2trash('bacon.txt')
```

In general, you should always use the send2trash.send2trash() function to delete files and folders. But while sending files to the recycle bin lets you recover them later, it will not free up disk space like permanently deleting them does. If you want your program to free up disk space, use the os and shutil functions for deleting files and folders. Note that the send2trash() function can only send files to the recycle bin; it cannot pull files out of it.

# Walking a Directory Tree

Say you want to rename every file in some folder and also every file in every subfolder of that folder. That is, you want to walk through the directory tree, touching each file as you go. Writing a program to do this could get tricky; fortunately, Python provides a function to handle this process for you.

Let's look at the *C:\delicious* folder with its contents, shown in Figure 10-1.

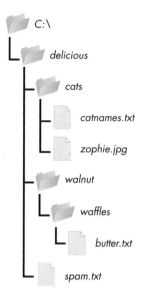

*Figure 10-1: An example folder that contains three folders and four files*

Here is an example program that uses the os.walk() function on the directory tree from Figure 10-1:

```
import os

for folderName, subfolders, filenames in os.walk('C:\\delicious'):
    print('The current folder is ' + folderName)

    for subfolder in subfolders:
        print('SUBFOLDER OF ' + folderName + ': ' + subfolder)

    for filename in filenames:
        print('FILE INSIDE ' + folderName + ': '+ filename)

    print('')
```

The os.walk() function is passed a single string value: the path of a folder. You can use os.walk() in a for loop statement to walk a directory tree, much like how you can use the range() function to walk over a range of numbers. Unlike range(), the os.walk() function will return three values on each iteration through the loop:

- A string of the current folder's name
- A list of strings of the folders in the current folder
- A list of strings of the files in the current folder

(By current folder, I mean the folder for the current iteration of the for loop. The current working directory of the program is *not* changed by os.walk().)

Just like you can choose the variable name i in the code for i in range(10):, you can also choose the variable names for the three values listed earlier. I usually use the names foldername, subfolders, and filenames.

When you run this program, it will output the following:

```
The current folder is C:\delicious
SUBFOLDER OF C:\delicious: cats
SUBFOLDER OF C:\delicious: walnut
FILE INSIDE C:\delicious: spam.txt

The current folder is C:\delicious\cats
FILE INSIDE C:\delicious\cats: catnames.txt
FILE INSIDE C:\delicious\cats: zophie.jpg

The current folder is C:\delicious\walnut
SUBFOLDER OF C:\delicious\walnut: waffles

The current folder is C:\delicious\walnut\waffles
FILE INSIDE C:\delicious\walnut\waffles: butter.txt.
```

Since os.walk() returns lists of strings for the subfolder and filename variables, you can use these lists in their own for loops. Replace the print() function calls with your own custom code. (Or if you don't need one or both of them, remove the for loops.)

## Compressing Files with the zipfile Module

You may be familiar with ZIP files (with the *.zip* file extension), which can hold the compressed contents of many other files. Compressing a file reduces its size, which is useful when transferring it over the internet. And since a ZIP file can also contain multiple files and subfolders, it's a handy way to package several files into one. This single file, called an *archive file*, can then be, say, attached to an email.

Your Python programs can create and open (or *extract*) ZIP files using functions in the zipfile module. Say you have a ZIP file named *example.zip* that has the contents shown in Figure 10-2.

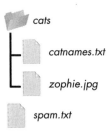

*cats*

*catnames.txt*

*zophie.jpg*

*spam.txt*

*Figure 10-2: The contents of* example.zip

You can download this ZIP file from *https://nostarch.com/automatestuff2/* or just follow along using a ZIP file already on your computer.

## Reading ZIP Files

To read the contents of a ZIP file, first you must create a ZipFile object (note the capital letters *Z* and *F*). ZipFile objects are conceptually similar to the File objects you saw returned by the open() function in the previous chapter: they are values through which the program interacts with the file. To create a ZipFile object, call the zipfile.ZipFile() function, passing it a string of the *.ZIP* file's filename. Note that zipfile is the name of the Python module, and ZipFile() is the name of the function.

For example, enter the following into the interactive shell:

```
>>> import zipfile, os

>>> from pathlib import Path
>>> p = Path.home()
>>> exampleZip = zipfile.ZipFile(p / 'example.zip')
>>> exampleZip.namelist()
['spam.txt', 'cats/', 'cats/catnames.txt', 'cats/zophie.jpg']
>>> spamInfo = exampleZip.getinfo('spam.txt')
>>> spamInfo.file_size
13908
>>> spamInfo.compress_size
3828
❶ >>> f'Compressed file is {round(spamInfo.file_size / spamInfo
.compress_size, 2)}x smaller!'
)
'Compressed file is 3.63x smaller!'
>>> exampleZip.close()
```

A ZipFile object has a namelist() method that returns a list of strings for all the files and folders contained in the ZIP file. These strings can be passed to the getinfo() ZipFile method to return a ZipInfo object about that particular file. ZipInfo objects have their own attributes, such as file_size and compress_size in bytes, which hold integers of the original file size and compressed file size, respectively. While a ZipFile object represents an entire archive file, a ZipInfo object holds useful information about a *single file* in the archive.

The command at ❶ calculates how efficiently *example.zip* is compressed by dividing the original file size by the compressed file size and prints this information.

## Extracting from ZIP Files

The extractall() method for ZipFile objects extracts all the files and folders from a ZIP file into the current working directory.

```
>>> import zipfile, os
>>> from pathlib import Path
>>> p = Path.home()
>>> exampleZip = zipfile.ZipFile(p / 'example.zip')
❶ >>> exampleZip.extractall()
>>> exampleZip.close()
```

After running this code, the contents of *example.zip* will be extracted to *C:\*. Optionally, you can pass a folder name to extractall() to have it extract the files into a folder other than the current working directory. If the folder passed to the extractall() method does not exist, it will be created. For instance, if you replaced the call at ❶ with exampleZip.extractall('C:\\ delicious'), the code would extract the files from *example.zip* into a newly created *C:\delicious* folder.

The extract() method for ZipFile objects will extract a single file from the ZIP file. Continue the interactive shell example:

```
>>> exampleZip.extract('spam.txt')
'C:\\spam.txt'
>>> exampleZip.extract('spam.txt', 'C:\\some\\new\\folders')
'C:\\some\\new\\folders\\spam.txt'
>>> exampleZip.close()
```

The string you pass to extract() must match one of the strings in the list returned by namelist(). Optionally, you can pass a second argument to extract() to extract the file into a folder other than the current working directory. If this second argument is a folder that doesn't yet exist, Python will create the folder. The value that extract() returns is the absolute path to which the file was extracted.

## Creating and Adding to ZIP Files

To create your own compressed ZIP files, you must open the ZipFile object in *write mode* by passing 'w' as the second argument. (This is similar to opening a text file in write mode by passing 'w' to the open() function.)

When you pass a path to the write() method of a ZipFile object, Python will compress the file at that path and add it into the ZIP file. The write() method's first argument is a string of the filename to add. The second argument is the *compression type* parameter, which tells the computer what algorithm it should use to compress the files; you can always just set this value to zipfile.ZIP_DEFLATED. (This specifies the *deflate* compression algorithm, which works well on all types of data.) Enter the following into the interactive shell:

```
>>> import zipfile
>>> newZip = zipfile.ZipFile('new.zip', 'w')
>>> newZip.write('spam.txt', compress_type=zipfile.ZIP_DEFLATED)
>>> newZip.close()
```

This code will create a new ZIP file named *new.zip* that has the compressed contents of *spam.txt*.

Keep in mind that, just as with writing to files, write mode will erase all existing contents of a ZIP file. If you want to simply add files to an existing ZIP file, pass 'a' as the second argument to zipfile.ZipFile() to open the ZIP file in *append mode*.

# Project: Renaming Files with American-Style Dates to European-Style Dates

Say your boss emails you thousands of files with American-style dates (MM-DD-YYYY) in their names and needs them renamed to European-style dates (DD-MM-YYYY). This boring task could take all day to do by hand! Let's write a program to do it instead.

Here's what the program does:

1. It searches all the filenames in the current working directory for American-style dates.

2. When one is found, it renames the file with the month and day swapped to make it European-style.

This means the code will need to do the following:

1. Create a regex that can identify the text pattern of American-style dates.

2. Call os.listdir() to find all the files in the working directory.

3. Loop over each filename, using the regex to check whether it has a date.

4. If it has a date, rename the file with shutil.move().

For this project, open a new file editor window and save your code as *renameDates.py*.

## Step 1: Create a Regex for American-Style Dates

The first part of the program will need to import the necessary modules and create a regex that can identify MM-DD-YYYY dates. The to-do comments will remind you what's left to write in this program. Typing them as TODO makes them easy to find using Mu editor's CTRL-F find feature. Make your code look like the following:

```
#! python3
# renameDates.py - Renames filenames with American MM-DD-YYYY date format
# to European DD-MM-YYYY.

❶ import shutil, os, re

# Create a regex that matches files with the American date format.
❷ datePattern = re.compile(r"""^(.*?) # all text before the date
    ((0|1)?\d)-                # one or two digits for the month
    ((0|1|2|3)?\d)-            # one or two digits for the day
    ((19|20)\d\d)             # four digits for the year
    (.*?)$                    # all text after the date
    """, re.VERBOSE❸)

# TODO: Loop over the files in the working directory.

# TODO: Skip files without a date.
```

```
# TODO: Get the different parts of the filename.

# TODO: Form the European-style filename.

# TODO: Get the full, absolute file paths.

# TODO: Rename the files.
```

From this chapter, you know the `shutil.move()` function can be used to rename files: its arguments are the name of the file to rename and the new filename. Because this function exists in the `shutil` module, you must import that module ❶.

But before renaming the files, you need to identify which files you want to rename. Filenames with dates such as *spam4-4-1984.txt* and *01-03-2014eggs.zip* should be renamed, while filenames without dates such as *littlebrother.epub* can be ignored.

You can use a regular expression to identify this pattern. After importing the `re` module at the top, call `re.compile()` to create a Regex object ❷. Passing `re.VERBOSE` for the second argument ❸ will allow whitespace and comments in the regex string to make it more readable.

The regular expression string begins with `^(.*?)` to match any text at the beginning of the filename that might come before the date. The `((0|1)?\d)` group matches the month. The first digit can be either 0 or 1, so the regex matches 12 for December but also 02 for February. This digit is also optional so that the month can be 04 or 4 for April. The group for the day is `((0|1|2|3)?\d)` and follows similar logic; 3, 03, and 31 are all valid numbers for days. (Yes, this regex will accept some invalid dates such as 4-31-2014, 2-29-2013, and 0-15-2014. Dates have a lot of thorny special cases that can be easy to miss. But for simplicity, the regex in this program works well enough.)

While 1885 is a valid year, you can just look for years in the 20th or 21st century. This will keep your program from accidentally matching nondate filenames with a date-like format, such as *10-10-1000.txt*.

The `(.*?)$` part of the regex will match any text that comes after the date.

## Step 2: Identify the Date Parts from the Filenames

Next, the program will have to loop over the list of filename strings returned from `os.listdir()` and match them against the regex. Any files that do not have a date in them should be skipped. For filenames that have a date, the matched text will be stored in several variables. Fill in the first three TODOs in your program with the following code:

```
#! python3
# renameDates.py - Renames filenames with American MM-DD-YYYY date format
# to European DD-MM-YYYY.

--snip--

# Loop over the files in the working directory.
for amerFilename in os.listdir('.'):
```

```
    mo = datePattern.search(amerFilename)

    # Skip files without a date.
❶ if mo == None:
    ❷ continue

❸ # Get the different parts of the filename.
    beforePart = mo.group(1)
    monthPart  = mo.group(2)
    dayPart    = mo.group(4)
    yearPart   = mo.group(6)
    afterPart  = mo.group(8)
```

*--snip--*

If the Match object returned from the search() method is None ❶, then the filename in amerFilename does not match the regular expression. The continue statement ❷ will skip the rest of the loop and move on to the next filename.

Otherwise, the various strings matched in the regular expression groups are stored in variables named beforePart, monthPart, dayPart, yearPart, and afterPart ❸. The strings in these variables will be used to form the European-style filename in the next step.

To keep the group numbers straight, try reading the regex from the beginning, and count up each time you encounter an opening parenthesis. Without thinking about the code, just write an outline of the regular expression. This can help you visualize the groups. Here's an example:

```
datePattern = re.compile(r"""^(1)   # all text before the date
    (2 (3) )-                        # one or two digits for the month
    (4 (5) )-                        # one or two digits for the day
    (6 (7) )                         # four digits for the year
    (8)$                             # all text after the date
    """, re.VERBOSE)
```

Here, the numbers **1** through **8** represent the groups in the regular expression you wrote. Making an outline of the regular expression, with just the parentheses and group numbers, can give you a clearer understanding of your regex before you move on with the rest of the program.

## Step 3: Form the New Filename and Rename the Files

As the final step, concatenate the strings in the variables made in the previous step with the European-style date: the date comes before the month. Fill in the three remaining TODOs in your program with the following code:

```
#! python3
# renameDates.py - Renames filenames with American MM-DD-YYYY date format
# to European DD-MM-YYYY.

--snip--
```

```
   # Form the European-style filename.
❶ euroFilename = beforePart + dayPart + '-' + monthPart + '-' + yearPart +
                 afterPart

   # Get the full, absolute file paths.
   absWorkingDir = os.path.abspath('.')
   amerFilename = os.path.join(absWorkingDir, amerFilename)
   euroFilename = os.path.join(absWorkingDir, euroFilename)

   # Rename the files.
❷ print(f'Renaming "{amerFilename}" to "{euroFilename}"...')
❸ #shutil.move(amerFilename, euroFilename)   # uncomment after testing
```

Store the concatenated string in a variable named euroFilename ❶. Then, pass the original filename in amerFilename and the new euroFilename variable to the shutil.move() function to rename the file ❸.

This program has the shutil.move() call commented out and instead prints the filenames that will be renamed ❷. Running the program like this first can let you double-check that the files are renamed correctly. Then you can uncomment the shutil.move() call and run the program again to actually rename the files.

### Ideas for Similar Programs

There are many other reasons you might want to rename a large number of files.

- To add a prefix to the start of the filename, such as adding *spam_* to rename *eggs.txt* to *spam_eggs.txt*
- To change filenames with European-style dates to American-style dates
- To remove the zeros from files such as *spam0042.txt*

## Project: Backing Up a Folder into a ZIP File

Say you're working on a project whose files you keep in a folder named *C:\AlsPythonBook*. You're worried about losing your work, so you'd like to create ZIP file "snapshots" of the entire folder. You'd like to keep different versions, so you want the ZIP file's filename to increment each time it is made; for example, *AlsPythonBook_1.zip*, *AlsPythonBook_2.zip*, *AlsPythonBook_3.zip*, and so on. You could do this by hand, but it is rather annoying, and you might accidentally misnumber the ZIP files' names. It would be much simpler to run a program that does this boring task for you.

For this project, open a new file editor window and save it as *backupToZip.py*.

### Step 1: Figure Out the ZIP File's Name

The code for this program will be placed into a function named backupToZip(). This will make it easy to copy and paste the function into other Python programs that need this functionality. At the end of the program, the function will be called to perform the backup. Make your program look like this:

```python
#! python3
# backupToZip.py - Copies an entire folder and its contents into
# a ZIP file whose filename increments.

❶ import zipfile, os

def backupToZip(folder):
    # Back up the entire contents of "folder" into a ZIP file.

    folder = os.path.abspath(folder)   # make sure folder is absolute

    # Figure out the filename this code should use based on
    # what files already exist.
❷   number = 1
❸   while True:
        zipFilename = os.path.basename(folder) + '_' + str(number) + '.zip'
        if not os.path.exists(zipFilename):
            break
        number = number + 1

❹   # TODO: Create the ZIP file.

    # TODO: Walk the entire folder tree and compress the files in each folder.
    print('Done.')

backupToZip('C:\\delicious')
```

Do the basics first: add the shebang (#!) line, describe what the program does, and import the zipfile and os modules ❶.

Define a backupToZip() function that takes just one parameter, folder. This parameter is a string path to the folder whose contents should be backed up. The function will determine what filename to use for the ZIP file it will create; then the function will create the file, walk the folder folder, and add each of the subfolders and files to the ZIP file. Write TODO comments for these steps in the source code to remind yourself to do them later ❹.

The first part, naming the ZIP file, uses the base name of the absolute path of folder. If the folder being backed up is *C:\delicious*, the ZIP file's name should be *delicious_N.zip*, where *N* = 1 is the first time you run the program, *N* = 2 is the second time, and so on.

You can determine what *N* should be by checking whether *delicious_1.zip* already exists, then checking whether *delicious_2.zip* already exists, and so on. Use a variable named number for *N* ❷, and keep incrementing it inside the loop that calls os.path.exists() to check whether the file exists ❸. The first nonexistent filename found will cause the loop to break, since it will have found the filename of the new zip.

## Step 2: Create the New ZIP File

Next let's create the ZIP file. Make your program look like the following:

```python
#! python3
# backupToZip.py - Copies an entire folder and its contents into
# a ZIP file whose filename increments.

--snip--
    while True:
        zipFilename = os.path.basename(folder) + '_' + str(number) + '.zip'
        if not os.path.exists(zipFilename):
            break
        number = number + 1

    # Create the ZIP file.
    print(f'Creating {zipFilename}...')
 ❶ backupZip = zipfile.ZipFile(zipFilename, 'w')

    # TODO: Walk the entire folder tree and compress the files in each folder.
    print('Done.')

backupToZip('C:\\delicious')
```

Now that the new ZIP file's name is stored in the zipFilename variable, you can call zipfile.ZipFile() to actually create the ZIP file ❶. Be sure to pass 'w' as the second argument so that the ZIP file is opened in write mode.

## Step 3: Walk the Directory Tree and Add to the ZIP File

Now you need to use the os.walk() function to do the work of listing every file in the folder and its subfolders. Make your program look like the following:

```python
#! python3
# backupToZip.py - Copies an entire folder and its contents into
# a ZIP file whose filename increments.

--snip--

    # Walk the entire folder tree and compress the files in each folder.
 ❶ for foldername, subfolders, filenames in os.walk(folder):
        print(f'Adding files in {foldername}...')
        # Add the current folder to the ZIP file.
      ❷ backupZip.write(foldername)

        # Add all the files in this folder to the ZIP file.
      ❸ for filename in filenames:
            newBase = os.path.basename(folder) + '_'
            if filename.startswith(newBase) and filename.endswith('.zip'):
                continue   # don't back up the backup ZIP files
            backupZip.write(os.path.join(foldername, filename))
    backupZip.close()
```

```
    print('Done.')


backupToZip('C:\\delicious')
```

You can use os.walk() in a for loop ❶, and on each iteration it will return the iteration's current folder name, the subfolders in that folder, and the filenames in that folder.

In the for loop, the folder is added to the ZIP file ❷. The nested for loop can go through each filename in the filenames list ❸. Each of these is added to the ZIP file, except for previously made backup ZIPs.

When you run this program, it will produce output that will look something like this:

```
Creating delicious_1.zip...
Adding files in C:\delicious...
Adding files in C:\delicious\cats...
Adding files in C:\delicious\waffles...
Adding files in C:\delicious\walnut...
Adding files in C:\delicious\walnut\waffles...
Done.
```

The second time you run it, it will put all the files in *C:\delicious* into a ZIP file named *delicious_2.zip*, and so on.

### Ideas for Similar Programs

You can walk a directory tree and add files to compressed ZIP archives in several other programs. For example, you can write programs that do the following:

- Walk a directory tree and archive just files with certain extensions, such as *.txt* or *.py*, and nothing else.
- Walk a directory tree and archive every file except the *.txt* and *.py* ones.
- Find the folder in a directory tree that has the greatest number of files or the folder that uses the most disk space.

## Summary

Even if you are an experienced computer user, you probably handle files manually with the mouse and keyboard. Modern file explorers make it easy to work with a few files. But sometimes you'll need to perform a task that would take hours using your computer's file explorer.

The os and shutil modules offer functions for copying, moving, renaming, and deleting files. When deleting files, you might want to use the send2trash module to move files to the recycle bin or trash rather than permanently deleting them. And when writing programs that handle files, it's a good idea to comment out the code that does the actual copy/move/rename/delete and add a print() call instead so you can run the program and verify exactly what it will do.

Often you will need to perform these operations not only on files in one folder but also on every folder in that folder, every folder in those folders, and so on. The os.walk() function handles this trek across the folders for you so that you can concentrate on what your program needs to do with the files in them.

The zipfile module gives you a way of compressing and extracting files in *.ZIP* archives through Python. Combined with the file-handling functions of os and shutil, zipfile makes it easy to package up several files from anywhere on your hard drive. These *.ZIP* files are much easier to upload to websites or send as email attachments than many separate files.

Previous chapters of this book have provided source code for you to copy. But when you write your own programs, they probably won't come out perfectly the first time. The next chapter focuses on some Python modules that will help you analyze and debug your programs so that you can quickly get them working correctly.

## Practice Questions

1. What is the difference between shutil.copy() and shutil.copytree()?
2. What function is used to rename files?
3. What is the difference between the delete functions in the send2trash and shutil modules?
4. ZipFile objects have a close() method just like File objects' close() method. What ZipFile method is equivalent to File objects' open() method?

## Practice Projects

For practice, write programs to do the following tasks.

### Selective Copy

Write a program that walks through a folder tree and searches for files with a certain file extension (such as *.pdf* or *.jpg*). Copy these files from whatever location they are in to a new folder.

### Deleting Unneeded Files

It's not uncommon for a few unneeded but humongous files or folders to take up the bulk of the space on your hard drive. If you're trying to free up room on your computer, you'll get the most bang for your buck by deleting the most massive of the unwanted files. But first you have to find them.

Write a program that walks through a folder tree and searches for exceptionally large files or folders—say, ones that have a file size of more than 100MB. (Remember that to get a file's size, you can use os.path.getsize() from the os module.) Print these files with their absolute path to the screen.

## Filling in the Gaps

Write a program that finds all files with a given prefix, such as *spam001.txt*, *spam002.txt*, and so on, in a single folder and locates any gaps in the numbering (such as if there is a *spam001.txt* and *spam003.txt* but no *spam002.txt*). Have the program rename all the later files to close this gap.

As an added challenge, write another program that can insert gaps into numbered files so that a new file can be added.

# 11

## DEBUGGING

Now that you know enough to write more complicated programs, you may start finding not-so-simple bugs in them. This chapter covers some tools and techniques for finding the root cause of bugs in your program to help you fix bugs faster and with less effort.

To paraphrase an old joke among programmers, writing code accounts for 90 percent of programming. Debugging code accounts for the other 90 percent.

Your computer will do only what you tell it to do; it won't read your mind and do what you *intended* it to do. Even professional programmers create bugs all the time, so don't feel discouraged if your program has a problem.

Fortunately, there are a few tools and techniques to identify what exactly your code is doing and where it's going wrong. First, you will look at logging and assertions, two features that can help you detect bugs early. In general, the earlier you catch bugs, the easier they will be to fix.

Second, you will look at how to use the debugger. The debugger is a feature of Mu that executes a program one instruction at a time, giving you a chance to inspect the values in variables while your code runs, and track how the values change over the course of your program. This is much slower than running the program at full speed, but it is helpful to see the actual values in a program while it runs, rather than deducing what the values might be from the source code.

## Raising Exceptions

Python raises an exception whenever it tries to execute invalid code. In Chapter 3, you read about how to handle Python's exceptions with try and except statements so that your program can recover from exceptions that you anticipated. But you can also raise your own exceptions in your code. Raising an exception is a way of saying, "Stop running the code in this function and move the program execution to the except statement."

Exceptions are raised with a raise statement. In code, a raise statement consists of the following:

- The raise keyword
- A call to the Exception() function
- A string with a helpful error message passed to the Exception() function

For example, enter the following into the interactive shell:

```
>>> raise Exception('This is the error message.')
Traceback (most recent call last):
  File "<pyshell#191>", line 1, in <module>
    raise Exception('This is the error message.')
Exception: This is the error message.
```

If there are no try and except statements covering the raise statement that raised the exception, the program simply crashes and displays the exception's error message.

Often it's the code that calls the function, rather than the function itself, that knows how to handle an exception. That means you will commonly see a raise statement inside a function and the try and except statements in the code calling the function. For example, open a new file editor tab, enter the following code, and save the program as *boxPrint.py*:

```
def boxPrint(symbol, width, height):
    if len(symbol) != 1:
      ❶ raise Exception('Symbol must be a single character string.')
    if width <= 2:
      ❷ raise Exception('Width must be greater than 2.')
```

```
        if height <= 2:
        ❸ raise Exception('Height must be greater than 2.')

        print(symbol * width)
        for i in range(height - 2):
            print(symbol + (' ' * (width - 2)) + symbol)
        print(symbol * width)

for sym, w, h in (('*', 4, 4), ('O', 20, 5), ('x', 1, 3), ('ZZ', 3, 3)):
    try:
        boxPrint(sym, w, h)
    ❹ except Exception as err:
        ❺ print('An exception happened: ' + str(err))
```

You can view the execution of this program at *https://autbor.com/boxprint*. Here we've defined a boxPrint() function that takes a character, a width, and a height, and uses the character to make a little picture of a box with that width and height. This box shape is printed to the screen.

Say we want the character to be a single character, and the width and height to be greater than 2. We add if statements to raise exceptions if these requirements aren't satisfied. Later, when we call boxPrint() with various arguments, our try/except will handle invalid arguments.

This program uses the except Exception as err form of the except statement ❹. If an Exception object is returned from boxPrint() ❶❷❸, this except statement will store it in a variable named err. We can then convert the Exception object to a string by passing it to str() to produce a user-friendly error message ❺. When you run this *boxPrint.py*, the output will look like this:

```
****
*  *
*  *
****
OOOOOOOOOOOOOOOOOOOO
O                  O
O                  O
O                  O
OOOOOOOOOOOOOOOOOOOO
An exception happened: Width must be greater than 2.
An exception happened: Symbol must be a single character string.
```

Using the try and except statements, you can handle errors more gracefully instead of letting the entire program crash.

## Getting the Traceback as a String

When Python encounters an error, it produces a treasure trove of error information called the *traceback*. The traceback includes the error message, the line number of the line that caused the error, and the sequence of the function calls that led to the error. This sequence of calls is called the *call stack*.

Open a new file editor tab in Mu, enter the following program, and save it as *errorExample.py*:

```
def spam():
    bacon()

def bacon():
    raise Exception('This is the error message.')

spam()
```

When you run *errorExample.py*, the output will look like this:

```
Traceback (most recent call last):
  File "errorExample.py", line 7, in <module>
    spam()
  File "errorExample.py", line 2, in spam
    bacon()
  File "errorExample.py", line 5, in bacon
    raise Exception('This is the error message.')
Exception: This is the error message.
```

From the traceback, you can see that the error happened on line 5, in the bacon() function. This particular call to bacon() came from line 2, in the spam() function, which in turn was called on line 7. In programs where functions can be called from multiple places, the call stack can help you determine which call led to the error.

Python displays the traceback whenever a raised exception goes unhandled. But you can also obtain it as a string by calling traceback.format_exc(). This function is useful if you want the information from an exception's traceback but also want an except statement to gracefully handle the exception. You will need to import Python's traceback module before calling this function.

For example, instead of crashing your program right when an exception occurs, you can write the traceback information to a text file and keep your program running. You can look at the text file later, when you're ready to debug your program. Enter the following into the interactive shell:

```
>>> import traceback
>>> try:
...         raise Exception('This is the error message.')
except:
...         errorFile = open('errorInfo.txt', 'w')
...         errorFile.write(traceback.format_exc())
...         errorFile.close()
...         print('The traceback info was written to errorInfo.txt.')

111
The traceback info was written to errorInfo.txt.
```

The 111 is the return value from the write() method, since 111 characters were written to the file. The traceback text was written to *errorInfo.txt*.

```
Traceback (most recent call last):
  File "<pyshell#28>", line 2, in <module>
Exception: This is the error message.
```

In "Logging" on page 255, you'll learn how to use the logging module, which is more effective than simply writing this error information to text files.

## Assertions

An *assertion* is a sanity check to make sure your code isn't doing something obviously wrong. These sanity checks are performed by assert statements. If the sanity check fails, then an AssertionError exception is raised. In code, an assert statement consists of the following:

- The assert keyword
- A condition (that is, an expression that evaluates to True or False)
- A comma
- A string to display when the condition is False

In plain English, an assert statement says, "I assert that the condition holds true, and if not, there is a bug somewhere, so immediately stop the program." For example, enter the following into the interactive shell:

```
>>> ages = [26, 57, 92, 54, 22, 15, 17, 80, 47, 73]
>>> ages.sort()
>>> ages
[15, 17, 22, 26, 47, 54, 57, 73, 80, 92]
>>> assert
ages[0] <= ages[-1] # Assert that the first age is <= the last age.
```

The assert statement here asserts that the first item in ages should be less than or equal to the last one. This is a sanity check; if the code in sort() is bug-free and did its job, then the assertion would be true.

Because the ages[0] <= ages[-1] expression evaluates to True, the assert statement does nothing.

However, let's pretend we had a bug in our code. Say we accidentally called the reverse() list method instead of the sort() list method. When we enter the following in the interactive shell, the assert statement raises an AssertionError:

```
>>> ages = [26, 57, 92, 54, 22, 15, 17, 80, 47, 73]
>>> ages.reverse()
>>> ages
[73, 47, 80, 17, 15, 22, 54, 92, 57, 26]
>>> assert ages[0] <= ages[-1] # Assert that the first age is <= the last age.
```

```
Traceback (most recent call last):
  File "<stdin>", line 1, in <module>
AssertionError
```

Unlike exceptions, your code should *not* handle assert statements with try and except; if an assert fails, your program *should* crash. By "failing fast" like this, you shorten the time between the original cause of the bug and when you first notice the bug. This will reduce the amount of code you will have to check before finding the bug's cause.

Assertions are for programmer errors, not user errors. Assertions should only fail while the program is under development; a user should never see an assertion error in a finished program. For errors that your program can run into as a normal part of its operation (such as a file not being found or the user entering invalid data), raise an exception instead of detecting it with an assert statement. You shouldn't use assert statements in place of raising exceptions, because users can choose to turn off assertions. If you run a Python script with python -O myscript.py instead of python myscript.py, Python will skip assert statements. Users might disable assertions when they're developing a program and need to run it in a production setting that requires peak performance. (Though, in many cases, they'll leave assertions enabled even then.)

Assertions also aren't a replacement for comprehensive testing. For instance, if the previous ages example was set to [10, 3, 2, 1, 20], then the assert ages[0] <= ages[-1] assertion wouldn't notice that the list was unsorted, because it just happened to have a first age that was less than or equal to the last age, which is the only thing the assertion checked for.

## *Using an Assertion in a Traffic Light Simulation*

Say you're building a traffic light simulation program. The data structure representing the stoplights at an intersection is a dictionary with keys 'ns' and 'ew', for the stoplights facing north-south and east-west, respectively. The values at these keys will be one of the strings 'green', 'yellow', or 'red'. The code would look something like this:

```
market_2nd = {'ns': 'green', 'ew': 'red'}
mission_16th = {'ns': 'red', 'ew': 'green'}
```

These two variables will be for the intersections of Market Street and 2nd Street, and Mission Street and 16th Street. To start the project, you want to write a switchLights() function, which will take an intersection dictionary as an argument and switch the lights.

At first, you might think that switchLights() should simply switch each light to the next color in the sequence: Any 'green' values should change to 'yellow', 'yellow' values should change to 'red', and 'red' values should change to 'green'. The code to implement this idea might look like this:

```
def switchLights(stoplight):
    for key in stoplight.keys():
```

```
        if stoplight[key] == 'green':
            stoplight[key] = 'yellow'
        elif stoplight[key] == 'yellow':
            stoplight[key] = 'red'
        elif stoplight[key] == 'red':
            stoplight[key] = 'green'

switchLights(market_2nd)
```

You may already see the problem with this code, but let's pretend you wrote the rest of the simulation code, thousands of lines long, without noticing it. When you finally do run the simulation, the program doesn't crash—but your virtual cars do!

Since you've already written the rest of the program, you have no idea where the bug could be. Maybe it's in the code simulating the cars or in the code simulating the virtual drivers. It could take hours to trace the bug back to the switchLights() function.

But if while writing switchLights() you had added an assertion to check that *at least one of the lights is always red*, you might have included the following at the bottom of the function:

```
assert 'red' in stoplight.values(), 'Neither light is red! ' + str(stoplight)
```

With this assertion in place, your program would crash with this error message:

```
Traceback (most recent call last):
  File "carSim.py", line 14, in <module>
    switchLights(market_2nd)
  File "carSim.py", line 13, in switchLights
    assert 'red' in stoplight.values(), 'Neither light is red! ' +
str(stoplight)
❶ AssertionError: Neither light is red! {'ns': 'yellow', 'ew': 'green'}
```

The important line here is the AssertionError ❶. While your program crashing is not ideal, it immediately points out that a sanity check failed: neither direction of traffic has a red light, meaning that traffic could be going both ways. By failing fast early in the program's execution, you can save yourself a lot of future debugging effort.

## Logging

If you've ever put a print() statement in your code to output some variable's value while your program is running, you've used a form of *logging* to debug your code. Logging is a great way to understand what's happening in your program and in what order it's happening. Python's logging module makes it easy to create a record of custom messages that you write. These log messages will describe when the program execution has reached the logging function call and list any variables you have specified at that point in time. On the other hand, a missing log message indicates a part of the code was skipped and never executed.

## Using the logging Module

To enable the logging module to display log messages on your screen as your program runs, copy the following to the top of your program (but under the #! python shebang line):

```
import logging
logging.basicConfig(level=logging.DEBUG, format=' %(asctime)s - %(levelname)
s - %(message)s')
```

You don't need to worry too much about how this works, but basically, when Python logs an event, it creates a LogRecord object that holds information about that event. The logging module's basicConfig() function lets you specify what details about the LogRecord object you want to see and how you want those details displayed.

Say you wrote a function to calculate the *factorial* of a number. In mathematics, factorial 4 is $1 \times 2 \times 3 \times 4$, or 24. Factorial 7 is $1 \times 2 \times 3 \times 4 \times 5 \times 6 \times 7$, or 5,040. Open a new file editor tab and enter the following code. It has a bug in it, but you will also enter several log messages to help yourself figure out what is going wrong. Save the program as *factorialLog.py*.

```
import logging
logging.basicConfig(level=logging.DEBUG, format='%(asctime)s - %(levelname)s
- %(message)s')
logging.debug('Start of program')

def factorial(n):
    logging.debug('Start of factorial(%s%%)' % (n))
    total = 1
    for i in range(n + 1):
        total *= i
        logging.debug('i is ' + str(i) + ', total is ' + str(total))
    logging.debug('End of factorial(%s%%)' % (n))
    return total

print(factorial(5))
logging.debug('End of program')
```

Here, we use the logging.debug() function when we want to print log information. This debug() function will call basicConfig(), and a line of information will be printed. This information will be in the format we specified in basicConfig() and will include the messages we passed to debug(). The print(factorial(5)) call is part of the original program, so the result is displayed even if logging messages are disabled.

The output of this program looks like this:

```
2019-05-23 16:20:12,664 - DEBUG - Start of program
2019-05-23 16:20:12,664 - DEBUG - Start of factorial(5)
2019-05-23 16:20:12,665 - DEBUG - i is 0, total is 0
2019-05-23 16:20:12,668 - DEBUG - i is 1, total is 0
2019-05-23 16:20:12,670 - DEBUG - i is 2, total is 0
2019-05-23 16:20:12,673 - DEBUG - i is 3, total is 0
2019-05-23 16:20:12,675 - DEBUG - i is 4, total is 0
2019-05-23 16:20:12,678 - DEBUG - i is 5, total is 0
2019-05-23 16:20:12,680 - DEBUG - End of factorial(5)
0
2019-05-23 16:20:12,684 - DEBUG - End of program
```

The factorial() function is returning 0 as the factorial of 5, which isn't right. The for loop should be multiplying the value in total by the numbers from 1 to 5. But the log messages displayed by logging.debug() show that the i variable is starting at 0 instead of 1. Since zero times anything is zero, the rest of the iterations also have the wrong value for total. Logging messages provide a trail of breadcrumbs that can help you figure out when things started to go wrong.

Change the for i in range(n + 1): line to for i in range(1, n + 1):, and run the program again. The output will look like this:

```
2019-05-23 17:13:40,650 - DEBUG - Start of program
2019-05-23 17:13:40,651 - DEBUG - Start of factorial(5)
2019-05-23 17:13:40,651 - DEBUG - i is 1, total is 1
2019-05-23 17:13:40,654 - DEBUG - i is 2, total is 2
2019-05-23 17:13:40,656 - DEBUG - i is 3, total is 6
2019-05-23 17:13:40,659 - DEBUG - i is 4, total is 24
2019-05-23 17:13:40,661 - DEBUG - i is 5, total is 120
2019-05-23 17:13:40,661 - DEBUG - End of factorial(5)
120
2019-05-23 17:13:40,666 - DEBUG - End of program
```

The factorial(5) call correctly returns 120. The log messages showed what was going on inside the loop, which led straight to the bug.

You can see that the logging.debug() calls printed out not just the strings passed to them but also a timestamp and the word *DEBUG*.

### Don't Debug with the print() Function

Typing import logging and logging.basicConfig(level=logging.DEBUG, format= '%(asctime)s - %(levelname)s - %(message)s') is somewhat unwieldy. You may want to use print() calls instead, but don't give in to this temptation! Once you're done debugging, you'll end up spending a lot of time removing print() calls from your code for each log message. You might even accidentally remove some print() calls that were being used for nonlog messages. The nice thing about log messages is that you're free to fill your program with as many as you like, and you can always disable them later by adding a single logging.disable(logging.CRITICAL) call. Unlike print(), the logging module makes it easy to switch between showing and hiding log messages.

Log messages are intended for the programmer, not the user. The user won't care about the contents of some dictionary value you need to see to help with debugging; use a log message for something like that. For messages that the user will want to see, like *File not found* or *Invalid input, please enter a number*, you should use a print() call. You don't want to deprive the user of useful information after you've disabled log messages.

## Logging Levels

*Logging levels* provide a way to categorize your log messages by importance. There are five logging levels, described in Table 11-1 from least to most important. Messages can be logged at each level using a different logging function.

**Table 11-1:** Logging Levels in Python

| Level | Logging function | Description |
|---|---|---|
| DEBUG | logging.debug() | The lowest level. Used for small details. Usually you care about these messages only when diagnosing problems. |
| INFO | logging.info() | Used to record information on general events in your program or confirm that things are working at their point in the program. |
| WARNING | logging.warning() | Used to indicate a potential problem that doesn't prevent the program from working but might do so in the future. |
| ERROR | logging.error() | Used to record an error that caused the program to fail to do something. |
| CRITICAL | logging.critical() | The highest level. Used to indicate a fatal error that has caused or is about to cause the program to stop running entirely. |

Your logging message is passed as a string to these functions. The logging levels are suggestions. Ultimately, it is up to you to decide which category your log message falls into. Enter the following into the interactive shell:

```
>>> import logging
>>> logging.basicConfig(level=logging.DEBUG, format=' %(asctime)s - %(levelname)s - %(message)s')
>>> logging.debug('Some debugging details.')
2019-05-18 19:04:26,901 - DEBUG - Some debugging details.
>>> logging.info('The logging module is working.')
2019-05-18 19:04:35,569 - INFO - The logging module is working.
>>> logging.warning('An error message is about to be logged.')
2019-05-18 19:04:56,843 - WARNING - An error message is about to be logged.
```

```
>>> logging.error('An error has occurred.')
2019-05-18 19:05:07,737 - ERROR - An error has occurred.
>>> logging.critical('The program is unable to recover!')
2019-05-18 19:05:45,794 - CRITICAL - The program is unable to recover!
```

The benefit of logging levels is that you can change what priority of logging message you want to see. Passing logging.DEBUG to the basicConfig() function's level keyword argument will show messages from all the logging levels (DEBUG being the lowest level). But after developing your program some more, you may be interested only in errors. In that case, you can set basicConfig()'s level argument to logging.ERROR. This will show only ERROR and CRITICAL messages and skip the DEBUG, INFO, and WARNING messages.

### Disabling Logging

After you've debugged your program, you probably don't want all these log messages cluttering the screen. The logging.disable() function disables these so that you don't have to go into your program and remove all the logging calls by hand. You simply pass logging.disable() a logging level, and it will suppress all log messages at that level or lower. So if you want to disable logging entirely, just add logging.disable(logging.CRITICAL) to your program. For example, enter the following into the interactive shell:

```
>>> import logging
>>> logging.basicConfig(level=logging.INFO, format=' %(asctime)s -
%(levelname)s -  %(message)s')
>>> logging.critical('Critical error! Critical error!')
2019-05-22 11:10:48,054 - CRITICAL - Critical error! Critical error!
>>> logging.disable(logging.CRITICAL)
>>> logging.critical('Critical error! Critical error!')
>>> logging.error('Error! Error!')
```

Since logging.disable() will disable all messages after it, you will probably want to add it near the import logging line of code in your program. This way, you can easily find it to comment out or uncomment that call to enable or disable logging messages as needed.

### Logging to a File

Instead of displaying the log messages to the screen, you can write them to a text file. The logging.basicConfig() function takes a filename keyword argument, like so:

```
import logging
logging.basicConfig(filename='myProgramLog.txt', level=logging.DEBUG, format='
%(asctime)s -  %(levelname)s -  %(message)s')
```

The log messages will be saved to *myProgramLog.txt*. While logging messages are helpful, they can clutter your screen and make it hard to read the program's output. Writing the logging messages to a file will keep your screen clear and store the messages so you can read them after running the program. You can open this text file in any text editor, such as Notepad or TextEdit.

## Mu's Debugger

The *debugger* is a feature of the Mu editor, IDLE, and other editor software that allows you to execute your program one line at a time. The debugger will run a single line of code and then wait for you to tell it to continue. By running your program "under the debugger" like this, you can take as much time as you want to examine the values in the variables at any given point during the program's lifetime. This is a valuable tool for tracking down bugs.

To run a program under Mu's debugger, click the **Debug** button in the top row of buttons, next to the Run button. Along with the usual output pane at the bottom, the Debug Inspector pane will open along the right side of the window. This pane lists the current value of variables in your program. In Figure 11-1, the debugger has paused the execution of the program just before it would have run the first line of code. You can see this line highlighted in the file editor.

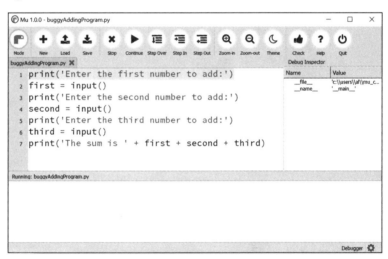

*Figure 11-1: Mu running a program under the debugger*

Debugging mode also adds the following new buttons to the top of the editor: Continue, Step Over, Step In, and Step Out. The usual Stop button is also available.

## Continue

Clicking the Continue button will cause the program to execute normally until it terminates or reaches a *breakpoint*. (I will describe breakpoints later in this chapter.) If you are done debugging and want the program to continue normally, click the Continue button.

## Step In

Clicking the Step In button will cause the debugger to execute the next line of code and then pause again. If the next line of code is a function call, the debugger will "step into" that function and jump to the first line of code of that function.

## Step Over

Clicking the Step Over button will execute the next line of code, similar to the Step In button. However, if the next line of code is a function call, the Step Over button will "step over" the code in the function. The function's code will be executed at full speed, and the debugger will pause as soon as the function call returns. For example, if the next line of code calls a spam() function but you don't really care about code inside this function, you can click Step Over to execute the code in the function at normal speed, and then pause when the function returns. For this reason, using the Over button is more common than using the Step In button.

## Step Out

Clicking the Step Out button will cause the debugger to execute lines of code at full speed until it returns from the current function. If you have stepped into a function call with the Step In button and now simply want to keep executing instructions until you get back out, click the Out button to "step out" of the current function call.

## Stop

If you want to stop debugging entirely and not bother to continue executing the rest of the program, click the Stop button. The Stop button will immediately terminate the program.

## Debugging a Number Adding Program

Open a new file editor tab and enter the following code:

```
print('Enter the first number to add:')
first = input()
print('Enter the second number to add:')
second = input()
print('Enter the third number to add:')
third = input()
print('The sum is ' + first + second + third)
```

Save it as *buggyAddingProgram.py* and run it first without the debugger enabled. The program will output something like this:

```
Enter the first number to add:
5
Enter the second number to add:
3
Enter the third number to add:
42
The sum is 5342
```

The program hasn't crashed, but the sum is obviously wrong. Run the program again, this time under the debugger.

When you click the Debug button, the program pauses on line 1, which is the line of code it is about to execute. Mu should look like Figure 10-1.

Click the **Step Over** button once to execute the first print() call. You should use Step Over instead of Step In here, since you don't want to step into the code for the print() function. (Although Mu should prevent the debugger from entering Python's built-in functions.) The debugger moves on to line 2, and highlights line 2 in the file editor, as shown in Figure 11-2. This shows you where the program execution currently is.

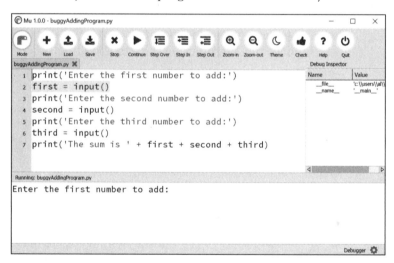

Figure 11-2: The Mu editor window after clicking Step Over

Click **Step Over** again to execute the input() function call. The highlighting will go away while Mu waits for you to type something for the input() call into the output pane. Enter 5 and press ENTER. The highlighting will return.

Keep clicking **Step Over**, and enter 3 and 42 as the next two numbers. When the debugger reaches line 7, the final print() call in the program, the Mu editor window should look like Figure 11-3.

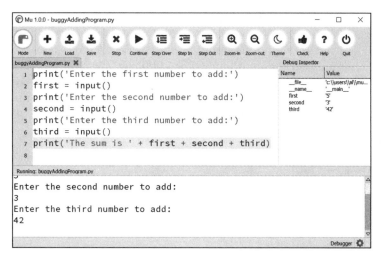

*Figure 11-3: The Debug Inspector pane on the right side shows that the variables are set to strings instead of integers, causing the bug.*

In the Debug Inspector pane, you should see that the first, second, and third variables are set to string values '5', '3', and '42' instead of integer values 5, 3, and 42. When the last line is executed, Python concatenates these strings instead of adding the numbers together, causing the bug.

Stepping through the program with the debugger is helpful but can also be slow. Often you'll want the program to run normally until it reaches a certain line of code. You can configure the debugger to do this with breakpoints.

### Breakpoints

A *breakpoint* can be set on a specific line of code and forces the debugger to pause whenever the program execution reaches that line. Open a new file editor tab and enter the following program, which simulates flipping a coin 1,000 times. Save it as *coinFlip.py*.

```
import random
heads = 0
for i in range(1, 1001):
❶   if random.randint(0, 1) == 1:
        heads = heads + 1
    if i == 500:
❷       print('Halfway done!')
print('Heads came up ' + str(heads) + ' times.')
```

The random.randint(0, 1) call ❶ will return 0 half of the time and 1 the other half of the time. This can be used to simulate a 50/50 coin flip where 1 represents heads. When you run this program without the debugger, it quickly outputs something like the following:

```
Halfway done!
Heads came up 490 times.
```

If you ran this program under the debugger, you would have to click the Step Over button thousands of times before the program terminated. If you were interested in the value of heads at the halfway point of the program's execution, when 500 of 1,000 coin flips have been completed, you could instead just set a breakpoint on the line print('Halfway done!') ❷. To set a breakpoint, click the line number in the file editor to cause a red dot to appear, marking the breakpoint like in Figure 11-4.

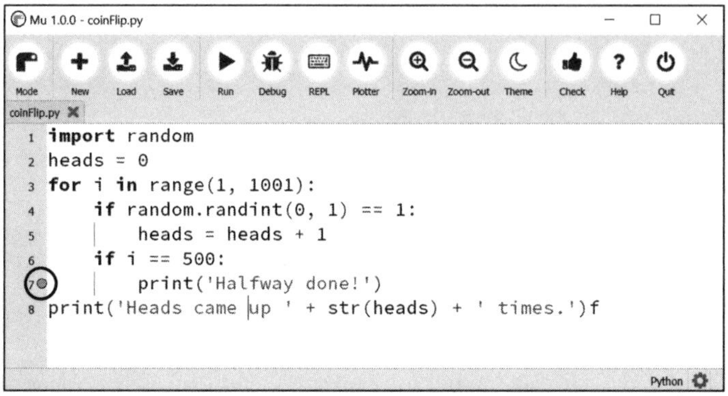

```
1  import random
2  heads = 0
3  for i in range(1, 1001):
4      if random.randint(0, 1) == 1:
5          heads = heads + 1
6      if i == 500:
7          print('Halfway done!')
8  print('Heads came up ' + str(heads) + ' times.')f
```

*Figure 11-4: Setting a breakpoint causes a red dot (circled) to appear next to the line number.*

You don't want to set a breakpoint on the if statement line, since the if statement is executed on every single iteration through the loop. When you set the breakpoint on the code in the if statement, the debugger breaks only when the execution enters the if clause.

The line with the breakpoint will have a red dot next to it. When you run the program under the debugger, it will start in a paused state at the first line, as usual. But if you click Continue, the program will run at full speed until it reaches the line with the breakpoint set on it. You can then click Continue, Step Over, Step In, or Step Out to continue as normal.

If you want to remove a breakpoint, click the line number again. The red dot will go away, and the debugger will not break on that line in the future.

## Summary

Assertions, exceptions, logging, and the debugger are all valuable tools to find and prevent bugs in your program. Assertions with the Python assert statement are a good way to implement "sanity checks" that give you an early warning when a necessary condition doesn't hold true. Assertions are only for errors that the program shouldn't try to recover from and should fail fast. Otherwise, you should raise an exception.

An exception can be caught and handled by the try and except statements. The logging module is a good way to look into your code while it's running and is much more convenient to use than the print() function because of its different logging levels and ability to log to a text file.

The debugger lets you step through your program one line at a time. Alternatively, you can run your program at normal speed and have the debugger pause execution whenever it reaches a line with a breakpoint set. Using the debugger, you can see the state of any variable's value at any point during the program's lifetime.

These debugging tools and techniques will help you write programs that work. Accidentally introducing bugs into your code is a fact of life, no matter how many years of coding experience you have.

## Practice Questions

1. Write an assert statement that triggers an AssertionError if the variable spam is an integer less than 10.

2. Write an assert statement that triggers an AssertionError if the variables eggs and bacon contain strings that are the same as each other, even if their cases are different (that is, 'hello' and 'hello' are considered the same, and 'goodbye' and 'GOODbye' are also considered the same).

3. Write an assert statement that *always* triggers an AssertionError.

4. What are the two lines that your program must have in order to be able to call logging.debug()?

5. What are the two lines that your program must have in order to have logging.debug() send a logging message to a file named *programLog.txt*?

6. What are the five logging levels?

7. What line of code can you add to disable all logging messages in your program?

8. Why is using logging messages better than using print() to display the same message?

9. What are the differences between the Step Over, Step In, and Step Out buttons in the debugger?

10. After you click Continue, when will the debugger stop?

11. What is a breakpoint?

12. How do you set a breakpoint on a line of code in Mu?

## Practice Project

For practice, write a program that does the following.

### Debugging Coin Toss

The following program is meant to be a simple coin toss guessing game. The player gets two guesses (it's an easy game). However, the program has several bugs in it. Run through the program a few times to find the bugs that keep the program from working correctly.

```
import random
guess = ''
while guess not in ('heads', 'tails'):
    print('Guess the coin toss! Enter heads or tails:')
    guess = input()
toss = random.randint(0, 1) # 0 is tails, 1 is heads
if toss == guess:
    print('You got it!')
else:
    print('Nope! Guess again!')
    guesss = input()
    if toss == guess:
        print('You got it!')
    else:
        print('Nope. You are really bad at this game.')
```

# 12

## WEB SCRAPING

In those rare, terrifying moments when I'm without Wi-Fi, I realize just how much of what I do on the computer is really what I do on the internet. Out of sheer habit I'll find myself trying to check email, read friends' Twitter feeds, or answer the question, "Did Kurtwood Smith have any major roles before he was in the original 1987 *RoboCop*?"[1]

Since so much work on a computer involves going on the internet, it'd be great if your programs could get online. *Web scraping* is the term for using a program to download and process content from the web. For example, Google runs many web scraping programs to index web pages for

---

1. The answer is no.

its search engine. In this chapter, you will learn about several modules that make it easy to scrape web pages in Python.

**webbrowser** Comes with Python and opens a browser to a specific page.

**requests** Downloads files and web pages from the internet.

**bs4** Parses HTML, the format that web pages are written in.

**selenium** Launches and controls a web browser. The selenium module is able to fill in forms and simulate mouse clicks in this browser.

# Project: mapIt.py with the webbrowser Module

The webbrowser module's open() function can launch a new browser to a specified URL. Enter the following into the interactive shell:

```
>>> import webbrowser
>>> webbrowser.open('https://inventwithpython.com/')
```

A web browser tab will open to the URL *https://inventwithpython.com/*. This is about the only thing the webbrowser module can do. Even so, the open() function does make some interesting things possible. For example, it's tedious to copy a street address to the clipboard and bring up a map of it on Google Maps. You could take a few steps out of this task by writing a simple script to automatically launch the map in your browser using the contents of your clipboard. This way, you only have to copy the address to a clipboard and run the script, and the map will be loaded for you.

This is what your program does:

1. Gets a street address from the command line arguments or clipboard
2. Opens the web browser to the Google Maps page for the address

This means your code will need to do the following:

1. Read the command line arguments from sys.argv.
2. Read the clipboard contents.
3. Call the webbrowser.open() function to open the web browser.

Open a new file editor tab and save it as *mapIt.py*.

## Step 1: Figure Out the URL

Based on the instructions in Appendix B, set up *mapIt.py* so that when you run it from the command line, like so . . .

```
C:\> mapit 870 Valencia St, San Francisco, CA 94110
```

. . . the script will use the command line arguments instead of the clipboard. If there are no command line arguments, then the program will know to use the contents of the clipboard.

First you need to figure out what URL to use for a given street address. When you load *https://maps.google.com/* in the browser and search for an address, the URL in the address bar looks something like this: *https://www .google.com/maps/place/870+Valencia+St/@37.7590311,-122.4215096,17z/data=!3 m1!4b1!4m2!3m1!1s0x808f7e3dadc07a37:0xc86b0b2bb93b73d8.*

The address is in the URL, but there's a lot of additional text there as well. Websites often add extra data to URLs to help track visitors or customize sites. But if you try just going to *https://www.google.com/maps/place/870+ Valencia+St+San+Francisco+CA/*, you'll find that it still brings up the correct page. So your program can be set to open a web browser to `'https://www .google.com/maps/place/your_address_string'` (where *your_address_string* is the address you want to map).

## Step 2: Handle the Command Line Arguments

Make your code look like this:

```
#! python3
# mapIt.py - Launches a map in the browser using an address from the
# command line or clipboard.

import webbrowser, sys
if len(sys.argv) > 1:
    # Get address from command line.
    address = ' '.join(sys.argv[1:])

# TODO: Get address from clipboard.
```

After the program's #! shebang line, you need to import the `webbrowser` module for launching the browser and import the sys module for reading the potential command line arguments. The `sys.argv` variable stores a list of the program's filename and command line arguments. If this list has more than just the filename in it, then `len(sys.argv)` evaluates to an integer greater than 1, meaning that command line arguments have indeed been provided.

Command line arguments are usually separated by spaces, but in this case, you want to interpret all of the arguments as a single string. Since `sys.argv` is a list of strings, you can pass it to the `join()` method, which returns a single string value. You don't want the program name in this string, so instead of `sys.argv`, you should pass `sys.argv[1:]` to chop off the first element of the array. The final string that this expression evaluates to is stored in the `address` variable.

If you run the program by entering this into the command line . . .

```
mapit 870 Valencia St, San Francisco, CA 94110
```

. . . the `sys.argv` variable will contain this list value:

```
['mapIt.py', '870', 'Valencia', 'St,', 'San', 'Francisco,', 'CA', '94110']
```

The address variable will contain the string `'870 Valencia St, San Francisco, CA 94110'`.

### Step 3: Handle the Clipboard Content and Launch the Browser

Make your code look like the following:

```
#! python3
# mapIt.py - Launches a map in the browser using an address from the
# command line or clipboard.

import webbrowser, sys, pyperclip
if len(sys.argv) > 1:
    # Get address from command line.
    address = ' '.join(sys.argv[1:])
else:
    # Get address from clipboard.
    address = pyperclip.paste()

webbrowser.open('https://www.google.com/maps/place/' + address)
```

If there are no command line arguments, the program will assume the address is stored on the clipboard. You can get the clipboard content with pyperclip.paste() and store it in a variable named address. Finally, to launch a web browser with the Google Maps URL, call webbrowser.open().

While some of the programs you write will perform huge tasks that save you hours, it can be just as satisfying to use a program that conveniently saves you a few seconds each time you perform a common task, such as getting a map of an address. Table 12-1 compares the steps needed to display a map with and without *mapIt.py*.

**Table 12-1:** Getting a Map with and Without *mapIt.py*

| Manually getting a map | Using *mapIt.py* |
| --- | --- |
| 1. Highlight the address. | 1. Highlight the address. |
| 2. Copy the address. | 2. Copy the address. |
| 3. Open the web browser. | 3. Run *mapIt.py*. |
| 4. Go to *https://maps.google.com/*. | |
| 5. Click the address text field. | |
| 6. Paste the address. | |
| 7. Press enter. | |

See how *mapIt.py* makes this task less tedious?

### Ideas for Similar Programs

As long as you have a URL, the webbrowser module lets users cut out the step of opening the browser and directing themselves to a website. Other programs could use this functionality to do the following:

- Open all links on a page in separate browser tabs.
- Open the browser to the URL for your local weather.
- Open several social network sites that you regularly check.

# Downloading Files from the Web with the requests Module

The requests module lets you easily download files from the web without having to worry about complicated issues such as network errors, connection problems, and data compression. The requests module doesn't come with Python, so you'll have to install it first. From the command line, run **pip install --user requests**. (Appendix A has additional details on how to install third-party modules.)

The requests module was written because Python's urllib2 module is too complicated to use. In fact, take a permanent marker and black out this entire paragraph. Forget I ever mentioned urllib2. If you need to download things from the web, just use the requests module.

Next, do a simple test to make sure the requests module installed itself correctly. Enter the following into the interactive shell:

```
>>> import requests
```

If no error messages show up, then the requests module has been successfully installed.

## Downloading a Web Page with the requests.get() Function

The requests.get() function takes a string of a URL to download. By calling type() on requests.get()'s return value, you can see that it returns a Response object, which contains the response that the web server gave for your request. I'll explain the Response object in more detail later, but for now, enter the following into the interactive shell while your computer is connected to the internet:

```
>>> import requests
❶ >>> res = requests.get('https://automatetheboringstuff.com/files/rj.txt')
>>> type(res)
<class 'requests.models.Response'>
❷ >>> res.status_code == requests.codes.ok
True
>>> len(res.text)
178981
>>> print(res.text[:250])
The Project Gutenberg EBook of Romeo and Juliet, by William Shakespeare

This eBook is for the use of anyone anywhere at no cost and with
almost no restrictions whatsoever.  You may copy it, give it away or
re-use it under the terms of the Proje
```

The URL goes to a text web page for the entire play of *Romeo and Juliet*, provided on this book's site ❶. You can tell that the request for this web page succeeded by checking the status_code attribute of the Response object. If it is equal to the value of requests.codes.ok, then everything went fine ❷. (Incidentally, the status code for "OK" in the HTTP protocol is 200. You may already be familiar with the 404 status code for "Not Found.")

You can find a complete list of HTTP status codes and their meanings at *https://en.wikipedia.org/wiki/List_of_HTTP_status_codes.*

If the request succeeded, the downloaded web page is stored as a string in the Response object's text variable. This variable holds a large string of the entire play; the call to len(res.text) shows you that it is more than 178,000 characters long. Finally, calling print(res.text[:250]) displays only the first 250 characters.

If the request failed and displayed an error message, like "Failed to establish a new connection" or "Max retries exceeded," then check your internet connection. Connecting to servers can be quite complicated, and I can't give a full list of possible problems here. You can find common causes of your error by doing a web search of the error message in quotes.

### Checking for Errors

As you've seen, the Response object has a status_code attribute that can be checked against requests.codes.ok (a variable that has the integer value 200) to see whether the download succeeded. A simpler way to check for success is to call the raise_for_status() method on the Response object. This will raise an exception if there was an error downloading the file and will do nothing if the download succeeded. Enter the following into the interactive shell:

```
>>> res = requests.get('https://inventwithpython.com/page_that_does_not_exist')
>>> res.raise_for_status()
Traceback (most recent call last):
  File "<stdin>", line 1, in <module>

  File "C:\Users\Al\AppData\Local\Programs\Python\Python37\lib\site-packages\requests\models
.py", line 940, in raise_for_status
    raise HTTPError(http_error_msg, response=self)
requests.exceptions.HTTPError: 404 Client Error: Not Found for url: https://inventwithpython
.com/page_that_does_not_exist.html
```

The raise_for_status() method is a good way to ensure that a program halts if a bad download occurs. This is a good thing: You want your program to stop as soon as some unexpected error happens. If a failed download *isn't* a deal breaker for your program, you can wrap the raise_for_status() line with try and except statements to handle this error case without crashing.

```
import requests
res = requests.get('https://inventwithpython.com/page_that_does_not_exist')
try:
    res.raise_for_status()
except Exception as exc:
    print('There was a problem: %s' % (exc))
```

This raise_for_status() method call causes the program to output the following:

```
There was a problem: 404 Client Error: Not Found for url: https://
inventwithpython.com/page_that_does_not_exist.html
```

Always call `raise_for_status()` after calling `requests.get()`. You want to be sure that the download has actually worked before your program continues.

## Saving Downloaded Files to the Hard Drive

From here, you can save the web page to a file on your hard drive with the standard `open()` function and `write()` method. There are some slight differences, though. First, you must open the file in *write binary* mode by passing the string `'wb'` as the second argument to `open()`. Even if the page is in plaintext (such as the *Romeo and Juliet* text you downloaded earlier), you need to write binary data instead of text data in order to maintain the *Unicode encoding* of the text.

To write the web page to a file, you can use a `for` loop with the `Response` object's `iter_content()` method.

```
>>> import requests
>>> res = requests.get('https://automatetheboringstuff.com/files/rj.txt')
>>> res.raise_for_status()
>>> playFile = open('RomeoAndJuliet.txt', 'wb')
>>> for chunk in res.iter_content(100000):
        playFile.write(chunk)

100000
78981
>>> playFile.close()
```

The `iter_content()` method returns "chunks" of the content on each iteration through the loop. Each chunk is of the *bytes* data type, and you get to specify how many bytes each chunk will contain. One hundred thousand bytes is generally a good size, so pass `100000` as the argument to `iter_content()`.

The file *RomeoAndJuliet.txt* will now exist in the current working directory. Note that while the filename on the website was *rj.txt*, the file on your hard drive has a different filename. The requests module simply handles downloading the contents of web pages. Once the page is downloaded, it is simply data in your program. Even if you were to lose your internet

---

**UNICODE ENCODINGS**

Unicode encodings are beyond the scope of this book, but you can learn more about them from these web pages:

- Joel on Software: The Absolute Minimum Every Software Developer Absolutely, Positively Must Know About Unicode and Character Sets (No Excuses!): *https://www.joelonsoftware.com/articles/Unicode.html*

- Pragmatic Unicode: *https://nedbatchelder.com/text/unipain.html*

---

connection after downloading the web page, all the page data would still be on your computer.

The write() method returns the number of bytes written to the file. In the previous example, there were 100,000 bytes in the first chunk, and the remaining part of the file needed only 78,981 bytes.

To review, here's the complete process for downloading and saving a file:

1. Call requests.get() to download the file.
2. Call open() with 'wb' to create a new file in write binary mode.
3. Loop over the Response object's iter_content() method.
4. Call write() on each iteration to write the content to the file.
5. Call close() to close the file.

That's all there is to the requests module! The for loop and iter_content() stuff may seem complicated compared to the open()/write()/close() workflow you've been using to write text files, but it's to ensure that the requests module doesn't eat up too much memory even if you download massive files. You can learn about the requests module's other features from *https://requests.readthedocs.org/.*

# HTML

Before you pick apart web pages, you'll learn some HTML basics. You'll also see how to access your web browser's powerful developer tools, which will make scraping information from the web much easier.

## Resources for Learning HTML

*Hypertext Markup Language (HTML)* is the format that web pages are written in. This chapter assumes you have some basic experience with HTML, but if you need a beginner tutorial, I suggest one of the following sites:

- *https://developer.mozilla.org/en-US/learn/html/*
- *https://htmldog.com/guides/html/beginner/*
- *https://www.codecademy.com/learn/learn-html*

## A Quick Refresher

In case it's been a while since you've looked at any HTML, here's a quick overview of the basics. An HTML file is a plaintext file with the *.html* file extension. The text in these files is surrounded by *tags*, which are words enclosed in angle brackets. The tags tell the browser how to format the web page. A starting tag and closing tag can enclose some text to form an *element*. The *text* (or *inner HTML*) is the content between the starting and closing tags. For example, the following HTML will display *Hello, world!* in the browser, with *Hello* in bold:

```
<strong>Hello</strong>, world!
```

This HTML will look like Figure 12-1 in a browser.

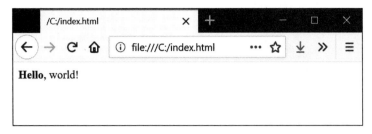

*Figure 12-1: Hello, world! rendered in the browser*

The opening `<strong>` tag says that the enclosed text will appear in bold. The closing `</strong>` tags tells the browser where the end of the bold text is.

There are many different tags in HTML. Some of these tags have extra properties in the form of *attributes* within the angle brackets. For example, the `<a>` tag encloses text that should be a link. The URL that the text links to is determined by the `href` attribute. Here's an example:

```
Al's free <a href="https://inventwithpython.com">Python books</a>.
```

This HTML will look like Figure 12-2 in a browser.

*Figure 12-2: The link rendered in the browser*

Some elements have an `id` attribute that is used to uniquely identify the element in the page. You will often instruct your programs to seek out an element by its `id` attribute, so figuring out an element's `id` attribute using the browser's developer tools is a common task in writing web scraping programs.

## Viewing the Source HTML of a Web Page

You'll need to look at the HTML source of the web pages that your programs will work with. To do this, right-click (or CTRL-click on macOS) any web page in your web browser, and select **View Source** or **View page source** to see the HTML text of the page (see Figure 12-3). This is the text your browser actually receives. The browser knows how to display, or *render*, the web page from this HTML.

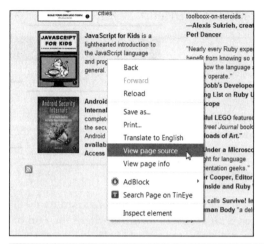

Figure 12-3: Viewing the source of a web page

I highly recommend viewing the source HTML of some of your favorite sites. It's fine if you don't fully understand what you are seeing when you look at the source. You won't need HTML mastery to write simple web scraping programs—after all, you won't be writing your own websites. You just need enough knowledge to pick out data from an existing site.

## Opening Your Browser's Developer Tools

In addition to viewing a web page's source, you can look through a page's HTML using your browser's developer tools. In Chrome and Internet Explorer for Windows, the developer tools are already installed, and you can press F12 to make them appear (see Figure 12-4). Pressing F12 again will make the developer tools disappear. In Chrome, you can also bring up the developer tools by selecting **View ▶ Developer ▶ Developer Tools**. In macOS, pressing ⌘-OPTION-I will open Chrome's Developer Tools.

*Figure 12-4: The Developer Tools window in the Chrome browser*

In Firefox, you can bring up the Web Developer Tools Inspector by pressing CTRL-SHIFT-C on Windows and Linux or by pressing ⌘-OPTION-C on macOS. The layout is almost identical to Chrome's developer tools.

In Safari, open the Preferences window, and on the Advanced pane check the **Show Develop menu in the menu bar** option. After it has been enabled, you can bring up the developer tools by pressing ⌘-OPTION-I.

After enabling or installing the developer tools in your browser, you can right-click any part of the web page and select **Inspect Element** from the context menu to bring up the HTML responsible for that part of the page. This will be helpful when you begin to parse HTML for your web scraping programs.

---

### DON'T USE REGULAR EXPRESSIONS TO PARSE HTML

Locating a specific piece of HTML in a string seems like a perfect case for regular expressions. However, I advise you against it. There are many different ways that HTML can be formatted and still be considered valid HTML, but trying to capture all these possible variations in a regular expression can be tedious and error prone. A module developed specifically for parsing HTML, such as bs4, will be less likely to result in bugs.

You can find an extended argument for why you shouldn't parse HTML with regular expressions at *https://stackoverflow.com/a/1732454/1893164/*.

## Using the Developer Tools to Find HTML Elements

Once your program has downloaded a web page using the requests module, you will have the page's HTML content as a single string value. Now you need to figure out which part of the HTML corresponds to the information on the web page you're interested in.

This is where the browser's developer tools can help. Say you want to write a program to pull weather forecast data from *https://weather.gov/*. Before writing any code, do a little research. If you visit the site and search for the 94105 ZIP code, the site will take you to a page showing the forecast for that area.

What if you're interested in scraping the weather information for that ZIP code? Right-click where it is on the page (or CONTROL-click on macOS) and select **Inspect Element** from the context menu that appears. This will bring up the Developer Tools window, which shows you the HTML that produces this particular part of the web page. Figure 12-5 shows the developer tools open to the HTML of the nearest forecast. Note that if the *https://weather.gov/* site changes the design of its web pages, you'll need to repeat this process to inspect the new elements.

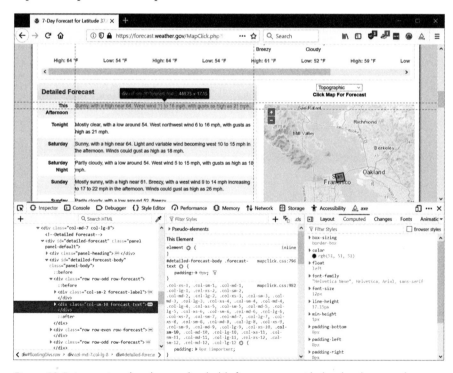

*Figure 12-5: Inspecting the element that holds forecast text with the developer tools*

From the developer tools, you can see that the HTML responsible for the forecast part of the web page is `<div class="col-sm-10 forecast-text">Sunny, with a high near 64. West wind 11 to 16 mph, with gusts as high as 21 mph .</div>`. This is exactly what you were looking for! It seems that the forecast information is contained inside a `<div>` element with the `forecast-text` CSS class. Right-click on this element in the browser's developer console, and from the context menu that appears, select **Copy ▶ CSS Selector**. This will copy a string such as `'div.row-odd:nth-child(1) > div:nth-child(2)'` to the clipboard. You can use this string for Beautiful Soup's select() or Selenium's find_element_by_css_selector() methods, as explained later in this chapter. Now that you know what you're looking for, the Beautiful Soup module will help you find it in the string.

## Parsing HTML with the bs4 Module

Beautiful Soup is a module for extracting information from an HTML page (and is much better for this purpose than regular expressions). The Beautiful Soup module's name is bs4 (for Beautiful Soup, version 4). To install it, you will need to run `pip install --user beautifulsoup4` from the command line. (Check out Appendix A for instructions on installing third-party modules.) While beautifulsoup4 is the name used for installation, to import Beautiful Soup you run `import bs4`.

For this chapter, the Beautiful Soup examples will *parse* (that is, analyze and identify the parts of) an HTML file on the hard drive. Open a new file editor tab in Mu, enter the following, and save it as *example.html*. Alternatively, download it from *https://nostarch.com/automatestuff2/*.

```
<!-- This is the example.html example file. -->

<html><head><title>The Website Title</title></head>
<body>
<p>Download my <strong>Python</strong> book from <a href="https://
inventwithpython.com">my website</a>.</p>
<p class="slogan">Learn Python the easy way!</p>
<p>By <span id="author">Al Sweigart</span></p>
</body></html>
```

As you can see, even a simple HTML file involves many different tags and attributes, and matters quickly get confusing with complex websites. Thankfully, Beautiful Soup makes working with HTML much easier.

## Creating a BeautifulSoup Object from HTML

The bs4.BeautifulSoup() function needs to be called with a string containing the HTML it will parse. The bs4.BeautifulSoup() function returns a BeautifulSoup object. Enter the following into the interactive shell while your computer is connected to the internet:

```
>>> import requests, bs4
>>> res = requests.get('https://nostarch.com')
>>> res.raise_for_status()
>>> noStarchSoup = bs4.BeautifulSoup(res.text, 'html.parser')
>>> type(noStarchSoup)
<class 'bs4.BeautifulSoup'>
```

This code uses requests.get() to download the main page from the No Starch Press website and then passes the text attribute of the response to bs4.BeautifulSoup(). The BeautifulSoup object that it returns is stored in a variable named noStarchSoup.

You can also load an HTML file from your hard drive by passing a File object to bs4.BeautifulSoup() along with a second argument that tells Beautiful Soup which parser to use to analyze the HTML.

Enter the following into the interactive shell (after making sure the *example.html* file is in the working directory):

```
>>> exampleFile = open('example.html')
>>> exampleSoup = bs4.BeautifulSoup(exampleFile, 'html.parser')
>>> type(exampleSoup)
<class 'bs4.BeautifulSoup'>
```

The 'html.parser' parser used here comes with Python. However, you can use the faster 'lxml' parser if you install the third-party lxml module. Follow the instructions in Appendix A to install this module by running pip install --user lxml. Forgetting to include this second argument will result in a UserWarning: No parser was explicitly specified warning.

Once you have a BeautifulSoup object, you can use its methods to locate specific parts of an HTML document.

## Finding an Element with the select() Method

You can retrieve a web page element from a BeautifulSoup object by calling the select()method and passing a string of a CSS *selector* for the element you are looking for. Selectors are like regular expressions: they specify a pattern to look for—in this case, in HTML pages instead of general text strings.

A full discussion of CSS selector syntax is beyond the scope of this book (there's a good selector tutorial in the resources at *https://nostarch.com /automatestuff2/*), but here's a short introduction to selectors. Table 12-2 shows examples of the most common CSS selector patterns.

**Table 12-2:** Examples of CSS Selectors

| Selector passed to the `select()` method | Will match . . . |
| --- | --- |
| `soup.select('div')` | All elements named `<div>` |
| `soup.select('#author')` | The element with an id attribute of author |
| `soup.select('.notice')` | All elements that use a CSS class attribute named notice |
| `soup.select('div span')` | All elements named `<span>` that are within an element named `<div>` |
| `soup.select('div > span')` | All elements named `<span>` that are *directly* within an element named `<div>`, with no other element in between |
| `soup.select('input[name]')` | All elements named `<input>` that have a name attribute with any value |
| `soup.select('input[type="button"]')` | All elements named `<input>` that have an attribute named type with value button |

The various selector patterns can be combined to make sophisticated matches. For example, soup.select('p #author') will match any element that has an id attribute of author, as long as it is also inside a <p> element. Instead of writing the selector yourself, you can also right-click on the element in your browser and select **Inspect Element**. When the browser's developer console opens, right-click on the element's HTML and select **Copy ▶ CSS Selector** to copy the selector string to the clipboard and paste it into your source code.

The select() method will return a list of Tag objects, which is how Beautiful Soup represents an HTML element. The list will contain one Tag object for every match in the BeautifulSoup object's HTML. Tag values can be passed to the str() function to show the HTML tags they represent. Tag values also have an attrs attribute that shows all the HTML attributes of the tag as a dictionary. Using the *example.html* file from earlier, enter the following into the interactive shell:

```
>>> import bs4
>>> exampleFile = open('example.html')
>>> exampleSoup = bs4.BeautifulSoup(exampleFile.read(), 'html.parser')
>>> elems = exampleSoup.select('#author')
>>> type(elems) # elems is a list of Tag objects.
<class 'list'>
>>> len(elems)
1
>>> type(elems[0])
<class 'bs4.element.Tag'>
>>> str(elems[0]) # The Tag object as a string.
'<span id="author">Al Sweigart</span>'
>>> elems[0].getText()
'Al Sweigart'
>>> elems[0].attrs
{'id': 'author'}
```

This code will pull the element with id="author" out of our example HTML. We use select('#author') to return a list of all the elements with id="author". We store this list of Tag objects in the variable elems, and len(elems) tells us there is one Tag object in the list; there was one match. Calling getText() on the element returns the element's text, or inner HTML. The text of an element is the content between the opening and closing tags: in this case, 'Al Sweigart'.

Passing the element to str() returns a string with the starting and closing tags and the element's text. Finally, attrs gives us a dictionary with the element's attribute, 'id', and the value of the id attribute, 'author'.

You can also pull all the \<p\> elements from the BeautifulSoup object. Enter this into the interactive shell:

```
>>> pElems = exampleSoup.select('p')
>>> str(pElems[0])
'<p>Download my <strong>Python</strong> book from <a href="https://
inventwithpython.com">my website</a>.</p>'
>>> pElems[0].getText()
'Download my Python book from my website.'
>>> str(pElems[1])
'<p class="slogan">Learn Python the easy way!</p>'
>>> pElems[1].getText()
'Learn Python the easy way!'
>>> str(pElems[2])
'<p>By <span id="author">Al Sweigart</span></p>'
>>> pElems[2].getText()
'By Al Sweigart'
```

This time, select() gives us a list of three matches, which we store in pElems. Using str() on pElems[0], pElems[1], and pElems[2] shows you each element as a string, and using getText() on each element shows you its text.

## Getting Data from an Element's Attributes

The get() method for Tag objects makes it simple to access attribute values from an element. The method is passed a string of an attribute name and returns that attribute's value. Using *example.html*, enter the following into the interactive shell:

```
>>> import bs4
>>> soup = bs4.BeautifulSoup(open('example.html'), 'html.parser')
>>> spanElem = soup.select('span')[0]
>>> str(spanElem)
'<span id="author">Al Sweigart</span>'
>>> spanElem.get('id')
'author'
>>> spanElem.get('some_nonexistent_addr') == None
True
>>> spanElem.attrs
{'id': 'author'}
```

Here we use select() to find any `<span>` elements and then store the first matched element in spanElem. Passing the attribute name 'id' to get() returns the attribute's value, 'author'.

# Project: Opening All Search Results

Whenever I search a topic on Google, I don't look at just one search result at a time. By middle-clicking a search result link (or clicking while holding CTRL), I open the first several links in a bunch of new tabs to read later. I search Google often enough that this workflow—opening my browser, searching for a topic, and middle-clicking several links one by one—is tedious. It would be nice if I could simply type a search term on the command line and have my computer automatically open a browser with all the top search results in new tabs. Let's write a script to do this with the search results page for the Python Package Index at *https://pypi.org/*. A program like this can be adapted to many other websites, although the Google and DuckDuckGo often employ measures that make scraping their search results pages difficult.

This is what your program does:

1. Gets search keywords from the command line arguments
2. Retrieves the search results page
3. Opens a browser tab for each result

This means your code will need to do the following:

1. Read the command line arguments from sys.argv.
2. Fetch the search result page with the requests module.
3. Find the links to each search result.
4. Call the webbrowser.open() function to open the web browser.

Open a new file editor tab and save it as *searchpypi.py*.

## Step 1: Get the Command Line Arguments and Request the Search Page

Before coding anything, you first need to know the URL of the search result page. By looking at the browser's address bar after doing a search, you can see that the result page has a URL like *https://pypi.org/search/?q=<SEARCH _TERM_HERE>*. The requests module can download this page and then you can use Beautiful Soup to find the search result links in the HTML. Finally, you'll use the webbrowser module to open those links in browser tabs.

Make your code look like the following:

```
#! python3
# searchpypi.py  - Opens several search results.

import requests, sys, webbrowser, bs4
```

```
print('Searching...')    # display text while downloading the search result page
res = requests.get('https://google.com/search?q=' 'https://pypi.org/search/?q='
+ ' '.join(sys.argv[1:]))
res.raise_for_status()

# TODO: Retrieve top search result links.

# TODO: Open a browser tab for each result.
```

The user will specify the search terms using command line arguments when they launch the program. These arguments will be stored as strings in a list in sys.argv.

## Step 2: Find All the Results

Now you need to use Beautiful Soup to extract the top search result links from your downloaded HTML. But how do you figure out the right selector for the job? For example, you can't just search for all <a> tags, because there are lots of links you don't care about in the HTML. Instead, you must inspect the search result page with the browser's developer tools to try to find a selector that will pick out only the links you want.

After doing a search for *Beautiful Soup*, you can open the browser's developer tools and inspect some of the link elements on the page. They can look complicated, something like pages of this: <a class="package -snippet" href="HYPERLINK "view-source:https://pypi.org/project/xml-parser /"/project/xml-parser/">.

It doesn't matter that the element looks incredibly complicated. You just need to find the pattern that all the search result links have.

Make your code look like the following:

```
#! python3
# searchpypi.py - Opens several google results.
import requests, sys, webbrowser, bs4
--snip--
# Retrieve top search result links.
soup = bs4.BeautifulSoup(res.text, 'html.parser')
# Open a browser tab for each result.
linkElems = soup.select('.package-snippet')
```

If you look at the <a> elements, though, the search result links all have class="package-snippet". Looking through the rest of the HTML source, it looks like the package-snippet class is used only for search result links. You don't have to know what the CSS class package-snippet is or what it does. You're just going to use it as a marker for the <a> element you are looking for. You can create a BeautifulSoup object from the downloaded page's HTML text and then use the selector '.package-snippet' to find all <a> elements that are within an element that has the package-snippet CSS class. Note that if the PyPI website changes its layout, you may need to update this program with a new CSS selector string to pass to soup.select(). The rest of the program will still be up to date.

## Step 3: Open Web Browsers for Each Result

Finally, we'll tell the program to open web browser tabs for our results. Add the following to the end of your program:

```
#! python3
# searchpypi.py - Opens several search results.
import requests, sys, webbrowser, bs4
--snip--
# Open a browser tab for each result.
linkElems = soup.select('.package-snippet')
numOpen = min(5, len(linkElems))
for i in range(numOpen):
    urlToOpen = 'https://pypi.org' + linkElems[i].get('href')
    print('Opening', urlToOpen)
    webbrowser.open(urlToOpen)
```

By default, you open the first five search results in new tabs using the webbrowser module. However, the user may have searched for something that turned up fewer than five results. The soup.select() call returns a list of all the elements that matched your '.package-snippet' selector, so the number of tabs you want to open is either 5 or the length of this list (whichever is smaller).

The built-in Python function min() returns the smallest of the integer or float arguments it is passed. (There is also a built-in max() function that returns the largest argument it is passed.) You can use min() to find out whether there are fewer than five links in the list and store the number of links to open in a variable named numOpen. Then you can run through a for loop by calling range(numOpen).

On each iteration of the loop, you use webbrowser.open() to open a new tab in the web browser. Note that the href attribute's value in the returned <a> elements do not have the initial https://pypi.org part, so you have to concatenate that to the href attribute's string value.

Now you can instantly open the first five PyPI search results for, say, *boring stuff* by running searchpypi boring stuff on the command line! (See Appendix B for how to easily run programs on your operating system.)

## Ideas for Similar Programs

The benefit of tabbed browsing is that you can easily open links in new tabs to peruse later. A program that automatically opens several links at once can be a nice shortcut to do the following:

- Open all the product pages after searching a shopping site such as Amazon.
- Open all the links to reviews for a single product.
- Open the result links to photos after performing a search on a photo site such as Flickr or Imgur.

# Project: Downloading All XKCD Comics

Blogs and other regularly updating websites usually have a front page with the most recent post as well as a Previous button on the page that takes you to the previous post. Then that post will also have a Previous button, and so on, creating a trail from the most recent page to the first post on the site. If you wanted a copy of the site's content to read when you're not online, you could manually navigate over every page and save each one. But this is pretty boring work, so let's write a program to do it instead.

XKCD is a popular geek webcomic with a website that fits this structure (see Figure 12-6). The front page at *https://xkcd.com/* has a Prev button that guides the user back through prior comics. Downloading each comic by hand would take forever, but you can write a script to do this in a couple of minutes.

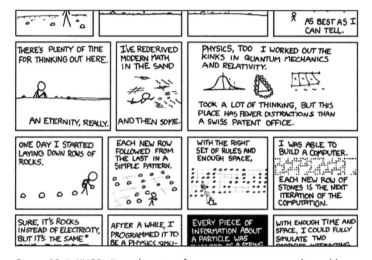

Figure 12-6: XKCD, "a webcomic of romance, sarcasm, math, and language"

Here's what your program does:

1. Loads the XKCD home page
2. Saves the comic image on that page
3. Follows the Previous Comic link
4. Repeats until it reaches the first comic

This means your code will need to do the following:

1. Download pages with the requests module.
2. Find the URL of the comic image for a page using Beautiful Soup.

3. Download and save the comic image to the hard drive with `iter_content()`.

4. Find the URL of the Previous Comic link, and repeat.

Open a new file editor tab and save it as *downloadXkcd.py*.

## Step 1: Design the Program

If you open the browser's developer tools and inspect the elements on the page, you'll find the following:

- The URL of the comic's image file is given by the `href` attribute of an `<img>` element.
- The `<img>` element is inside a `<div id="comic">` element.
- The Prev button has a `rel` HTML attribute with the value `prev`.
- The first comic's Prev button links to the *https://xkcd.com/#* URL, indicating that there are no more previous pages.

Make your code look like the following:

```python
#! python3
# downloadXkcd.py - Downloads every single XKCD comic.

import requests, os, bs4

url = 'https://xkcd.com'              # starting url
os.makedirs('xkcd', exist_ok=True)    # store comics in ./xkcd
while not url.endswith('#'):
    # TODO: Download the page.

    # TODO: Find the URL of the comic image.

    # TODO: Download the image.

    # TODO: Save the image to ./xkcd.

    # TODO: Get the Prev button's url.

print('Done.')
```

You'll have a url variable that starts with the value `'https://xkcd.com'` and repeatedly update it (in a for loop) with the URL of the current page's Prev link. At every step in the loop, you'll download the comic at url. You'll know to end the loop when url ends with `'#'`.

You will download the image files to a folder in the current working directory named *xkcd*. The call `os.makedirs()` ensures that this folder exists, and the `exist_ok=True` keyword argument prevents the function from throwing an exception if this folder already exists. The remaining code is just comments that outline the rest of your program.

## Step 2: Download the Web Page

Let's implement the code for downloading the page. Make your code look like the following:

```python
#! python3
# downloadXkcd.py - Downloads every single XKCD comic.

import requests, os, bs4

url = 'https://xkcd.com'                 # starting url
os.makedirs('xkcd', exist_ok=True)       # store comics in ./xkcd
while not url.endswith('#'):
    # Download the page.
    print('Downloading page %s...' % url)
    res = requests.get(url)
    res.raise_for_status()

    soup = bs4.BeautifulSoup(res.text, 'html.parser')

    # TODO: Find the URL of the comic image.

    # TODO: Download the image.

    # TODO: Save the image to ./xkcd.

    # TODO: Get the Prev button's url.

print('Done.')
```

First, print url so that the user knows which URL the program is about to download; then use the requests module's request.get() function to download it. As always, you immediately call the Response object's raise_for_status() method to throw an exception and end the program if something went wrong with the download. Otherwise, you create a BeautifulSoup object from the text of the downloaded page.

## Step 3: Find and Download the Comic Image

Make your code look like the following:

```python
#! python3
# downloadXkcd.py - Downloads every single XKCD comic.

import requests, os, bs4

--snip--

    # Find the URL of the comic image.
    comicElem = soup.select('#comic img')
```

```
    if comicElem == []:
        print('Could not find comic image.')
    else:
        comicUrl = 'https:' + comicElem[0].get('src')
        # Download the image.
        print('Downloading image %s...' % (comicUrl))
        res = requests.get(comicUrl)
        res.raise_for_status()

    # TODO: Save the image to ./xkcd.

    # TODO: Get the Prev button's url.

print('Done.')
```

From inspecting the XKCD home page with your developer tools, you know that the <img> element for the comic image is inside a <div> element with the id attribute set to comic, so the selector '#comic img' will get you the correct <img> element from the BeautifulSoup object.

A few XKCD pages have special content that isn't a simple image file. That's fine; you'll just skip those. If your selector doesn't find any elements, then soup.select('#comic img') will return a blank list. When that happens, the program can just print an error message and move on without downloading the image.

Otherwise, the selector will return a list containing one <img> element. You can get the src attribute from this <img> element and pass it to requests .get() to download the comic's image file.

### Step 4: Save the Image and Find the Previous Comic

Make your code look like the following:

```
#! python3
# downloadXkcd.py - Downloads every single XKCD comic.

import requests, os, bs4

--snip--

        # Save the image to ./xkcd.
        imageFile = open(os.path.join('xkcd', os.path.basename(comicUrl)),
'wb')
        for chunk in res.iter_content(100000):
            imageFile.write(chunk)
        imageFile.close()

    # Get the Prev button's url.
    prevLink = soup.select('a[rel="prev"]')[0]
    url = 'https://xkcd.com' + prevLink.get('href')

print('Done.')
```

At this point, the image file of the comic is stored in the res variable. You need to write this image data to a file on the hard drive.

You'll need a filename for the local image file to pass to open(). The comicUrl will have a value like 'https://imgs.xkcd.com/comics/heartbleed _explanation.png'—which you might have noticed looks a lot like a file path. And in fact, you can call os.path.basename() with comicUrl, and it will return just the last part of the URL, 'heartbleed_explanation.png'. You can use this as the filename when saving the image to your hard drive. You join this name with the name of your xkcd folder using os.path.join() so that your program uses backslashes (\) on Windows and forward slashes (/) on macOS and Linux. Now that you finally have the filename, you can call open() to open a new file in 'wb' "write binary" mode.

Remember from earlier in this chapter that to save files you've downloaded using requests, you need to loop over the return value of the iter _content() method. The code in the for loop writes out chunks of the image data (at most 100,000 bytes each) to the file and then you close the file. The image is now saved to your hard drive.

Afterward, the selector 'a[rel="prev"]' identifies the <a> element with the rel attribute set to prev, and you can use this <a> element's href attribute to get the previous comic's URL, which gets stored in url. Then the while loop begins the entire download process again for this comic.

The output of this program will look like this:

```
Downloading page https://xkcd.com...
Downloading image https://imgs.xkcd.com/comics/phone_alarm.png...
Downloading page https://xkcd.com/1358/...
Downloading image https://imgs.xkcd.com/comics/nro.png...
Downloading page https://xkcd.com/1357/...
Downloading image https://imgs.xkcd.com/comics/free_speech.png...
Downloading page https://xkcd.com/1356/...
Downloading image https://imgs.xkcd.com/comics/orbital_mechanics.png...
Downloading page https://xkcd.com/1355/...
Downloading image https://imgs.xkcd.com/comics/airplane_message.png...
Downloading page https://xkcd.com/1354/...
Downloading image https://imgs.xkcd.com/comics/heartbleed_explanation.png...
--snip--
```

This project is a good example of a program that can automatically follow links in order to scrape large amounts of data from the web. You can learn about Beautiful Soup's other features from its documentation at *https://www.crummy.com/software/BeautifulSoup/bs4/doc/*.

## *Ideas for Similar Programs*

Downloading pages and following links are the basis of many web crawling programs. Similar programs could also do the following:

- Back up an entire site by following all of its links.
- Copy all the messages off a web forum.
- Duplicate the catalog of items for sale on an online store.

The requests and bs4 modules are great as long as you can figure out the URL you need to pass to requests.get(). However, sometimes this isn't so easy to find. Or perhaps the website you want your program to navigate requires you to log in first. The selenium module will give your programs the power to perform such sophisticated tasks.

## Controlling the Browser with the selenium Module

The selenium module lets Python directly control the browser by programmatically clicking links and filling in login information, almost as though there were a human user interacting with the page. Using selenium, you can interact with web pages in a much more advanced way than with requests and bs4; but because it launches a web browser, it is a bit slower and hard to run in the background if, say, you just need to download some files from the web.

Still, if you need to interact with a web page in a way that, say, depends on the JavaScript code that updates the page, you'll need to use selenium instead of requests. That's because major ecommerce websites such as Amazon almost certainly have software systems to recognize traffic that they suspect is a script harvesting their info or signing up for multiple free accounts. These sites may refuse to serve pages to you after a while, breaking any scripts you've made. The selenium module is much more likely to function on these sites long-term than requests.

A major "tell" to websites that you're using a script is the *user-agent* string, which identifies the web browser and is included in all HTTP requests. For example, the user-agent string for the requests module is something like 'python-requests/2.21.0'. You can visit a site such as *https:// www.whatsmyua.info/* to see your user-agent string. Using selenium, you're much more likely to "pass for human" because not only is Selenium's user-agent is the same as a regular browser (for instance, 'Mozilla/5.0 (Windows NT 10.0; Win64; x64; rv:65.0) Gecko/20100101 Firefox/65.0'), but it has the same traffic patterns: a selenium-controlled browser will download images, advertisements, cookies, and privacy-invading trackers just like a regular browser. However, selenium can still be detected by websites, and major ticketing and ecommerce websites often block browsers controlled by selenium to prevent web scraping of their pages.

### Starting a selenium-Controlled Browser

The following examples will show you how to control Firefox's web browser. If you don't already have Firefox, you can download it for free from *https:// getfirefox.com/*. You can install selenium by running pip install --user selenium from a command line terminal. More information is available in Appendix A.

Importing the modules for selenium is slightly tricky. Instead of import selenium, you need to run from selenium import webdriver. (The exact reason why the selenium module is set up this way is beyond the scope of this book.)

After that, you can launch the Firefox browser with selenium. Enter the following into the interactive shell:

```
>>> from selenium import webdriver
>>> browser = webdriver.Firefox()
>>> type(browser)
<class 'selenium.webdriver.firefox.webdriver.WebDriver'>
>>> browser.get('https://inventwithpython.com')
```

You'll notice when webdriver.Firefox() is called, the Firefox web browser starts up. Calling type() on the value webdriver.Firefox() reveals it's of the WebDriver data type. And calling browser.get('https://inventwithpython.com') directs the browser to *https://inventwithpython.com/*. Your browser should look something like Figure 12-7.

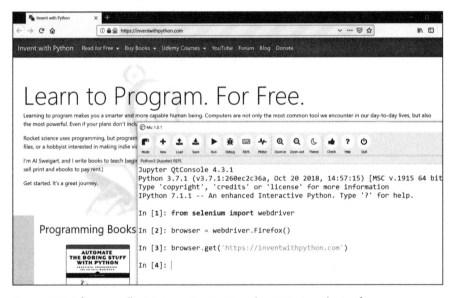

*Figure 12-7: After we call webdriver.Firefox() and get() in Mu, the Firefox browser appears.*

If you encounter the error message "'geckodriver' executable needs to be in PATH.", then you need to manually download the webdriver for Firefox before you can use selenium to control it. You can also control browsers other than Firefox if you install the webdriver for them.

For Firefox, go to *https://github.com/mozilla/geckodriver/releases* and download the geckodriver for your operating system. ("Gecko" is the name of the browser engine used in Firefox.) For example, on Windows you'll want to download the *geckodriver-v0.24.0-win64.zip* link, and on macOS, you'll want the *geckodriver-v0.24.0-macos.tar.gz* link. Newer versions will have slightly different links. The downloaded ZIP file will contain a *geckodriver.exe* (on Windows) or *geckodriver* (on macOS and Linux) file that you can put on your system PATH. Appendix B has information about the system PATH, or you can learn more at *https://stackoverflow.com/q/40208051/1893164*.

For Chrome, go to *https://sites.google.com/a/chromium.org/chromedriver/downloads* and download the ZIP file for your operating system. This ZIP file will contain a *chromedriver.exe* (on Windows) or *chromedriver* (on macOS or Linux) file that you can put on your system PATH.

Other major web browsers also have webdrivers available, and you can often find these by performing an internet search for "<browser name> webdriver".

If you still have problems opening up a new browser under the control of selenium, it may be because the current version of the browser is incompatible with the selenium module. One workaround is to install an older version of the web browser—or, more simply, an older version of the selenium module. You can find the list of selenium version numbers at *https://pypi.org/project/selenium/#history*. Unfortunately, the compatibility between versions of selenium and a browser sometimes breaks, and you may need to search the web for possible solutions. Appendix A has more information about running pip to install a specific version of selenium. (For example, you might run pip install --user -U selenium==3.14.1.)

### Finding Elements on the Page

WebDriver objects have quite a few methods for finding elements on a page. They are divided into the find_element_* and find_elements_* methods. The find_element_* methods return a single WebElement object, representing the first element on the page that matches your query. The find_elements_* methods return a list of WebElement_* objects for *every* matching element on the page.

Table 12-3 shows several examples of find_element_* and find_elements_* methods being called on a WebDriver object that's stored in the variable browser.

**Table 12-3:** Selenium's WebDriver Methods for Finding Elements

| Method name | WebElement object/list returned |
| --- | --- |
| browser.find_element_by_class_name(*name*)<br>browser.find_elements_by_class_name(*name*) | Elements that use the CSS class *name* |
| browser.find_element_by_css_selector(*selector*)<br>browser.find_elements_by_css_selector(*selector*) | Elements that match the CSS *selector* |
| browser.find_element_by_id(*id*)<br>browser.find_elements_by_id(*id*) | Elements with a matching *id* attribute value |
| browser.find_element_by_link_text(*text*)<br>browser.find_elements_by_link_text(*text*) | <a> elements that completely match the *text* provided |
| browser.find_element_by_partial_link_text(*text*)<br>browser.find_elements_by_partial_link_text(*text*) | <a> elements that contain the *text* provided |
| browser.find_element_by_name(*name*)<br>browser.find_elements_by_name(*name*) | Elements with a matching *name* attribute value |
| browser.find_element_by_tag_name(*name*)<br>browser.find_elements_by_tag_name(*name*) | Elements with a matching tag *name* (case-insensitive; an <a> element is matched by 'a' and 'A') |

Except for the *_by_tag_name() methods, the arguments to all the methods are case sensitive. If no elements exist on the page that match what the method is looking for, the selenium module raises a NoSuchElement exception. If you do not want this exception to crash your program, add try and except statements to your code.

Once you have the WebElement object, you can find out more about it by reading the attributes or calling the methods in Table 12-4.

**Table 12-4:** WebElement Attributes and Methods

| Attribute or method | Description |
| --- | --- |
| tag_name | The tag name, such as 'a' for an <a> element |
| get_attribute(*name*) | The value for the element's name attribute |
| text | The text within the element, such as 'hello' in hello |
| clear() | For text field or text area elements, clears the text typed into it |
| is_displayed() | Returns True if the element is visible; otherwise returns False |
| is_enabled() | For input elements, returns True if the element is enabled; otherwise returns False |
| is_selected() | For checkbox or radio button elements, returns True if the element is selected; otherwise returns False |
| location | A dictionary with keys 'x' and 'y' for the position of the element in the page |

For example, open a new file editor tab and enter the following program:

```
from selenium import webdriver
browser = webdriver.Firefox()
browser.get('https://inventwithpython.com')
try:
    elem = browser.find_element_by_class_name(' cover-thumb')
    print('Found <%s> element with that class name!' % (elem.tag_name))
except:
    print('Was not able to find an element with that name.')
```

Here we open Firefox and direct it to a URL. On this page, we try to find elements with the class name 'bookcover', and if such an element is found, we print its tag name using the tag_name attribute. If no such element was found, we print a different message.

This program will output the following:

```
Found <img> element with that class name!
```

We found an element with the class name 'bookcover' and the tag name 'img'.

### Clicking the Page

WebElement objects returned from the find_element_* and find_elements_* methods have a click() method that simulates a mouse click on that element. This method can be used to follow a link, make a selection on a radio button, click a Submit button, or trigger whatever else might happen when the element is clicked by the mouse. For example, enter the following into the interactive shell:

```
>>> from selenium import webdriver
>>> browser = webdriver.Firefox()
>>> browser.get('https://inventwithpython.com')
>>> linkElem = browser.find_element_by_link_text('Read Online for Free')
>>> type(linkElem)
<class 'selenium.webdriver.remote.webelement.FirefoxWebElement'>
>>> linkElem.click() # follows the "Read Online for Free" link
```

This opens Firefox to *https://inventwithpython.com/*, gets the WebElement object for the <a> element with the text *Read It Online*, and then simulates clicking that <a> element. It's just like if you clicked the link yourself; the browser then follows that link.

### Filling Out and Submitting Forms

Sending keystrokes to text fields on a web page is a matter of finding the <input> or <textarea> element for that text field and then calling the send_keys()method. For example, enter the following into the interactive shell:

```
>>> from selenium import webdriver
>>> browser = webdriver.Firefox()
>>> browser.get('https://login.metafilter.com')
>>> userElem = browser.find_element_by_id('user_name)
>>> userElem.send_keys('your_real_username_here')


>>> passwordElem = browser.find_element_by_id('user_pass')
>>> passwordElem.send_keys('your_real_password_here')
>>> passwordElem.submit()
```

As long as login page for MetaFilter hasn't changed the id of the Username and Password text fields since this book was published, the previous code will fill in those text fields with the provided text. (You can always use the browser's inspector to verify the id.) Calling the submit() method on any element will have the same result as clicking the Submit button for the form that element is in. (You could have just as easily called emailElem.submit(), and the code would have done the same thing.)

**WARNING** *Avoid putting your passwords in source code whenever possible. It's easy to accidentally leak your passwords to others when they are left unencrypted on your hard drive. If possible, have your program prompt users to enter their passwords from the keyboard using the* pyinputplus.inputPassword() *function described in Chapter 8.*

## Sending Special Keys

The selenium module has a module for keyboard keys that are impossible to type into a string value, which function much like escape characters. These values are stored in attributes in the selenium.webdriver.common.keys module. Since that is such a long module name, it's much easier to run from selenium .webdriver.common.keys import Keys at the top of your program; if you do, then you can simply write Keys anywhere you'd normally have to write selenium .webdriver.common.keys. Table 12-5 lists the commonly used Keys variables.

**Table 12-5:** Commonly Used Variables in the selenium.webdriver.common.keys Module

| Attributes | Meanings |
|---|---|
| Keys.DOWN, Keys.UP, Keys.LEFT, Keys.RIGHT | The keyboard arrow keys |
| Keys.ENTER, Keys.RETURN | The ENTER and RETURN keys |
| Keys.HOME, Keys.END, Keys.PAGE_DOWN, Keys.PAGE_UP | The HOME, END, PAGEDOWN, and PAGEUP keys |
| Keys.ESCAPE, Keys.BACK_SPACE, Keys.DELETE | The ESC, BACKSPACE, and DELETE keys |
| Keys.F1, Keys.F2, . . . , Keys.F12 | The F1 to F12 keys at the top of the keyboard |
| Keys.TAB | The TAB key |

For example, if the cursor is not currently in a text field, pressing the HOME and END keys will scroll the browser to the top and bottom of the page, respectively. Enter the following into the interactive shell, and notice how the send_keys() calls scroll the page:

```
>>> from selenium import webdriver
>>> from selenium.webdriver.common.keys import Keys
>>> browser = webdriver.Firefox()
>>> browser.get('https://nostarch.com')
>>> htmlElem = browser.find_element_by_tag_name('html')
>>> htmlElem.send_keys(Keys.END)     # scrolls to bottom
>>> htmlElem.send_keys(Keys.HOME)    # scrolls to top
```

The <html> tag is the base tag in HTML files: the full content of the HTML file is enclosed within the <html> and </html> tags. Calling browser .find_element_by_tag_name('html') is a good place to send keys to the general web page. This would be useful if, for example, new content is loaded once you've scrolled to the bottom of the page.

### Clicking Browser Buttons

The selenium module can simulate clicks on various browser buttons as well through the following methods:

browser.back()   Clicks the Back button.

browser.forward()   Clicks the Forward button.

browser.refresh()   Clicks the Refresh/Reload button.

browser.quit()   Clicks the Close Window button.

### More Information on Selenium

Selenium can do much more beyond the functions described here. It can modify your browser's cookies, take screenshots of web pages, and run custom JavaScript. To learn more about these features, you can visit the selenium documentation at *https://selenium-python.readthedocs.org/*.

## Summary

Most boring tasks aren't limited to the files on your computer. Being able to programmatically download web pages will extend your programs to the internet. The requests module makes downloading straightforward, and with some basic knowledge of HTML concepts and selectors, you can utilize the BeautifulSoup module to parse the pages you download.

But to fully automate any web-based tasks, you need direct control of your web browser through the selenium module. The selenium module will allow you to log in to websites and fill out forms automatically. Since a web browser is the most common way to send and receive information over the internet, this is a great ability to have in your programmer toolkit.

## Practice Questions

1. Briefly describe the differences between the webbrowser, requests, bs4, and selenium modules.
2. What type of object is returned by requests.get()? How can you access the downloaded content as a string value?
3. What requests method checks that the download worked?
4. How can you get the HTTP status code of a requests response?
5. How do you save a requests response to a file?
6. What is the keyboard shortcut for opening a browser's developer tools?
7. How can you view (in the developer tools) the HTML of a specific element on a web page?
8. What is the CSS selector string that would find the element with an id attribute of main?

9. What is the CSS selector string that would find the elements with a CSS class of highlight?

10. What is the CSS selector string that would find all the <div> elements inside another <div> element?

11. What is the CSS selector string that would find the <button> element with a value attribute set to favorite?

12. Say you have a Beautiful Soup Tag object stored in the variable spam for the element <div>Hello, world!</div>. How could you get a string 'Hello, world!' from the Tag object?

13. How would you store all the attributes of a Beautiful Soup Tag object in a variable named linkElem?

14. Running import selenium doesn't work. How do you properly import the selenium module?

15. What's the difference between the find_element_* and find_elements_* methods?

16. What methods do Selenium's WebElement objects have for simulating mouse clicks and keyboard keys?

17. You could call send_keys(Keys.ENTER) on the Submit button's WebElement object, but what is an easier way to submit a form with selenium?

18. How can you simulate clicking a browser's Forward, Back, and Refresh buttons with selenium?

## Practice Projects

For practice, write programs to do the following tasks.

### Command Line Emailer

Write a program that takes an email address and string of text on the command line and then, using selenium, logs in to your email account and sends an email of the string to the provided address. (You might want to set up a separate email account for this program.)

This would be a nice way to add a notification feature to your programs. You could also write a similar program to send messages from a Facebook or Twitter account.

### Image Site Downloader

Write a program that goes to a photo-sharing site like Flickr or Imgur, searches for a category of photos, and then downloads all the resulting images. You could write a program that works with any photo site that has a search feature.

## 2048

*2048* is a simple game where you combine tiles by sliding them up, down, left, or right with the arrow keys. You can actually get a fairly high score by repeatedly sliding in an up, right, down, and left pattern over and over again. Write a program that will open the game at *https://gabrielecirulli .github.io/2048/* and keep sending up, right, down, and left keystrokes to automatically play the game.

## Link Verification

Write a program that, given the URL of a web page, will attempt to download every linked page on the page. The program should flag any pages that have a 404 "Not Found" status code and print them out as broken links.

# 13

## WORKING WITH EXCEL
## SPREADSHEETS

Although we don't often think of spreadsheets as programming tools, almost
everyone uses them to organize information into two-dimensional data structures, perform calculations with formulas, and produce output
as charts. In the next two chapters, we'll integrate
Python into two popular spreadsheet applications:
Microsoft Excel and Google Sheets.

Excel is a popular and powerful spreadsheet application for Windows.
The openpyxl module allows your Python programs to read and modify
Excel spreadsheet files. For example, you might have the boring task of
copying certain data from one spreadsheet and pasting it into another one.
Or you might have to go through thousands of rows and pick out just a
handful of them to make small edits based on some criteria. Or you might
have to look through hundreds of spreadsheets of department budgets,
searching for any that are in the red. These are exactly the sort of boring,
mindless spreadsheet tasks that Python can do for you.

Although Excel is proprietary software from Microsoft, there are free alternatives that run on Windows, macOS, and Linux. Both LibreOffice Calc and OpenOffice Calc work with Excel's *.xlsx* file format for spreadsheets, which means the openpyxl module can work on spreadsheets from these applications as well. You can download the software from *https://www.libreoffice.org/* and *https://www.openoffice.org/*, respectively. Even if you already have Excel installed on your computer, you may find these programs easier to use. The screenshots in this chapter, however, are all from Excel 2010 on Windows 10.

## Excel Documents

First, let's go over some basic definitions: an Excel spreadsheet document is called a *workbook*. A single workbook is saved in a file with the *.xlsx* extension. Each workbook can contain multiple *sheets* (also called *worksheets*). The sheet the user is currently viewing (or last viewed before closing Excel) is called the *active sheet*.

Each sheet has *columns* (addressed by letters starting at *A*) and *rows* (addressed by numbers starting at 1). A box at a particular column and row is called a *cell*. Each cell can contain a number or text value. The grid of cells with data makes up a sheet.

## Installing the openpyxl Module

Python does not come with OpenPyXL, so you'll have to install it. Follow the instructions for installing third-party modules in Appendix A; the name of the module is openpyxl.

This book uses version 2.6.2 of OpenPyXL. It's important that you install this version by running pip install --user -U openpyxl==2.6.2 because newer versions of OpenPyXL are incompatible with the information in this book. To test whether it is installed correctly, enter the following into the interactive shell:

```
>>> import openpyxl
```

If the module was correctly installed, this should produce no error messages. Remember to import the openpyxl module before running the interactive shell examples in this chapter, or you'll get a NameError: name 'openpyxl' is not defined error.

You can find the full documentation for OpenPyXL at *https://openpyxl.readthedocs.org/*.

## Reading Excel Documents

The examples in this chapter will use a spreadsheet named *example.xlsx* stored in the root folder. You can either create the spreadsheet yourself or download it from *https://nostarch.com/automatestuff2/*. Figure 13-1 shows the tabs for the three default sheets named *Sheet1*, *Sheet2*, and *Sheet3* that Excel automatically provides for new workbooks. (The number of default sheets created may vary between operating systems and spreadsheet programs.)

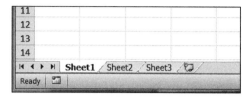

Figure 13-1: The tabs for a workbook's sheets are in the lower-left corner of Excel.

Sheet 1 in the example file should look like Table 13-1. (If you didn't download *example.xlsx* from the website, you should enter this data into the sheet yourself.)

**Table 13-1:** The *example.xlsx* Spreadsheet

|   | A | B | C |
|---|---|---|---|
| 1 | 4/5/2015  1:34:02 PM | Apples | 73 |
| 2 | 4/5/2015  3:41:23 AM | Cherries | 85 |
| 3 | 4/6/2015  12:46:51 PM | Pears | 14 |
| 4 | 4/8/2015  8:59:43 AM | Oranges | 52 |
| 5 | 4/10/2015  2:07:00 AM | Apples | 152 |
| 6 | 4/10/2015  6:10:37 PM | Bananas | 23 |
| 7 | 4/10/2015  2:40:46 AM | Strawberries | 98 |

Now that we have our example spreadsheet, let's see how we can manipulate it with the openpyxl module.

## Opening Excel Documents with OpenPyXL

Once you've imported the openpyxl module, you'll be able to use the openpyxl .load_workbook() function. Enter the following into the interactive shell:

```
>>> import openpyxl
>>> wb = openpyxl.load_workbook('example.xlsx')
>>> type(wb)
<class 'openpyxl.workbook.workbook.Workbook'>
```

The openpyxl.load_workbook() function takes in the filename and returns a value of the workbook data type. This Workbook object represents the Excel file, a bit like how a File object represents an opened text file.

Remember that *example.xlsx* needs to be in the current working directory in order for you to work with it. You can find out what the current working directory is by importing os and using os.getcwd(), and you can change the current working directory using os.chdir().

## Getting Sheets from the Workbook

You can get a list of all the sheet names in the workbook by accessing the sheetnames attribute. Enter the following into the interactive shell:

```
>>> import openpyxl
>>> wb = openpyxl.load_workbook('example.xlsx')
>>> wb.sheetnames # The workbook's sheets' names.
['Sheet1', 'Sheet2', 'Sheet3']
>>> sheet = wb['Sheet3'] # Get a sheet from the workbook.
>>> sheet
<Worksheet "Sheet3">
>>> type(sheet)
<class 'openpyxl.worksheet.worksheet.Worksheet'>
>>> sheet.title # Get the sheet's title as a string.
'Sheet3'
>>> anotherSheet = wb.active # Get the active sheet.
>>> anotherSheet
<Worksheet "Sheet1">
```

Each sheet is represented by a Worksheet object, which you can obtain by using the square brackets with the sheet name string like a dictionary key. Finally, you can use the active attribute of a Workbook object to get the workbook's active sheet. The active sheet is the sheet that's on top when the workbook is opened in Excel. Once you have the Worksheet object, you can get its name from the title attribute.

## Getting Cells from the Sheets

Once you have a Worksheet object, you can access a Cell object by its name. Enter the following into the interactive shell:

```
>>> import openpyxl
>>> wb = openpyxl.load_workbook('example.xlsx')
>>> sheet = wb['Sheet1'] # Get a sheet from the workbook.
>>> sheet['A1'] # Get a cell from the sheet.
<Cell 'Sheet1'.A1>
>>> sheet['A1'].value # Get the value from the cell.
datetime.datetime(2015, 4, 5, 13, 34, 2)
>>> c = sheet['B1'] # Get another cell from the sheet.
>>> c.value
'Apples'
>>> # Get the row, column, and value from the cell.
>>> 'Row %s, Column %s is %s' % (c.row, c.column, c.value)
'Row 1, Column B is Apples'
>>> 'Cell %s is %s' % (c.coordinate, c.value)
'Cell B1 is Apples'
>>> sheet['C1'].value
73
```

The Cell object has a value attribute that contains, unsurprisingly, the value stored in that cell. Cell objects also have row, column, and coordinate attributes that provide location information for the cell.

Here, accessing the value attribute of our `Cell` object for cell B1 gives us the string `'Apples'`. The `row` attribute gives us the integer `1`, the `column` attribute gives us `'B'`, and the `coordinate` attribute gives us `'B1'`.

OpenPyXL will automatically interpret the dates in column A and return them as `datetime` values rather than strings. The `datetime` data type is explained further in Chapter 17.

Specifying a column by letter can be tricky to program, especially because after column Z, the columns start by using two letters: AA, AB, AC, and so on. As an alternative, you can also get a cell using the sheet's `cell()` method and passing integers for its `row` and `column` keyword arguments. The first row or column integer is `1`, not `0`. Continue the interactive shell example by entering the following:

```
>>> sheet.cell(row=1, column=2)
<Cell 'Sheet1'.B1>
>>> sheet.cell(row=1, column=2).value
'Apples'
>>> for i in range(1, 8, 2): # Go through every other row:
...     print(i, sheet.cell(row=i, column=2).value)
...
1 Apples
3 Pears
5 Apples
7 Strawberries
```

As you can see, using the sheet's `cell()` method and passing it `row=1` and `column=2` gets you a `Cell` object for cell B1, just like specifying `sheet['B1']` did. Then, using the `cell()` method and its keyword arguments, you can write a for loop to print the values of a series of cells.

Say you want to go down column B and print the value in every cell with an odd row number. By passing `2` for the `range()` function's "step" parameter, you can get cells from every second row (in this case, all the odd-numbered rows). The for loop's `i` variable is passed for the `row` keyword argument to the `cell()` method, while `2` is always passed for the `column` keyword argument. Note that the integer `2`, not the string `'B'`, is passed.

You can determine the size of the sheet with the `Worksheet` object's `max_row` and `max_column` attributes. Enter the following into the interactive shell:

```
>>> import openpyxl
>>> wb = openpyxl.load_workbook('example.xlsx')
>>> sheet = wb['Sheet1']
>>> sheet.max_row # Get the highest row number.
7
>>> sheet.max_column # Get the highest column number.
3
```

Note that the `max_column` attribute is an integer rather than the letter that appears in Excel.

## Converting Between Column Letters and Numbers

To convert from letters to numbers, call the openpyxl.utils.column_index_from _string() function. To convert from numbers to letters, call the openpyxl.utils .get_column_letter() function. Enter the following into the interactive shell:

```
>>> import openpyxl
>>> from openpyxl.utils import get_column_letter, column_index_from_string
>>> get_column_letter(1) # Translate column 1 to a letter.
'A'
>>> get_column_letter(2)
'B'
>>> get_column_letter(27)
'AA'
>>> get_column_letter(900)
'AHP'
>>> wb = openpyxl.load_workbook('example.xlsx')
>>> sheet = wb['Sheet1']
>>> get_column_letter(sheet.max_column)
'C'
>>> column_index_from_string('A') # Get A's number.
1
>>> column_index_from_string('AA')
27
```

After you import these two functions from the openpyxl.utils module, you can call get_column_letter() and pass it an integer like 27 to figure out what the letter name of the 27th column is. The function column_index_string() does the reverse: you pass it the letter name of a column, and it tells you what number that column is. You don't need to have a workbook loaded to use these functions. If you want, you can load a workbook, get a Worksheet object, and use a Worksheet attribute like max_column to get an integer. Then, you can pass that integer to get_column_letter().

## Getting Rows and Columns from the Sheets

You can slice Worksheet objects to get all the Cell objects in a row, column, or rectangular area of the spreadsheet. Then you can loop over all the cells in the slice. Enter the following into the interactive shell:

```
>>> import openpyxl
>>> wb = openpyxl.load_workbook('example.xlsx')
>>> sheet = wb['Sheet1']
>>> tuple(sheet['A1':'C3']) # Get all cells from A1 to C3.
((<Cell 'Sheet1'.A1>, <Cell 'Sheet1'.B1>, <Cell 'Sheet1'.C1>), (<Cell
'Sheet1'.A2>, <Cell 'Sheet1'.B2>, <Cell 'Sheet1'.C2>), (<Cell 'Sheet1'.A3>,
<Cell 'Sheet1'.B3>, <Cell 'Sheet1'.C3>))
```
❶ ```
>>> for rowOfCellObjects in sheet['A1':'C3']:
```
❷ ```
...     for cellObj in rowOfCellObjects:
...         print(cellObj.coordinate, cellObj.value)
...     print('--- END OF ROW ---')
```

A1 2015-04-05 13:34:02

```
B1 Apples
C1 73
--- END OF ROW ---
A2 2015-04-05 03:41:23
B2 Cherries
C2 85
--- END OF ROW ---
A3 2015-04-06 12:46:51
B3 Pears
C3 14
--- END OF ROW ---
```

Here, we specify that we want the Cell objects in the rectangular area from A1 to C3, and we get a Generator object containing the Cell objects in that area. To help us visualize this Generator object, we can use tuple() on it to display its Cell objects in a tuple.

This tuple contains three tuples: one for each row, from the top of the desired area to the bottom. Each of these three inner tuples contains the Cell objects in one row of our desired area, from the leftmost cell to the right. So overall, our slice of the sheet contains all the Cell objects in the area from A1 to C3, starting from the top-left cell and ending with the bottom-right cell.

To print the values of each cell in the area, we use two for loops. The outer for loop goes over each row in the slice ❶. Then, for each row, the nested for loop goes through each cell in that row ❷.

To access the values of cells in a particular row or column, you can also use a Worksheet object's rows and columns attribute. These attributes must be converted to lists with the list() function before you can use the square brackets and an index with them. Enter the following into the interactive shell:

```
>>> import openpyxl
>>> wb = openpyxl.load_workbook('example.xlsx')
>>> sheet = wb.active
>>> list(sheet.columns)[1] # Get second column's cells.
(<Cell 'Sheet1'.B1>, <Cell 'Sheet1'.B2>, <Cell 'Sheet1'.B3>, <Cell 'Sheet1'.
B4>, <Cell 'Sheet1'.B5>, <Cell 'Sheet1'.B6>, <Cell 'Sheet1'.B7>)
>>> for cellObj in list(sheet.columns)[1]:
        print(cellObj.value)

Apples
Cherries
Pears
Oranges
Apples
Bananas
Strawberries
```

Using the rows attribute on a Worksheet object will give you a tuple of tuples. Each of these inner tuples represents a row, and contains the Cell objects in that row. The columns attribute also gives you a tuple of tuples, with each of the inner tuples containing the Cell objects in a particular column. For *example.xlsx*, since there are 7 rows and 3 columns, rows gives us a tuple of 7 tuples (each containing 3 Cell objects), and columns gives us a tuple of 3 tuples (each containing 7 Cell objects).

To access one particular tuple, you can refer to it by its index in the larger tuple. For example, to get the tuple that represents column B, you use list(sheet.columns)[1]. To get the tuple containing the Cell objects in column A, you'd use list(sheet.columns)[0]. Once you have a tuple representing one row or column, you can loop through its Cell objects and print their values.

### Workbooks, Sheets, Cells

As a quick review, here's a rundown of all the functions, methods, and data types involved in reading a cell out of a spreadsheet file:

1. Import the openpyxl module.
2. Call the openpyxl.load_workbook() function.
3. Get a Workbook object.
4. Use the active or sheetnames attributes.
5. Get a Worksheet object.
6. Use indexing or the cell() sheet method with row and column keyword arguments.
7. Get a Cell object.
8. Read the Cell object's value attribute.

# Project: Reading Data from a Spreadsheet

Say you have a spreadsheet of data from the 2010 US Census and you have the boring task of going through its thousands of rows to count both the total population and the number of census tracts for each county. (A census tract is simply a geographic area defined for the purposes of the census.) Each row represents a single census tract. We'll name the spreadsheet file *censuspopdata.xlsx*, and you can download it from *https://nostarch.com/automatestuff2/*. Its contents look like Figure 13-2.

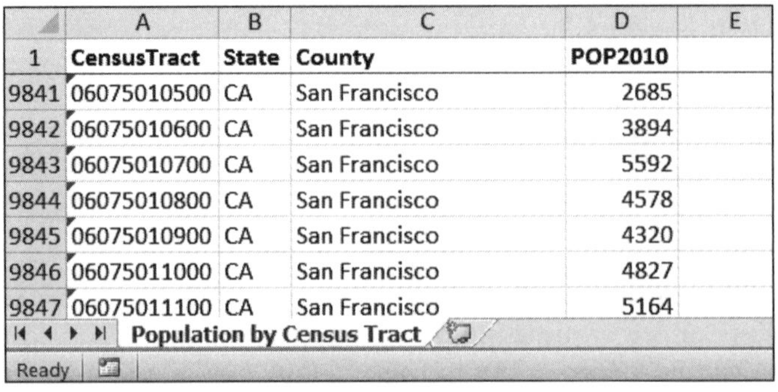

| | A | B | C | D | E |
|---|---|---|---|---|---|
| 1 | CensusTract | State | County | POP2010 | |
| 9841 | 06075010500 | CA | San Francisco | 2685 | |
| 9842 | 06075010600 | CA | San Francisco | 3894 | |
| 9843 | 06075010700 | CA | San Francisco | 5592 | |
| 9844 | 06075010800 | CA | San Francisco | 4578 | |
| 9845 | 06075010900 | CA | San Francisco | 4320 | |
| 9846 | 06075011000 | CA | San Francisco | 4827 | |
| 9847 | 06075011100 | CA | San Francisco | 5164 | |

| ◄ ◄ ► ► | Population by Census Tract |
|---|---|
| Ready | |

*Figure 13-2: The* censuspopdata.xlsx *spreadsheet*

Even though Excel can calculate the sum of multiple selected cells, you'd still have to select the cells for each of the 3,000-plus counties. Even if it takes just a few seconds to calculate a county's population by hand, this would take hours to do for the whole spreadsheet.

In this project, you'll write a script that can read from the census spreadsheet file and calculate statistics for each county in a matter of seconds.

This is what your program does:

1. Reads the data from the Excel spreadsheet
2. Counts the number of census tracts in each county
3. Counts the total population of each county
4. Prints the results

This means your code will need to do the following:

1. Open and read the cells of an Excel document with the openpyxl module.
2. Calculate all the tract and population data and store it in a data structure.
3. Write the data structure to a text file with the *.py* extension using the pprint module.

### Step 1: Read the Spreadsheet Data

There is just one sheet in the *censuspopdata.xlsx* spreadsheet, named 'Population by Census Tract', and each row holds the data for a single census tract. The columns are the tract number (A), the state abbreviation (B), the county name (C), and the population of the tract (D).

Open a new file editor tab and enter the following code. Save the file as *readCensusExcel.py*.

```
#! python3
# readCensusExcel.py - Tabulates population and number of census tracts for
# each county.

❶ import openpyxl, pprint
  print('Opening workbook...')
❷ wb = openpyxl.load_workbook('censuspopdata.xlsx')
❸ sheet = wb['Population by Census Tract']
  countyData = {}

  # TODO: Fill in countyData with each county's population and tracts.
  print('Reading rows...')
❹ for row in range(2, sheet.max_row + 1):
      # Each row in the spreadsheet has data for one census tract.
      state  = sheet['B' + str(row)].value
      county = sheet['C' + str(row)].value
      pop    = sheet['D' + str(row)].value

  # TODO: Open a new text file and write the contents of countyData to it.
```

This code imports the openpyxl module, as well as the pprint module that you'll use to print the final county data ❶. Then it opens the *censuspopdata .xlsx* file ❷, gets the sheet with the census data ❸, and begins iterating over its rows ❹.

Note that you've also created a variable named countyData, which will contain the populations and number of tracts you calculate for each county. Before you can store anything in it, though, you should determine exactly how you'll structure the data inside it.

## Step 2: Populate the Data Structure

The data structure stored in countyData will be a dictionary with state abbreviations as its keys. Each state abbreviation will map to another dictionary, whose keys are strings of the county names in that state. Each county name will in turn map to a dictionary with just two keys, 'tracts' and 'pop'. These keys map to the number of census tracts and population for the county. For example, the dictionary will look similar to this:

```
{'AK': {'Aleutians East': {'pop': 3141, 'tracts': 1},
        'Aleutians West': {'pop': 5561, 'tracts': 2},
        'Anchorage': {'pop': 291826, 'tracts': 55},
        'Bethel': {'pop': 17013, 'tracts': 3},
        'Bristol Bay': {'pop': 997, 'tracts': 1},
        --snip--
```

If the previous dictionary were stored in countyData, the following expressions would evaluate like this:

```
>>> countyData['AK']['Anchorage']['pop']
291826
>>> countyData['AK']['Anchorage']['tracts']
55
```

More generally, the countyData dictionary's keys will look like this:

```
countyData[state abbrev][county]['tracts']
countyData[state abbrev][county]['pop']
```

Now that you know how countyData will be structured, you can write the code that will fill it with the county data. Add the following code to the bottom of your program:

```
#! python 3
# readCensusExcel.py - Tabulates population and number of census tracts for
# each county.

--snip--
```

```
for row in range(2, sheet.max_row + 1):
    # Each row in the spreadsheet has data for one census tract.
    state  = sheet['B' + str(row)].value
    county = sheet['C' + str(row)].value
    pop    = sheet['D' + str(row)].value

    # Make sure the key for this state exists.
❶  countyData.setdefault(state, {})
    # Make sure the key for this county in this state exists.
❷  countyData[state].setdefault(county, {'tracts': 0, 'pop': 0})

    # Each row represents one census tract, so increment by one.
❸  countyData[state][county]['tracts'] += 1
    # Increase the county pop by the pop in this census tract.
❹  countyData[state][county]['pop'] += int(pop)

# TODO: Open a new text file and write the contents of countyData to it.
```

The last two lines of code perform the actual calculation work, incrementing the value for tracts ❸ and increasing the value for pop ❹ for the current county on each iteration of the for loop.

The other code is there because you cannot add a county dictionary as the value for a state abbreviation key until the key itself exists in countyData. (That is, countyData['AK']['Anchorage']['tracts'] += 1 will cause an error if the 'AK' key doesn't exist yet.) To make sure the state abbreviation key exists in your data structure, you need to call the setdefault() method to set a value if one does not already exist for state ❶.

Just as the countyData dictionary needs a dictionary as the value for each state abbreviation key, each of *those* dictionaries will need its own dictionary as the value for each county key ❷. And each of *those* dictionaries in turn will need keys 'tracts' and 'pop' that start with the integer value 0. (If you ever lose track of the dictionary structure, look back at the example dictionary at the start of this section.)

Since setdefault() will do nothing if the key already exists, you can call it on every iteration of the for loop without a problem.

## Step 3: Write the Results to a File

After the for loop has finished, the countyData dictionary will contain all of the population and tract information keyed by county and state. At this point, you could program more code to write this to a text file or another Excel spreadsheet. For now, let's just use the pprint.pformat() function to write the countyData dictionary value as a massive string to a file named *census2010.py*. Add the following code to the bottom of your program (making sure to keep it unindented so that it stays outside the for loop):

```
#! python 3
# readCensusExcel.py - Tabulates population and number of census tracts for
# each county.
```

```
--snip--

for row in range(2, sheet.max_row + 1):
--snip--

# Open a new text file and write the contents of countyData to it.
print('Writing results...')
resultFile = open('census2010.py', 'w')
resultFile.write('allData = ' + pprint.pformat(countyData))
resultFile.close()
print('Done.')
```

The `pprint.pformat()` function produces a string that itself is formatted as valid Python code. By outputting it to a text file named *census2010.py*, you've generated a Python program from your Python program! This may seem complicated, but the advantage is that you can now import *census2010.py* just like any other Python module. In the interactive shell, change the current working directory to the folder with your newly created *census2010.py* file and then import it:

```
>>> import os

>>> import census2010
>>> census2010.allData['AK']['Anchorage']
{'pop': 291826, 'tracts': 55}
>>> anchoragePop = census2010.allData['AK']['Anchorage']['pop']
>>> print('The 2010 population of Anchorage was ' + str(anchoragePop))
The 2010 population of Anchorage was 291826
```

The *readCensusExcel.py* program was throwaway code: once you have its results saved to *census2010.py*, you won't need to run the program again. Whenever you need the county data, you can just run `import census2010`.

Calculating this data by hand would have taken hours; this program did it in a few seconds. Using OpenPyXL, you will have no trouble extracting information that is saved to an Excel spreadsheet and performing calculations on it. You can download the complete program from *https://nostarch.com/automatestuff2/*.

## Ideas for Similar Programs

Many businesses and offices use Excel to store various types of data, and it's not uncommon for spreadsheets to become large and unwieldy. Any program that parses an Excel spreadsheet has a similar structure: it loads the spreadsheet file, preps some variables or data structures, and then loops through each of the rows in the spreadsheet. Such a program could do the following:

- Compare data across multiple rows in a spreadsheet.
- Open multiple Excel files and compare data between spreadsheets.

- Check whether a spreadsheet has blank rows or invalid data in any cells and alert the user if it does.

- Read data from a spreadsheet and use it as the input for your Python programs.

# Writing Excel Documents

OpenPyXL also provides ways of writing data, meaning that your programs can create and edit spreadsheet files. With Python, it's simple to create spreadsheets with thousands of rows of data.

## Creating and Saving Excel Documents

Call the openpyxl.Workbook() function to create a new, blank Workbook object. Enter the following into the interactive shell:

```
>>> import openpyxl
>>> wb = openpyxl.Workbook() # Create a blank workbook.
>>> wb.sheetnames # It starts with one sheet.
['Sheet']
>>> sheet = wb.active
>>> sheet.title
'Sheet'
>>> sheet.title = 'Spam Bacon Eggs Sheet' # Change title.
>>> wb.sheetnames
['Spam Bacon Eggs Sheet']
```

The workbook will start off with a single sheet named *Sheet*. You can change the name of the sheet by storing a new string in its title attribute.

Any time you modify the Workbook object or its sheets and cells, the spreadsheet file will not be saved until you call the save() workbook method. Enter the following into the interactive shell (with *example.xlsx* in the current working directory):

```
>>> import openpyxl
>>> wb = openpyxl.load_workbook('example.xlsx')
>>> sheet = wb.active
>>> sheet.title = 'Spam Spam Spam'
>>> wb.save('example_copy.xlsx') # Save the workbook.
```

Here, we change the name of our sheet. To save our changes, we pass a filename as a string to the save() method. Passing a different filename than the original, such as 'example_copy.xlsx', saves the changes to a copy of the spreadsheet.

Whenever you edit a spreadsheet you've loaded from a file, you should always save the new, edited spreadsheet to a different filename than the original. That way, you'll still have the original spreadsheet file to work with in case a bug in your code caused the new, saved file to have incorrect or corrupt data.

## Creating and Removing Sheets

Sheets can be added to and removed from a workbook with the create_sheet()
method and del operator. Enter the following into the interactive shell:

```
>>> import openpyxl
>>> wb = openpyxl.Workbook()
>>> wb.sheetnames
['Sheet']
>>> wb.create_sheet() # Add a new sheet.
<Worksheet "Sheet1">
>>> wb.sheetnames
['Sheet', 'Sheet1']
>>> # Create a new sheet at index 0.
>>> wb.create_sheet(index=0, title='First Sheet')
<Worksheet "First Sheet">
>>> wb.sheetnames
['First Sheet', 'Sheet', 'Sheet1']
>>> wb.create_sheet(index=2, title='Middle Sheet')
<Worksheet "Middle Sheet">
>>> wb.sheetnames
['First Sheet', 'Sheet', 'Middle Sheet', 'Sheet1']
```

The create_sheet() method returns a new Worksheet object named Sheet*X*,
which by default is set to be the last sheet in the workbook. Optionally, the
index and name of the new sheet can be specified with the index and title
keyword arguments.

Continue the previous example by entering the following:

```
>>> wb.sheetnames
['First Sheet', 'Sheet', 'Middle Sheet', 'Sheet1']
>>> del wb['Middle Sheet']
>>> del wb['Sheet1']
>>> wb.sheetnames
['First Sheet', 'Sheet']
```

You can use the del operator to delete a sheet from a workbook, just like
you can use it to delete a key-value pair from a dictionary.

Remember to call the save() method to save the changes after adding
sheets to or removing sheets from the workbook.

## Writing Values to Cells

Writing values to cells is much like writing values to keys in a dictionary.
Enter this into the interactive shell:

```
>>> import openpyxl
>>> wb = openpyxl.Workbook()
>>> sheet = wb['Sheet']
>>> sheet['A1'] = 'Hello, world!' # Edit the cell's value.
>>> sheet['A1'].value
'Hello, world!'
```

If you have the cell's coordinate as a string, you can use it just like a dictionary key on the Worksheet object to specify which cell to write to.

## Project: Updating a Spreadsheet

In this project, you'll write a program to update cells in a spreadsheet of produce sales. Your program will look through the spreadsheet, find specific kinds of produce, and update their prices. Download this spreadsheet from *https://nostarch.com/automatestuff2/*. Figure 13-3 shows what the spreadsheet looks like.

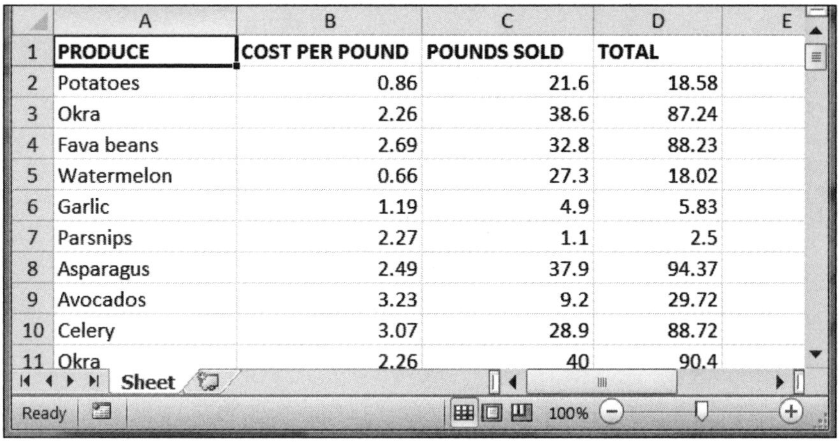

| ▲ | A | B | C | D | E |
|---|---|---|---|---|---|
| 1 | PRODUCE | COST PER POUND | POUNDS SOLD | TOTAL | |
| 2 | Potatoes | 0.86 | 21.6 | 18.58 | |
| 3 | Okra | 2.26 | 38.6 | 87.24 | |
| 4 | Fava beans | 2.69 | 32.8 | 88.23 | |
| 5 | Watermelon | 0.66 | 27.3 | 18.02 | |
| 6 | Garlic | 1.19 | 4.9 | 5.83 | |
| 7 | Parsnips | 2.27 | 1.1 | 2.5 | |
| 8 | Asparagus | 2.49 | 37.9 | 94.37 | |
| 9 | Avocados | 3.23 | 9.2 | 29.72 | |
| 10 | Celery | 3.07 | 28.9 | 88.72 | |
| 11 | Okra | 2.26 | 40 | 90.4 | |

*Figure 13-3: A spreadsheet of produce sales*

Each row represents an individual sale. The columns are the type of produce sold (A), the cost per pound of that produce (B), the number of pounds sold (C), and the total revenue from the sale (D). The TOTAL column is set to the Excel formula *=ROUND(B3\*C3, 2)*, which multiplies the cost per pound by the number of pounds sold and rounds the result to the nearest cent. With this formula, the cells in the TOTAL column will automatically update themselves if there is a change in column B or C.

Now imagine that the prices of garlic, celery, and lemons were entered incorrectly, leaving you with the boring task of going through thousands of rows in this spreadsheet to update the cost per pound for any garlic, celery, and lemon rows. You can't do a simple find-and-replace for the price, because there might be other items with the same price that you don't want to mistakenly "correct." For thousands of rows, this would take hours to do by hand. But you can write a program that can accomplish this in seconds.

Your program does the following:

1. Loops over all the rows
2. If the row is for garlic, celery, or lemons, changes the price

This means your code will need to do the following:

1. Open the spreadsheet file.
2. For each row, check whether the value in column A is Celery, Garlic, or Lemon.
3. If it is, update the price in column B.
4. Save the spreadsheet to a new file (so that you don't lose the old spreadsheet, just in case).

## Step 1: Set Up a Data Structure with the Update Information

The prices that you need to update are as follows:

| Celery | 1.19 |
|--------|------|
| Garlic | 3.07 |
| Lemon  | 1.27 |

You could write code like this:

```python
if produceName == 'Celery':
    cellObj = 1.19
if produceName == 'Garlic':
    cellObj = 3.07
if produceName == 'Lemon':
    cellObj = 1.27
```

Having the produce and updated price data hardcoded like this is a bit inelegant. If you needed to update the spreadsheet again with different prices or different produce, you would have to change a lot of the code. Every time you change code, you risk introducing bugs.

A more flexible solution is to store the corrected price information in a dictionary and write your code to use this data structure. In a new file editor tab, enter the following code:

```python
#! python3
# updateProduce.py - Corrects costs in produce sales spreadsheet.

import openpyxl

wb = openpyxl.load_workbook('produceSales.xlsx')
sheet = wb['Sheet']

# The produce types and their updated prices
PRICE_UPDATES = {'Garlic': 3.07,
                 'Celery': 1.19,
                 'Lemon': 1.27}

# TODO: Loop through the rows and update the prices.
```

Save this as *updateProduce.py*. If you need to update the spreadsheet again, you'll need to update only the PRICE_UPDATES dictionary, not any other code.

## Step 2: Check All Rows and Update Incorrect Prices

The next part of the program will loop through all the rows in the spreadsheet. Add the following code to the bottom of *updateProduce.py*:

```
#! python3
# updateProduce.py - Corrects costs in produce sales spreadsheet.

--snip--

# Loop through the rows and update the prices.
❶ for rowNum in range(2, sheet.max_row):     # skip the first row
❷     produceName = sheet.cell(row=rowNum, column=1).value
❸     if produceName in PRICE_UPDATES:
            sheet.cell(row=rowNum, column=2).value = PRICE_UPDATES[produceName]

❹ wb.save('updatedProduceSales.xlsx')
```

We loop through the rows starting at row 2, since row 1 is just the header ❶. The cell in column 1 (that is, column A) will be stored in the variable produceName ❷. If produceName exists as a key in the PRICE_UPDATES dictionary ❸, then you know this is a row that must have its price corrected. The correct price will be in PRICE_UPDATES[produceName].

Notice how clean using PRICE_UPDATES makes the code. Only one if statement, rather than code like if produceName == 'Garlic': , is necessary for every type of produce to update. And since the code uses the PRICE_UPDATES dictionary instead of hardcoding the produce names and updated costs into the for loop, you modify only the PRICE_UPDATES dictionary and not the code if the produce sales spreadsheet needs additional changes.

After going through the entire spreadsheet and making changes, the code saves the Workbook object to *updatedProduceSales.xlsx* ❹. It doesn't overwrite the old spreadsheet just in case there's a bug in your program and the updated spreadsheet is wrong. After checking that the updated spreadsheet looks right, you can delete the old spreadsheet.

You can download the complete source code for this program from *https://nostarch.com/automatestuff2/*.

## Ideas for Similar Programs

Since many office workers use Excel spreadsheets all the time, a program that can automatically edit and write Excel files could be really useful. Such a program could do the following:

- Read data from one spreadsheet and write it to parts of other spreadsheets.
- Read data from websites, text files, or the clipboard and write it to a spreadsheet.
- Automatically "clean up" data in spreadsheets. For example, it could use regular expressions to read multiple formats of phone numbers and edit them to a single, standard format.

## Setting the Font Style of Cells

Styling certain cells, rows, or columns can help you emphasize important areas in your spreadsheet. In the produce spreadsheet, for example, your program could apply bold text to the potato, garlic, and parsnip rows. Or perhaps you want to italicize every row with a cost per pound greater than $5. Styling parts of a large spreadsheet by hand would be tedious, but your programs can do it instantly.

To customize font styles in cells, important, import the `Font()` function from the `openpyxl.styles` module.

```
from openpyxl.styles import Font
```

This allows you to type `Font()` instead of `openpyxl.styles.Font()`. (See "Importing Modules" on page 47 to review this style of `import` statement.)

Here's an example that creates a new workbook and sets cell A1 to have a 24-point, italicized font. Enter the following into the interactive shell:

```
>>> import openpyxl
>>> from openpyxl.styles import Font
>>> wb = openpyxl.Workbook()
>>> sheet = wb['Sheet']
❶ >>> italic24Font = Font(size=24, italic=True) # Create a font.
❷ >>> sheet['A1'].font = italic24Font # Apply the font to A1.
>>> sheet['A1'] = 'Hello, world!'
>>> wb.save('styles.xlsx')
```

In this example, `Font(size=24, italic=True)` returns a `Font` object, which is stored in `italic24Font` ❶. The keyword arguments to `Font()`, `size` and `italic`, configure the `Font` object's styling information. And when `sheet['A1'].font` is assigned the `italic24Font` object ❷, all that font styling information gets applied to cell A1.

## Font Objects

To set font attributes, you pass keyword arguments to `Font()`. Table 13-2 shows the possible keyword arguments for the `Font()` function.

**Table 13-2:** Keyword Arguments for Font Objects

| Keyword argument | Data type | Description |
| --- | --- | --- |
| name | String | The font name, such as 'Calibri' or 'Times New Roman' |
| size | Integer | The point size |
| bold | Boolean | True, for bold font |
| italic | Boolean | True, for italic font |

You can call Font() to create a Font object and store that Font object in a variable. You then assign that variable to a Cell object's font attribute. For example, this code creates various font styles:

```
>>> import openpyxl
>>> from openpyxl.styles import Font
>>> wb = openpyxl.Workbook()
>>> sheet = wb['Sheet']

>>> fontObj1 = Font(name='Times New Roman', bold=True)
>>> sheet['A1'].font = fontObj1
>>> sheet['A1'] = 'Bold Times New Roman'

>>> fontObj2 = Font(size=24, italic=True)
>>> sheet['B3'].font = fontObj2
>>> sheet['B3'] = '24 pt Italic'

>>> wb.save('styles.xlsx')
```

Here, we store a Font object in fontObj1 and then set the A1 Cell object's font attribute to fontObj1. We repeat the process with another Font object to set the font of a second cell. After you run this code, the styles of the A1 and B3 cells in the spreadsheet will be set to custom font styles, as shown in Figure 13-4.

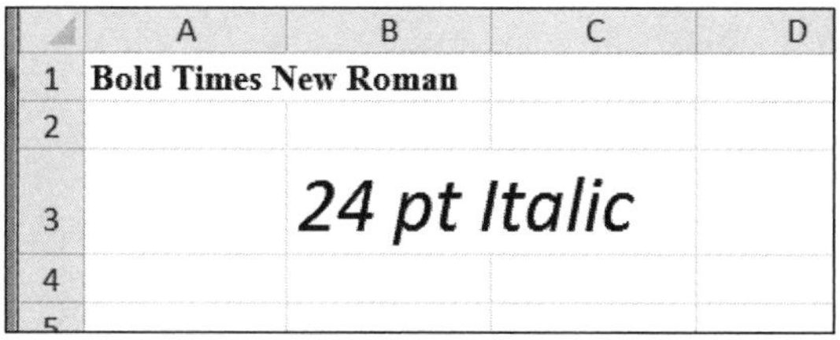

Figure 13-4: A spreadsheet with custom font styles

For cell A1, we set the font name to 'Times New Roman' and set bold to true, so our text appears in bold Times New Roman. We didn't specify a size, so the openpyxl default, 11, is used. In cell B3, our text is italic, with a size of 24; we didn't specify a font name, so the openpyxl default, Calibri, is used.

## Formulas

Excel formulas, which begin with an equal sign, can configure cells to contain values calculated from other cells. In this section, you'll use the openpyxl module to programmatically add formulas to cells, just like any normal value. For example:

```
>>> sheet['B9'] = '=SUM(B1:B8)'
```

This will store *=SUM(B1:B8)* as the value in cell B9. This sets the B9 cell to a formula that calculates the sum of values in cells B1 to B8. You can see this in action in Figure 13-5.

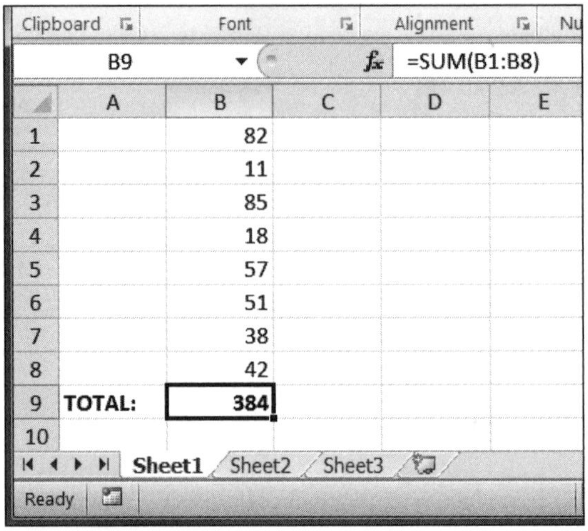

Figure 13-5: Cell B9 contains the formula =SUM(B1:B8), which adds the cells B1 to B8.

An Excel formula is set just like any other text value in a cell. Enter the following into the interactive shell:

```
>>> import openpyxl
>>> wb = openpyxl.Workbook()
>>> sheet = wb.active
>>> sheet['A1'] = 200
>>> sheet['A2'] = 300
>>> sheet['A3'] = '=SUM(A1:A2)' # Set the formula.
>>> wb.save('writeFormula.xlsx')
```

The cells in A1 and A2 are set to 200 and 300, respectively. The value in cell A3 is set to a formula that sums the values in A1 and A2. When the spreadsheet is opened in Excel, A3 will display its value as 500.

Excel formulas offer a level of programmability for spreadsheets but can quickly become unmanageable for complicated tasks. For example, even if you're deeply familiar with Excel formulas, it's a headache to try to decipher what *=IFERROR(TRIM(IF(LEN(VLOOKUP(F7, Sheet2!$A$1:$B$10000, 2, FALSE))>0,SUBSTITUTE(VLOOKUP(F7, Sheet2!$A$1:$B$10000, 2, FALSE), " ", "")))), "")* actually does. Python code is much more readable.

## Adjusting Rows and Columns

In Excel, adjusting the sizes of rows and columns is as easy as clicking and dragging the edges of a row or column header. But if you need to set a row or column's size based on its cells' contents or if you want to set sizes in a large number of spreadsheet files, it will be much quicker to write a Python program to do it.

Rows and columns can also be hidden entirely from view. Or they can be "frozen" in place so that they are always visible on the screen and appear on every page when the spreadsheet is printed (which is handy for headers).

### Setting Row Height and Column Width

Worksheet objects have row_dimensions and column_dimensions attributes that control row heights and column widths. Enter this into the interactive shell:

```
>>> import openpyxl
>>> wb = openpyxl.Workbook()
>>> sheet = wb.active
>>> sheet['A1'] = 'Tall row'
>>> sheet['B2'] = 'Wide column'
>>> # Set the height and width:
>>> sheet.row_dimensions[1].height = 70
>>> sheet.column_dimensions['B'].width = 20
>>> wb.save('dimensions.xlsx')
```

A sheet's row_dimensions and column_dimensions are dictionary-like values; row_dimensions contains RowDimension objects and column_dimensions contains ColumnDimension objects. In row_dimensions, you can access one of the objects using the number of the row (in this case, 1 or 2). In column_dimensions, you can access one of the objects using the letter of the column (in this case, A or B).

The *dimensions.xlsx* spreadsheet looks like Figure 13-6.

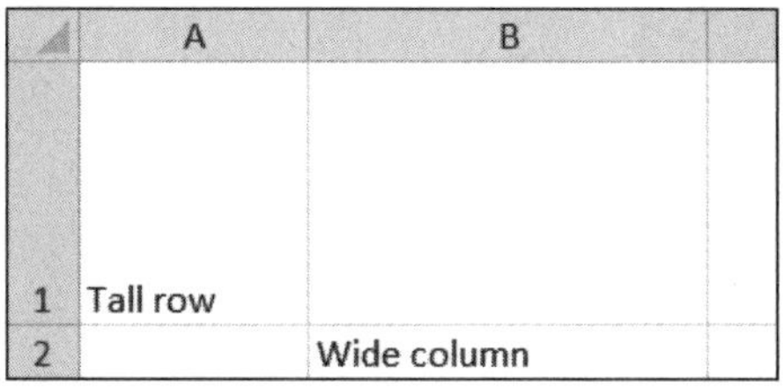

*Figure 13-6: Row 1 and column B set to larger heights and widths*

Once you have the RowDimension object, you can set its height. Once you have the ColumnDimension object, you can set its width. The row height can be set to an integer or float value between 0 and 409. This value represents the height measured in *points*, where one point equals 1/72 of an inch. The default row height is 12.75. The column width can be set to an integer or float value between 0 and 255. This value represents the number of characters at the default font size (11 point) that can be displayed in the cell. The default column width is 8.43 characters. Columns with widths of 0 or rows with heights of 0 are hidden from the user.

## Merging and Unmerging Cells

A rectangular area of cells can be merged into a single cell with the merge_cells() sheet method. Enter the following into the interactive shell:

```
>>> import openpyxl
>>> wb = openpyxl.Workbook()
>>> sheet = wb.active
>>> sheet.merge_cells('A1:D3') # Merge all these cells.
>>> sheet['A1'] = 'Twelve cells merged together.'
>>> sheet.merge_cells('C5:D5') # Merge these two cells.
>>> sheet['C5'] = 'Two merged cells.'
>>> wb.save('merged.xlsx')
```

The argument to merge_cells() is a single string of the top-left and bottom-right cells of the rectangular area to be merged: 'A1:D3' merges 12 cells into a single cell. To set the value of these merged cells, simply set the value of the top-left cell of the merged group.

When you run this code, *merged.xlsx* will look like Figure 13-7.

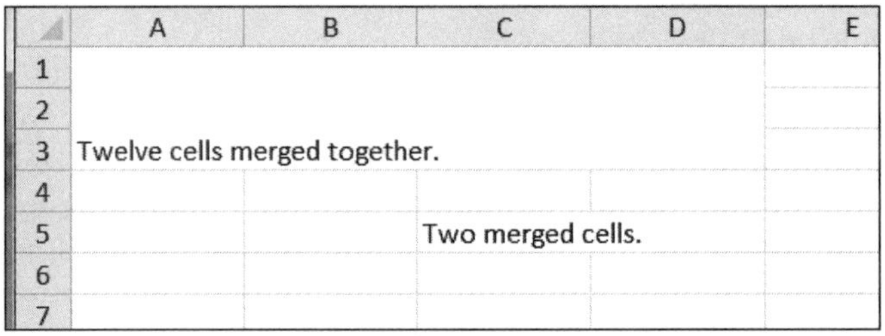

*Figure 13-7: Merged cells in a spreadsheet*

To unmerge cells, call the unmerge_cells() sheet method. Enter this into the interactive shell:

```
>>> import openpyxl
>>> wb = openpyxl.load_workbook('merged.xlsx')
>>> sheet = wb.active
>>> sheet.unmerge_cells('A1:D3') # Split these cells up.
>>> sheet.unmerge_cells('C5:D5')
>>> wb.save('merged.xlsx')
```

If you save your changes and then take a look at the spreadsheet, you'll see that the merged cells have gone back to being individual cells.

## Freezing Panes

For spreadsheets too large to be displayed all at once, it's helpful to "freeze" a few of the top rows or leftmost columns onscreen. Frozen column or row headers, for example, are always visible to the user even as they scroll through the spreadsheet. These are known as *freeze panes*. In OpenPyXL, each Worksheet object has a freeze_panes attribute that can be set to a Cell object or a string of a cell's coordinates. Note that all rows above and all columns to the left of this cell will be frozen, but the row and column of the cell itself will not be frozen.

To unfreeze all panes, set freeze_panes to None or 'A1'. Table 13-3 shows which rows and columns will be frozen for some example settings of freeze_panes.

**Table 13-3:** Frozen Pane Examples

| freeze_panes setting | Rows and columns frozen |
|---|---|
| sheet.freeze_panes = 'A2' | Row 1 |
| sheet.freeze_panes = 'B1' | Column A |
| sheet.freeze_panes = 'C1' | Columns A and B |
| sheet.freeze_panes = 'C2' | Row 1 and columns A and B |
| sheet.freeze_panes = 'A1' or sheet.freeze_panes = None | No frozen panes |

Make sure you have the produce sales spreadsheet from *https://nostarch.com/automatestuff2/*. Then enter the following into the interactive shell:

```
>>> import openpyxl
>>> wb = openpyxl.load_workbook('produceSales.xlsx')
>>> sheet = wb.active
>>> sheet.freeze_panes = 'A2' # Freeze the rows above A2.
>>> wb.save('freezeExample.xlsx')
```

If you set the freeze_panes attribute to 'A2', row 1 will always be viewable, no matter where the user scrolls in the spreadsheet. You can see this in Figure 13-8.

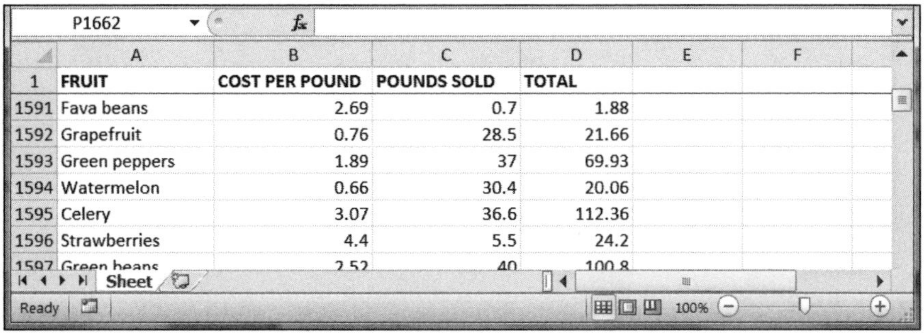

*Figure 13-8: With freeze_panes set to 'A2', row 1 is always visible, even as the user scrolls down.*

# Charts

OpenPyXL supports creating bar, line, scatter, and pie charts using the data in a sheet's cells. To make a chart, you need to do the following:

1. Create a `Reference` object from a rectangular selection of cells.
2. Create a `Series` object by passing in the `Reference` object.
3. Create a `Chart` object.
4. Append the `Series` object to the `Chart` object.
5. Add the `Chart` object to the `Worksheet` object, optionally specifying which cell should be the top-left corner of the chart.

The `Reference` object requires some explaining. You create `Reference` objects by calling the `openpyxl.chart.Reference()` function and passing three arguments:

1. The `Worksheet` object containing your chart data.
2. A tuple of two integers, representing the top-left cell of the rectangular selection of cells containing your chart data: the first integer in the tuple is the row, and the second is the column. Note that 1 is the first row, not 0.
3. A tuple of two integers, representing the bottom-right cell of the rectangular selection of cells containing your chart data: the first integer in the tuple is the row, and the second is the column.

Figure 13-9 shows some sample coordinate arguments.

*Figure 13-9: From left to right: (1, 1), (10, 1); (3, 2), (6, 4); (5, 3), (5, 3)*

Enter this interactive shell example to create a bar chart and add it to the spreadsheet:

```
>>> import openpyxl
>>> wb = openpyxl.Workbook()
>>> sheet = wb.active
>>> for i in range(1, 11): # create some data in column A
...     sheet['A' + str(i)] = i
...
>>> refObj = openpyxl.chart.Reference(sheet, min_col=1, min_row=1, max_col=1,
max_row=10)
```

```
>>> seriesObj = openpyxl.chart.Series(refObj, title='First series')

>>> chartObj = openpyxl.chart.BarChart()
>>> chartObj.title = 'My Chart'
>>> chartObj.append(seriesObj)

>>> sheet.add_chart(chartObj, 'C5')
>>> wb.save('sampleChart.xlsx')
```

This produces a spreadsheet that looks like Figure 13-10.

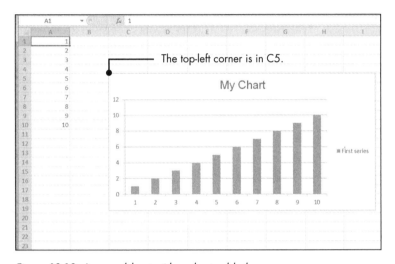

*Figure 13-10: A spreadsheet with a chart added*

We've created a bar chart by calling openpyxl.chart.BarChart(). You can also create line charts, scatter charts, and pie charts by calling openpyxl.charts .LineChart(), openpyxl.chart.ScatterChart(), and openpyxl.chart.PieChart().

## Summary

Often the hard part of processing information isn't the processing itself but simply getting the data in the right format for your program. But once you have your spreadsheet loaded into Python, you can extract and manipulate its data much faster than you could by hand.

You can also generate spreadsheets as output from your programs. So if colleagues need your text file or PDF of thousands of sales contacts transferred to a spreadsheet file, you won't have to tediously copy and paste it all into Excel.

Equipped with the openpyxl module and some programming knowledge, you'll find processing even the biggest spreadsheets a piece of cake.

In the next chapter, we'll take a look at using Python to interact with another spreadsheet program: the popular online Google Sheets application.

# Practice Questions

For the following questions, imagine you have a Workbook object in the variable wb, a Worksheet object in sheet, a Cell object in cell, a Comment object in comm, and an Image object in img.

1. What does the openpyxl.load_workbook() function return?
2. What does the wb.sheetnames workbook attribute contain?
3. How would you retrieve the Worksheet object for a sheet named 'Sheet1'?
4. How would you retrieve the Worksheet object for the workbook's active sheet?
5. How would you retrieve the value in the cell C5?
6. How would you set the value in the cell C5 to "Hello"?
7. How would you retrieve the cell's row and column as integers?
8. What do the sheet.max_column and sheet.max_row sheet attributes hold, and what is the data type of these attributes?
9. If you needed to get the integer index for column 'M', what function would you need to call?
10. If you needed to get the string name for column 14, what function would you need to call?
11. How can you retrieve a tuple of all the Cell objects from A1 to F1?
12. How would you save the workbook to the filename *example.xlsx*?
13. How do you set a formula in a cell?
14. If you want to retrieve the result of a cell's formula instead of the cell's formula itself, what must you do first?
15. How would you set the height of row 5 to 100?
16. How would you hide column C?
17. What is a freeze pane?
18. What five functions and methods do you have to call to create a bar chart?

# Practice Projects

For practice, write programs that perform the following tasks.

## Multiplication Table Maker

Create a program *multiplicationTable.py* that takes a number *N* from the command line and creates an *N×N* multiplication table in an Excel spreadsheet. For example, when the program is run like this:

```
py multiplicationTable.py 6
```

. . . it should create a spreadsheet that looks like Figure 13-11.

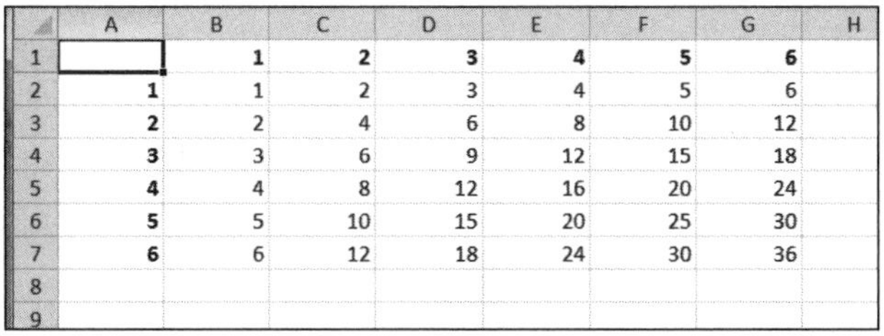

| | A | B | C | D | E | F | G | H |
|---|---|---|---|---|---|---|---|---|
| 1 | | 1 | 2 | 3 | 4 | 5 | 6 | |
| 2 | 1 | 1 | 2 | 3 | 4 | 5 | 6 | |
| 3 | 2 | 2 | 4 | 6 | 8 | 10 | 12 | |
| 4 | 3 | 3 | 6 | 9 | 12 | 15 | 18 | |
| 5 | 4 | 4 | 8 | 12 | 16 | 20 | 24 | |
| 6 | 5 | 5 | 10 | 15 | 20 | 25 | 30 | |
| 7 | 6 | 6 | 12 | 18 | 24 | 30 | 36 | |
| 8 | | | | | | | | |
| 9 | | | | | | | | |

*Figure 13-11: A multiplication table generated in a spreadsheet*

Row 1 and column A should be used for labels and should be in bold.

## Blank Row Inserter

Create a program *blankRowInserter.py* that takes two integers and a filename string as command line arguments. Let's call the first integer *N* and the second integer *M*. Starting at row *N*, the program should insert *M* blank rows into the spreadsheet. For example, when the program is run like this:

```
python blankRowInserter.py 3 2 myProduce.xlsx
```

. . . the "before" and "after" spreadsheets should look like Figure 13-12.

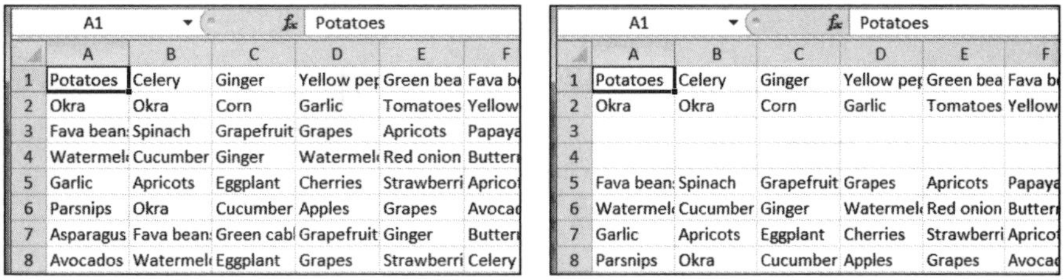

A1 Potatoes

| | A | B | C | D | E | F |
|---|---|---|---|---|---|---|
| 1 | Potatoes | Celery | Ginger | Yellow pep | Green bea | Fava b |
| 2 | Okra | Okra | Corn | Garlic | Tomatoes | Yellow |
| 3 | Fava bean: | Spinach | Grapefruit | Grapes | Apricots | Papaya |
| 4 | Watermel | Cucumber | Ginger | Watermel | Red onion | Butter |
| 5 | Garlic | Apricots | Eggplant | Cherries | Strawberri | Aprico |
| 6 | Parsnips | Okra | Cucumber | Apples | Grapes | Avoca |
| 7 | Asparagus | Fava bean: | Green cab | Grapefruit | Ginger | Butter |
| 8 | Avocados | Watermel | Eggplant | Grapes | Strawberri | Celery |

A1 Potatoes

| | A | B | C | D | E | F |
|---|---|---|---|---|---|---|
| 1 | Potatoes | Celery | Ginger | Yellow pep | Green bea | Fava b |
| 2 | Okra | Okra | Corn | Garlic | Tomatoes | Yellow |
| 3 | | | | | | |
| 4 | | | | | | |
| 5 | Fava bean: | Spinach | Grapefruit | Grapes | Apricots | Papaya |
| 6 | Watermel | Cucumber | Ginger | Watermel | Red onion | Butter |
| 7 | Garlic | Apricots | Eggplant | Cherries | Strawberri | Aprico |
| 8 | Parsnips | Okra | Cucumber | Apples | Grapes | Avoca |

*Figure 13-12: Before (left) and after (right) the two blank rows are inserted at row 3*

You can write this program by reading in the contents of the spreadsheet. Then, when writing out the new spreadsheet, use a `for` loop to copy the first *N* lines. For the remaining lines, add *M* to the row number in the output spreadsheet.

## Spreadsheet Cell Inverter

Write a program to invert the row and column of the cells in the spreadsheet. For example, the value at row 5, column 3 will be at row 3, column 5 (and vice versa). This should be done for all cells in the spreadsheet. For example, the "before" and "after" spreadsheets would look something like Figure 13-13.

Spreadsheet before (top):

| | A | B | C | D | E | F | G | H | I | J |
|---|---|---|---|---|---|---|---|---|---|---|
| | ITEM | | | | | | | | | |
| 1 | ITEM | SOLD | | | | | | | | |
| 2 | Eggplant | 334 | | | | | | | | |
| 3 | Cucumber | 252 | | | | | | | | |
| 4 | Green cabl | 238 | | | | | | | | |
| 5 | Eggplant | 516 | | | | | | | | |
| 6 | Garlic | 98 | | | | | | | | |
| 7 | Parsnips | 16 | | | | | | | | |
| 8 | Asparagus | 335 | | | | | | | | |
| 9 | Avocados | 84 | | | | | | | | |
| 10 | | | | | | | | | | |

Spreadsheet after (bottom):

| | A | B | C | D | E | F | G | H | I | J |
|---|---|---|---|---|---|---|---|---|---|---|
| | ITEM | | | | | | | | | |
| 1 | ITEM | Eggplant | Cucumber | Green cabl | Eggplant | Garlic | Parsnips | Asparagus | Avocados | |
| 2 | SOLD | 334 | 252 | 238 | 516 | 98 | 16 | 335 | 84 | |
| 3 | | | | | | | | | | |
| 4 | | | | | | | | | | |
| 5 | | | | | | | | | | |
| 6 | | | | | | | | | | |
| 7 | | | | | | | | | | |
| 8 | | | | | | | | | | |
| 9 | | | | | | | | | | |
| 10 | | | | | | | | | | |

*Figure 13-13: The spreadsheet before (top) and after (bottom) inversion*

You can write this program by using nested for loops to read the spreadsheet's data into a list of lists data structure. This data structure could have sheetData[x][y] for the cell at column x and row y. Then, when writing out the new spreadsheet, use sheetData[y][x] for the cell at column x and row y.

### Text Files to Spreadsheet

Write a program to read in the contents of several text files (you can make the text files yourself) and insert those contents into a spreadsheet, with one line of text per row. The lines of the first text file will be in the cells of column A, the lines of the second text file will be in the cells of column B, and so on.

Use the readlines() File object method to return a list of strings, one string per line in the file. For the first file, output the first line to column 1, row 1. The second line should be written to column 1, row 2, and so on. The next file that is read with readlines() will be written to column 2, the next file to column 3, and so on.

### Spreadsheet to Text Files

Write a program that performs the tasks of the previous program in reverse order: the program should open a spreadsheet and write the cells of column A into one text file, the cells of column B into another text file, and so on.

# 14

## WORKING WITH GOOGLE SHEETS

Google Sheets, the free, web-based spreadsheet application available to anyone with a Google account or Gmail address, has become a useful, feature-rich competitor to Excel. Google Sheets has its own API, but this API can be confusing to learn and use. This chapter covers the EZSheets third-party module, documented at *https://ezsheets.readthedocs.io/*. While not as full featured as the official Google Sheets API, EZSheets makes common spreadsheet tasks easy to perform.

## Installing and Setting Up EZSheets

You can install EZSheets by opening a new terminal window and running pip install --user ezsheets. As part of this installation, EZSheets will also install the google-api-python-client, google-auth-httplib2, and `google-auth-oauthlib` modules. These modules allow your program to log in to Google's servers and make API requests. EZSheets handles the interaction with these modules, so you don't need to concern yourself with how they work.

### Obtaining Credentials and Token Files

Before you can use EZSheets, you need to enable the Google Sheets and Google Drive APIs for your Google account. Visit the following web pages and click the **Enable API** buttons at the top of each:

- *https://console.developers.google.com/apis/library/sheets.googleapis.com/*
- *https://console.developers.google.com/apis/library/drive.googleapis.com/*

You'll also need to obtain three files, which you should save in the same folder as your *.py* Python script that uses EZSheets:

- A credentials file named *credentials-sheets.json*
- A token for Google Sheets named *token-sheets.pickle*
- A token for Google Drive named *token-drive.pickle*

The credentials file will generate the token files. The easiest way to obtain a credentials file is to go to the Google Sheets Python Quickstart page at *https://developers.google.com/sheets/api/quickstart/python/* and click the blue **Enable the Google Sheets API** button, as shown in Figure 14-1. You'll need to log in to your Google account to view this page.

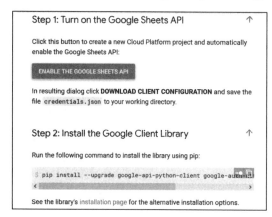

*Figure 14-1: Obtaining a* credentials.json *file.*

Clicking this button will bring up a window with a **Download Client Configuration** link that lets you download a *credentials.json* file. Rename this file to *credentials-sheets.json* and place it in the same folder as your Python scripts.

Once you have a *credentials-sheets.json* file, run the `import ezsheets` module. The first time you import the EZSheets module, it will open a new browser window for you to log in to your Google account. Click **Allow**, as shown in Figure 14-2.

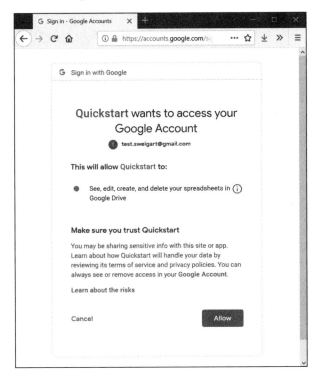

Figure 14-2: Allowing Quickstart to access your Google account

The message about Quickstart comes from the fact that you downloaded the credentials file from the Google Sheets Python Quickstart page. Note that this window will open *twice*: first for Google Sheets access and second for Google Drive access. EZSheets uses Google Drive access to upload, download, and delete spreadsheets.

After you log in, the browser window will prompt you to close it, and the *token-sheets.pickle* and *token-drive.pickle* files will appear in the same folder as *credentials-sheets.json*. You only need to go through this process the first time you run `import ezsheets`.

If you encounter an error after clicking Allow and the page seems to hang, make sure you have first enabled the Google Sheets and Drive APIs from the links at the start of this section. It may take a few minutes for Google's servers to register this change, so you may have to wait before you can use EZSheets.

Don't share the credential or token files with anyone—treat them like passwords.

### Revoking the Credentials File

If you accidentally share the credential or token files with someone, they won't be able to change your Google account password, but they will have access to your spreadsheets. You can revoke these files by going to the Google Cloud Platform developer's console page at *https://console.developers.google.com/*. You'll need to log in to your Google account to view this page. Click the **Credentials** link on the sidebar. Then click the trash can icon next to the credentials file you've accidentally shared, as shown in Figure 14-3.

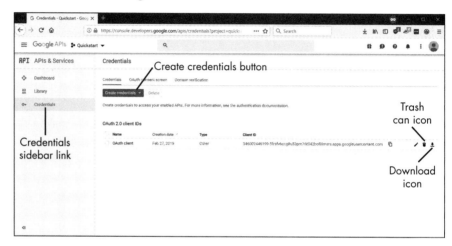

*Figure 14-3: The Credentials page in the Google Cloud Platform developer's console*

To generate a new credentials file from this page, click the **Create Credentials** button and select **OAuth client ID**, also shown in Figure 14-3. Next, for Application Type, select **Other** and give the file any name you like. This new credentials file will then be listed on the page, and you can click on the download icon to download it. The downloaded file will have a long, complicated filename, so you should rename it to the default filename that EZSheets attempts to load: *credentials-sheets.json*. You can also generate a new credential file by clicking the Enable the Google Sheets API button mentioned in the previous section.

## Spreadsheet Objects

In Google Sheets, a *spreadsheet* can contain multiple *sheets* (also called *worksheets*), and each sheet contains columns and rows of values. Figure 14-4 shows a spreadsheet titled "Education Data" containing three sheets titled "Students," "Classes," and "Resources." The first column of each sheet is labeled A, and the first row is labeled 1.

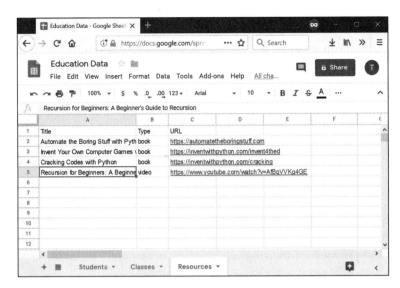

Figure 14-4: A spreadsheet titled "Education Data" with three sheets

While most of your work will involve modifying the Sheet objects, you can also modify Spreadsheet objects, as you'll see in the next section.

## Creating, Uploading, and Listing Spreadsheets

You can make a new Spreadsheet object from an existing spreadsheet, a blank spreadsheet, or an uploaded spreadsheet. To make a Spreadsheet object from an existing Google Sheets spreadsheet, you'll need the spreadsheet's ID string. The unique ID for a Google Sheets spreadsheet can be found in the URL, after the *spreadsheets/d/* part and before the */edit* part. For example, the spreadsheet featured in Figure 14-4 is located at the URL *https://docs .google.com/spreadsheets/d/1J-Jx6Ne2K_vqI9J2SO-TAXOFbxx_9tUjwnkPC22LjeU /edit#gid=151537240/*, so its ID is 1J-Jx6Ne2K_vqI9J2SO-TAXOFbxx_9tUjwnkPC22LjeU.

**NOTE** *The specific spreadsheet IDs used in this chapter are for my Google account's spreadsheets. They won't work if you enter them into your interactive shell. Go to* https://sheets.google.com/ *to create spreadsheets under your account and then get the IDs from the address bar.*

Pass your spreadsheet's ID as a string to the ezsheets.Spreadsheet() function to obtain a Spreadsheet object for its spreadsheet:

```
>>> import ezsheets
>>> ss = ezsheets.Spreadsheet('1J-Jx6Ne2K_vqI9J2SO-TAXOFbxx_9tUjwnkPC22LjeU')
>>> ss
Spreadsheet(spreadsheetId='1J-Jx6Ne2K_vqI9J2SO-TAXOFbxx_9tUjwnkPC22LjeU')
>>> ss.title
'Education Data'
```

For convenience, you can also obtain a Spreadsheet object of an existing spreadsheet by passing the spreadsheet's full URL to the function. Or, if there is only one spreadsheet in your Google account with that title, you can pass the title of the spreadsheet as a string.

To make a new, blank spreadsheet, call the ezsheets.createSpreadsheet() function and pass it a string for the new spreadsheet's title. For example, enter the following into the interactive shell:

```
>>> import ezsheets
>>> ss = ezsheets.createSpreadsheet('Title of My New Spreadsheet')
>>> ss.title
'Title of My New Spreadsheet'
```

To upload an existing Excel, OpenOffice, CSV, or TSV spreadsheet to Google Sheets, pass the filename of the spreadsheet to ezsheets.upload(). Enter the following into the interactive shell, replacing *my_spreadsheet.xlsx* with a spreadsheet file of your own:

```
>>> import ezsheets
>>> ss = ezsheets.upload('my_spreadsheet.xlsx')
>>> ss.title
'my_spreadsheet'
```

You can list the spreadsheets in your Google account by calling the listSpreadsheets() function. Enter the following into the interactive shell after uploading a spreadsheet:

```
>>> ezsheets.listSpreadsheets()
{'1J-Jx6Ne2K_vqI9J2SO-TAXOFbxx_9tUjwnkPC22LjeU': 'Education Data'}
```

The listSpreadsheets() function returns a dictionary where the keys are spreadsheet IDs and the values are the titles of each spreadsheet.

Once you've obtained a Spreadsheet object, you can use its attributes and methods to manipulate the online spreadsheet hosted on Google Sheets.

## Spreadsheet Attributes

While the actual data lives in a spreadsheet's individual sheets, the Spreadsheet object has the following attributes for manipulating the spreadsheet itself: title, spreadsheetId, url, sheetTitles, and sheets. Enter the following into the interactive shell:

```
>>> import ezsheets
>>> ss = ezsheets.Spreadsheet('1J-Jx6Ne2K_vqI9J2SO-TAXOFbxx_9tUjwnkPC22LjeU')
>>> ss.title          # The title of the spreadsheet.
'Education Data'
>>> ss.title = 'Class Data' # Change the title.
>>> ss.spreadsheetId # The unique ID (this is a read-only attribute).
'1J-Jx6Ne2K_vqI9J2SO-TAXOFbxx_9tUjwnkPC22LjeU'
>>> ss.url            # The original URL (this is a read-only attribute).
```

```
'https://docs.google.com/spreadsheets/d/1J-Jx6Ne2K_vqI9J2SO-
TAXOFbxx_9tUjwnkPC22LjeU/'
>>> ss.sheetTitles   # The titles of all the Sheet objects
('Students', 'Classes', 'Resources')
>>> ss.sheets        # The Sheet objects in this Spreadsheet, in order.
(<Sheet sheetId=0, title='Students', rowCount=1000, columnCount=26>, <Sheet
sheetId=1669384683, title='Classes', rowCount=1000, columnCount=26>, <Sheet
sheetId=151537240, title='Resources', rowCount=1000, columnCount=26>)
>>> ss[0]            # The first Sheet object in this Spreadsheet.
<Sheet sheetId=0, title='Students', rowCount=1000, columnCount=26>
>>> ss['Students']   # Sheets can also be accessed by title.
<Sheet sheetId=0, title='Students', rowCount=1000, columnCount=26>
>>> del ss[0]        # Delete the first Sheet object in this Spreadsheet.
>>> ss.sheetTitles   # The "Students" Sheet object has been deleted:
('Classes', 'Resources')
```

If someone changes the spreadsheet through the Google Sheets website, your script can update the Spreadsheet object to match the online data by calling the refresh() method:

```
>>> ss.refresh()
```

This will refresh not only the Spreadsheet object's attributes but also the data in the Sheet objects it contains. The changes you make to the Spreadsheet object will be reflected in the online spreadsheet in real time.

## Downloading and Uploading Spreadsheets

You can download a Google Sheets spreadsheet in a number of formats: Excel, OpenOffice, CSV, TSV, and PDF. You can also download it as a ZIP file containing HTML files of the spreadsheet's data. EZSheets contains functions for each of these options:

```
>>> import ezsheets
>>> ss = ezsheets.Spreadsheet('1J-Jx6Ne2K_vqI9J2SO-TAXOFbxx_9tUjwnkPC22LjeU')
>>> ss.title
'Class Data'
>>> ss.downloadAsExcel() # Downloads the spreadsheet as an Excel file.
'Class_Data.xlsx'
>>> ss.downloadAsODS() # Downloads the spreadsheet as an OpenOffice file.
'Class_Data.ods'
>>> ss.downloadAsCSV() # Only downloads the first sheet as a CSV file.
'Class_Data.csv'
>>> ss.downloadAsTSV() # Only downloads the first sheet as a TSV file.
'Class_Data.tsv'
>>> ss.downloadAsPDF() # Downloads the spreadsheet as a PDF.
'Class_Data.pdf'
>>> ss.downloadAsHTML() # Downloads the spreadsheet as a ZIP of HTML files.
'Class_Data.zip'
```

Note that files in the CSV and TSV formats can contain only one sheet; therefore, if you download a Google Sheets spreadsheet in this format, you will get the first sheet only. To download other sheets, you'll need to change the Sheet object's index attribute to 0. See "Creating and Deleting Sheets" on page 341 for information on how to do this.

The download functions all return a string of the downloaded file's filename. You can also specify your own filename for the spreadsheet by passing the new filename to the download function:

```
>>> ss.downloadAsExcel('a_different_filename.xlsx')
'a_different_filename.xlsx'
```

The function should return the updated filename.

## Deleting Spreadsheets

To delete a spreadsheet, call the delete() method:

```
>>> import ezsheets
>>> ss = ezsheets.createSpreadsheet('Delete me') # Create the spreadsheet.
>>> ezsheets.listSpreadsheets() # Confirm that we've created a spreadsheet.
{'1aCw2NNJSZblDbhygVv77kPsL3djmgV5zJZllSOZ_mRk': 'Delete me'}
>>> ss.delete() # Delete the spreadsheet.
>>> ezsheets.listSpreadsheets()
{}
```

The delete() method will move your spreadsheet to the Trash folder on your Google Drive. You can view the contents of your Trash folder at *https://drive.google.com/drive/trash*. To permanently delete your spreadsheet, pass True for the permanent keyword argument:

```
>>> ss.delete(permanent=True)
```

In general, permanently deleting your spreadsheets is not a good idea, because it would be impossible to recover a spreadsheet that a bug in your script accidentally deleted. Even free Google Drive accounts have gigabytes of storage available, so you most likely don't need to worry about freeing up space.

# Sheet Objects

A Spreadsheet object will have one or more Sheet objects. The Sheet objects represent the rows and columns of data in each sheet. You can access these sheets using the square brackets operator and an integer index. The Spreadsheet object's sheets attribute holds a tuple of Sheet objects in the order in which they appear in the spreadsheet. To access the Sheet objects in a spreadsheet, enter the following into the interactive shell:

```
>>> import ezsheets
>>> ss = ezsheets.Spreadsheet('1J-Jx6Ne2K_vqI9J2SO-TAXOFbxx_9tUjwnkPC22LjeU')
>>> ss.sheets      # The Sheet objects in this Spreadsheet, in order.
```

```
(<Sheet sheetId=1669384683, title='Classes', rowCount=1000, columnCount=26>,
<Sheet sheetId=151537240, title='Resources', rowCount=1000, columnCount=26>)
>>> ss.sheets[0] # Gets the first Sheet object in this Spreadsheet.
<Sheet sheetId=1669384683, title='Classes', rowCount=1000, columnCount=26>
>>> ss[0]        # Also gets the first Sheet object in this Spreadsheet.
<Sheet sheetId=1669384683, title='Classes', rowCount=1000, columnCount=26>
```

You can also obtain a Sheet object with the square brackets operator and a string of the sheet's name. The Spreadsheet object's sheetTitles attribute holds a tuple of all the sheet titles. For example, enter the following into the interactive shell:

```
>>> ss.sheetTitles # The titles of all the Sheet objects in this Spreadsheet.
('Classes', 'Resources')
>>> ss['Classes'] # Sheets can also be accessed by title.
<Sheet sheetId=1669384683, title='Classes', rowCount=1000, columnCount=26>
```

Once you have a Sheet object, you can read data from and write data to it using the Sheet object's methods, as explained in the next section.

## Reading and Writing Data

Just as in Excel, Google Sheets worksheets have columns and rows of cells containing data. You can use the square brackets operator to read and write data from and to these cells. For example, to create a new spreadsheet and add data to it, enter the following into the interactive shell:

```
>>> import ezsheets
>>> ss = ezsheets.createSpreadsheet('My Spreadsheet')
>>> sheet = ss[0] # Get the first sheet in this spreadsheet.
>>> sheet.title
'Sheet1'
>>> sheet = ss[0]
>>> sheet['A1'] = 'Name' # Set the value in cell A1.
>>> sheet['B1'] = 'Age'
>>> sheet['C1'] = 'Favorite Movie'
>>> sheet['A1'] # Read the value in cell A1.
'Name'
>>> sheet['A2'] # Empty cells return a blank string.
''
>>> sheet[2, 1] # Column 2, Row 1 is the same address as B1.
'Age'
>>> sheet['A2'] = 'Alice'
>>> sheet['B2'] = 30
>>> sheet['C2'] = 'RoboCop'
```

These instructions should produce a Google Sheets spreadsheet that looks like Figure 14-5.

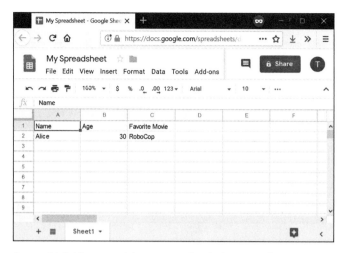

Figure 14-5: The spreadsheet created with the example instructions

Multiple users can update a sheet simultaneously. To refresh the local data in the Sheet object, call its refresh() method:

```
>>> sheet.refresh()
```

All of the data in the Sheet object is loaded when the Spreadsheet object is first loaded, so the data is read instantly. However, writing values to the online spreadsheet requires a network connection and can take about a second. If you have thousands of cells to update, updating them one at a time might be quite slow.

### Column and Row Addressing

Cell addressing works in Google Sheets just like in Excel. The only difference is that, unlike Python's 0-based list indexes, Google Sheets have 1-based columns and rows: the first column or row is at index 1, not 0. You can convert the 'A2' string-style address to the (column, row) tuple-style address (and vice versa) with the convertAddress() function. The getColumnLetterOf() and getColumnNumberOf() functions will also convert a column address between letters and numbers. Enter the following into the interactive shell:

```
>>> import ezsheets
>>> ezsheets.convertAddress('A2') # Converts addresses...
(1, 2)
>>> ezsheets.convertAddress(1, 2) # ...and converts them back, too.
'A2'
>>> ezsheets.getColumnLetterOf(2)
'B'
>>> ezsheets.getColumnNumberOf('B')
2
>>> ezsheets.getColumnLetterOf(999)
'ALK'
```

```
>>> ezsheets.getColumnNumberOf('ZZZ')
18278
```

The 'A2' string-style addresses are convenient if you're typing addresses into your source code. But the (column, row) tuple-style addresses are convenient if you're looping over a range of addresses and need a numeric form for the column. The convertAddress(), getColumnLetterOf(), and getColumnNumberOf() functions are helpful when you need to convert between the two formats.

## Reading and Writing Entire Columns and Rows

As mentioned, writing data one cell at a time can often take too long. Fortunately, EZSheets has Sheet methods for reading and writing entire columns and rows at the same time. The getColumn(), getRow(), updateColumn(), and updateRow() methods will, respectively, read and write columns and rows. These methods make requests to the Google Sheets servers to update the spreadsheet, so they require that you be connected to the internet. In this section's example, we'll upload *produceSales.xlsx* from the last chapter to Google Sheets. The first eight rows look like Table 14-1.

**Table 14-1:** The First Eight Rows of the *produceSales.xlsx* Spreadsheet

|   | A | B | C | D |
|---|---|---|---|---|
| 1 | PRODUCE | COST PER POUND | POUNDS SOLD | TOTAL |
| 2 | Potatoes | 0.86 | 21.6 | 18.58 |
| 3 | Okra | 2.26 | 38.6 | 87.24 |
| 4 | Fava beans | 2.69 | 32.8 | 88.23 |
| 5 | Watermelon | 0.66 | 27.3 | 18.02 |
| 6 | Garlic | 1.19 | 4.9 | 5.83 |
| 7 | Parsnips | 2.27 | 1.1 | 2.5 |
| 8 | Asparagus | 2.49 | 37.9 | 94.37 |

To upload this spreadsheet, enter the following into the interactive shell:

```
>>> import ezsheets
>>> ss = ezsheets.upload('produceSales.xlsx')
>>> sheet = ss[0]
>>> sheet.getRow(1) # The first row is row 1, not row 0.
['PRODUCE', 'COST PER POUND', 'POUNDS SOLD', 'TOTAL', '', '']
>>> sheet.getRow(2)
['Potatoes', '0.86', '21.6', '18.58', '', '']
>>> columnOne = sheet.getColumn(1)
>>> sheet.getColumn(1)
['PRODUCE', 'Potatoes', 'Okra', 'Fava beans', 'Watermelon', 'Garlic',
--snip--
>>> sheet.getColumn('A') # Same result as getColumn(1)
['PRODUCE', 'Potatoes', 'Okra', 'Fava beans', 'Watermelon', 'Garlic',
--snip--
>>> sheet.getRow(3)
```

```
['Okra', '2.26', '38.6', '87.24', '', '']
>>> sheet.updateRow(3, ['Pumpkin', '11.50', '20', '230'])
>>> sheet.getRow(3)
['Pumpkin', '11.50', '20', '230', '', '']
>>> columnOne = sheet.getColumn(1)
>>> for i, value in enumerate(columnOne):
...     # Make the Python list contain uppercase strings:
...     columnOne[i] = value.upper()
...
>>> sheet.updateColumn(1, columnOne) # Update the entire column in one
request.
```

The getRow() and getColumn() functions retrieve the data from every cell in a specific row or column as a list of values. Note that empty cells become blank string values in the list. You can pass getColumn() either a column number or letter to tell it to retrieve a specific column's data. The previous example shows that getColumn(1) and getColumn('A') return the same list.

The updateRow() and updateColumn() functions will overwrite all the data in the row or column, respectively, with the list of values passed to the function. In this example, the third row initially contains information about okra, but the updateRow() call replaces it with data about pumpkin. Call sheet.getRow(3) again to view the new values in the third row.

Next, let's update the "produceSales" spreadsheet. Updating cells one at a time is slow if you have many cells to update. Getting a column or row as a list, updating the list, and then updating the entire column or row with the list is much faster, since all the changes can be made in one request.

To get all of the rows at once, call the getRows() method to return a list of lists. The inner lists inside the outer list each represent a single row of the sheet. You can modify the values in this data structure to change the produce name, pounds sold, and total cost of some of the rows. Then you pass it to the updateRows() method by entering the following into the interactive shell:

```
>>> rows = sheet.getRows() # Get every row in the spreadsheet.
>>> rows[0] # Examine the values in the first row.
['PRODUCE', 'COST PER POUND', 'POUNDS SOLD', 'TOTAL', '', '']
>>> rows[1]
['POTATOES', '0.86', '21.6', '18.58', '', '']
>>> rows[1][0] = 'PUMPKIN' # Change the produce name.
>>> rows[1]
['PUMPKIN', '0.86', '21.6', '18.58', '', '']
>>> rows[10]
['OKRA', '2.26', '40', '90.4', '', '']
>>> rows[10][2] = '400' # Change the pounds sold.
>>> rows[10][3] = '904' # Change the total.
>>> rows[10]
['OKRA', '2.26', '400', '904', '', '']
>>> sheet.updateRows(rows) # Update the online spreadsheet with the changes.
```

You can update the entire sheet in a single request by passing updateRows() the list of lists returned from getRows(), amended with the changes made to rows 1 and 10.

Note that the rows in the Google Sheet have empty strings at the end. This is because the uploaded sheet has a column count of 6, but we have only 4 columns of data. You can read the number of rows and columns in a sheet with the `rowCount` and `columnCount` attributes. Then by setting these values, you can change the size of the sheet.

```
>>> sheet.rowCount          # The number of rows in the sheet.
23758
>>> sheet.columnCount       # The number of columns in the sheet.
6
>>> sheet.columnCount = 4   # Change the number of columns to 4.
>>> sheet.columnCount       # Now the number of columns in the sheet is 4.
4
```

These instructions should delete the fifth and sixth columns of the "produceSales" spreadsheet, as shown in Figure 14-6.

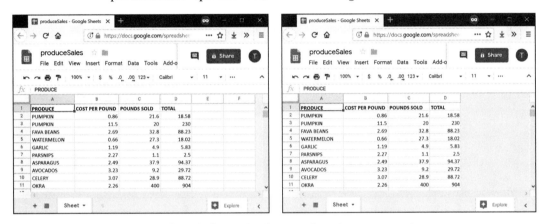

Figure 14-6: The sheet before (left) and after (right) changing the column count to 4

According to *https://support.google.com/drive/answer/37603?hl=en/*, Google Sheets spreadsheets can have up to 5 million cells in them. However, it's a good idea to make sheets only as big as you need to minimize the time it takes to update and refresh the data.

## Creating and Deleting Sheets

All Google Sheets spreadsheets start with a single sheet named "Sheet1." You can add additional sheets to the end of the list of sheets with the `createSheet()` method, to which you pass a string to use as the new sheet's title. An optional second argument can specify the integer index of the new sheet. To create a spreadsheet and then add new sheets to it, enter the following into the interactive shell:

```
>>> import ezsheets
>>> ss = ezsheets.createSpreadsheet('Multiple Sheets')
>>> ss.sheetTitles
('Sheet1',)
```

```
>>> ss.createSheet('Spam') # Create a new sheet at the end of the list of
sheets.
<Sheet sheetId=2032744541, title='Spam', rowCount=1000, columnCount=26>
>>> ss.createSheet('Eggs') # Create another new sheet.
<Sheet sheetId=417452987, title='Eggs', rowCount=1000, columnCount=26>
>>> ss.sheetTitles
('Sheet1', 'Spam', 'Eggs')
>>> ss.createSheet('Bacon', 0) # Create a sheet at index 0 in the list of
sheets.
<Sheet sheetId=814694991, title='Bacon', rowCount=1000, columnCount=26>
>>> ss.sheetTitles
('Bacon', 'Sheet1', 'Spam', 'Eggs')
```

These instructions add three new sheets to the spreadsheet: "Bacon," "Spam," and "Eggs" (in addition to the default "Sheet1"). The sheets in a spreadsheet are ordered, and new sheets go to the end of the list unless you pass a second argument to createSheet() specifying the sheet's index. Here, you create the sheet titled "Bacon" at index 0, making "Bacon" the first sheet in the spreadsheet and displacing the other three sheets by one position. This is similar to the behavior of the insert() list method.

You can see the new sheets on the tabs at the bottom of the screen, as shown in Figure 14-7.

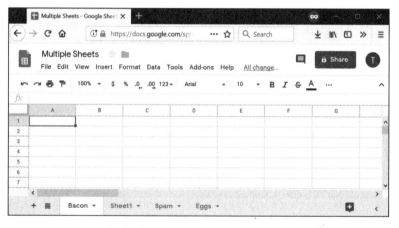

Figure 14-7: The "Multiple Sheets" spreadsheet after adding sheets "Spam," "Eggs," and "Bacon"

The Sheet object's delete() method will delete the sheet from the spreadsheet. If you want to keep the sheet but delete the data it contains, call the clear() method to clear all the cells and make it a blank sheet. Enter the following into the interactive shell:

```
>>> ss.sheetTitles
('Bacon', 'Sheet1', 'Spam', 'Eggs')
>>> ss[0].delete()        # Delete the sheet at index 0: the "Bacon" sheet.
>>> ss.sheetTitles
('Sheet1', 'Spam', 'Eggs')
```

```
>>> ss['Spam'].delete()  # Delete the "Spam" sheet.
>>> ss.sheetTitles
('Sheet1', 'Eggs')
>>> sheet = ss['Eggs']   # Assign a variable to the "Eggs" sheet.
>>> sheet.delete()       # Delete the "Eggs" sheet.
>>> ss.sheetTitles
('Sheet1',)
>>> ss[0].clear()        # Clear all the cells on the "Sheet1" sheet.
>>> ss.sheetTitles       # The "Sheet1" sheet is empty but still exists.
('Sheet1',)
```

Deleting sheets is permanent; there's no way to recover the data. However, you can back up sheets by copying them to another spreadsheet with the copyTo() method, as explained in the next section.

### Copying Sheets

Every Spreadsheet object has an ordered list of the Sheet objects it contains, and you can use this list to reorder the sheets (as shown in the previous section) or copy them to other spreadsheets. To copy a Sheet object to another Spreadsheet object, call the copyTo() method. Pass it the destination Spreadsheet object as an argument. To create two spreadsheets and copy the first spreadsheet's data to the other sheet, enter the following into the interactive shell:

```
>>> import ezsheets
>>> ss1 = ezsheets.createSpreadsheet('First Spreadsheet')
>>> ss2 = ezsheets.createSpreadsheet('Second Spreadsheet')
>>> ss1[0]
<Sheet sheetId=0, title='Sheet1', rowCount=1000, columnCount=26>
>>> ss1[0].updateRow(1, ['Some', 'data', 'in', 'the', 'first', 'row'])
>>> ss1[0].copyTo(ss2)  # Copy the ss1's Sheet1 to the ss2 spreadsheet.
>>> ss2.sheetTitles     # ss2 now contains a copy of ss1's Sheet1.
('Sheet1', 'Copy of Sheet1')
```

Note that since the destination spreadsheet (ss2 in the previous example) already had a sheet named Sheet1, the copied sheet will be named Copy of Sheet1. Copied sheets appear at the end of the list of the destination spreadsheet's sheets. If you wish, you can change their index attribute to reorder them in the new spreadsheet.

## Working with Google Sheets Quotas

Because Google Sheets is online, it's easy to share sheets among multiple users who can all access the sheets simultaneously. However, this also means that reading and updating the sheets will be slower than reading and updating Excel files stored locally on your hard drive. In addition, Google Sheets has limits on how many read and write operations you can perform.

According to Google's developer guidelines, users are restricted to creating 250 new spreadsheets a day, and free Google accounts can perform 100 read and 100 write requests per 100 seconds. Attempting to exceed this quota will raise the googleapiclient.errors.HttpError "Quota exceeded for quota group" exception. EZSheets will automatically catch this exception and retry the request. When this happens, the function calls to read or write data will take several seconds (or even a full minute or two) before they return. If the request continues to fail (which is possible if another script using the same credentials is also making requests), EZSheets will re-raise this exception.

This means that, on occasion, your EZSheets method calls may take several seconds before they return. If you want to view your API usage or increase your quota, go to the IAM & Admin Quotas page at *https://console .developers.google.com/quotas/* to learn about paying for increased usage. If you'd rather just deal with the HttpError exceptions yourself, you can set ezsheets.IGNORE_QUOTA to True, and EZSheet's methods will raise these exceptions when it encounters them.

## Summary

Google Sheets is a popular online spreadsheet application that runs in your browser. Using the EZSheets third-party module, you can download, create, read, and modify spreadsheets. EZSheets represents spreadsheets as Spreadsheet objects, each of which contains an ordered list of Sheet objects. Each sheet has columns and rows of data that you can read and update in several ways.

While Google Sheets makes sharing data and cooperative editing easy, its main disadvantage is speed: you must update spreadsheets with web requests, which can take a few seconds to execute. But for most purposes, this speed restriction won't affect Python scripts using EZSheets. Google Sheets also limits how often you can make changes.

For complete documentation of EZSheet's features, visit *https://ezsheets .readthedocs.io/*.

## Practice Questions

1. What three files do you need for EZSheets to access Google Sheets?
2. What two types of objects does EZSheets have?
3. How can you create an Excel file from a Google Sheet spreadsheet?
4. How can you create a Google Sheet spreadsheet from an Excel file?
5. The ss variable contains a Spreadsheet object. What code will read data from the cell B2 in a sheet titled "Students"?
6. How can you find the column letters for column 999?
7. How can you find out how many rows and columns a sheet has?
8. How do you delete a spreadsheet? Is this deletion permanent?

9. What functions will create a new `Spreadsheet` object and a new `Sheet` object, respectively?

10. What will happen if, by making frequent read and write requests with EZSheets, you exceed your Google account's quota?

# Practice Projects

For practice, write programs to do the following tasks.

## Downloading Google Forms Data

Google Forms allows you to create simple online forms that make it easy to collect information from people. The information they enter into the form is stored in a Google Sheet. For this project, write a program that can automatically download the form information that users have submitted. Go to *https://docs.google.com/forms/* and start a new form; it will be blank. Add fields to the form that ask the user for a name and email address. Then click the **Send** button in the upper right to get a link to your new form, such as *https://goo.gl/forms/QZsq5sC2Qe4fYO592/*. Try to enter a few example responses into this form.

On the "Responses" tab of your form, click the green **Create Spreadsheet** button to create a Google Sheets spreadsheet that will hold the responses that users submit. You should see your example responses in the first rows of this spreadsheet. Then write a Python script using EZSheets to collect a list of the email addresses on this spreadsheet.

## Converting Spreadsheets to Other Formats

You can use Google Sheets to convert a spreadsheet file into other formats. Write a script that passes a submitted file to `upload()`. Once the spreadsheet has uploaded to Google Sheets, download it using `downloadAsExcel()`, `downloadAsODS()`, and other such functions to create a copy of the spreadsheet in these other formats.

## Finding Mistakes in a Spreadsheet

After a long day at the bean-counting office, I've finished a spreadsheet with all the bean totals and uploaded them to Google Sheets. The spreadsheet is publicly viewable (but not editable). You can get this spreadsheet with the following code:

```
>>> import ezsheets
>>> ss = ezsheets.Spreadsheet('1jDZEdvSIh4TmZxccyyOZXrH-ELlrwq8_YYiZrEOB4jg')
```

You can look at this spreadsheet in your browser by going to *https://docs .google.com/spreadsheets/d/1jDZEdvSIh4TmZxccyyOZXrH-ELlrwq8_YYiZrEOB4jg /edit?usp=sharing/*. The columns of the first sheet in this spreadsheet are "Beans per Jar," "Jars," and "Total Beans." The "Total Beans" column is the product of the numbers in the "Beans per Jar" and "Jars" columns. However, there is a mistake in one of the 15,000 rows in this sheet. That's too many rows to check by hand. Luckily, you can write a script that checks the totals.

As a hint, you can access the individual cells in a row with `ss[0].getRow` (*rowNum*), where *ss* is the Spreadsheet object and *rowNum* is the row number. Remember that row numbers in Google Sheets begin at 1, not 0. The cell values will be strings, so you'll need to convert them to integers so your program can work with them. The expression `int(ss[0].getRow(2)[0]) * int(ss[0].getRow(2)[1]) == int(ss[0].getRow(2)[2])` evaluates to `True` if the row has the correct total. Put this code in a loop to identify which row in the sheet has the incorrect total.

# 15

## WORKING WITH PDF AND WORD DOCUMENTS

PDF and Word documents are binary files, which makes them much more complex than plaintext files. In addition to text, they store lots of font, color, and layout information. If you want your programs to read or write to PDFs or Word documents, you'll need to do more than simply pass their filenames to open().

Fortunately, there are Python modules that make it easy for you to interact with PDFs and Word documents. This chapter will cover two such modules: PyPDF2 and Python-Docx.

## PDF Documents

*PDF* stands for *Portable Document Format* and uses the *.pdf* file extension. Although PDFs support many features, this chapter will focus on the two things you'll be doing most often with them: reading text content from PDFs and crafting new PDFs from existing documents.

The module you'll use to work with PDFs is PyPDF2 version 1.26.0. It's important that you install this version because future versions of PyPDF2 may be incompatible with the code. To install it, run `pip install --user PyPDF2==1.26.0` from the command line. This module name is case sensitive, so make sure the *y* is lowercase and everything else is uppercase. (Check out Appendix A for full details about installing third-party modules.) If the module was installed correctly, running `import PyPDF2` in the interactive shell shouldn't display any errors.

---

**THE PROBLEMATIC PDF FORMAT**

While PDF files are great for laying out text in a way that's easy for people to print and read, they're not straightforward for software to parse into plaintext. As a result, PyPDF2 might make mistakes when extracting text from a PDF and may even be unable to open some PDFs at all. There isn't much you can do about this, unfortunately. PyPDF2 may simply be unable to work with some of your particular PDF files. That said, I haven't found any PDF files so far that can't be opened with PyPDF2.

---

## Extracting Text from PDFs

PyPDF2 does not have a way to extract images, charts, or other media from PDF documents, but it can extract text and return it as a Python string. To start learning how PyPDF2 works, we'll use it on the example PDF shown in Figure 15-1.

*Figure 15-1: The PDF page that we will be extracting text from*

Download this PDF from *https://nostarch.com/automatestuff2/* and enter the following into the interactive shell:

```
>>> import PyPDF2
>>> pdfFileObj = open('meetingminutes.pdf', 'rb')
>>> pdfReader = PyPDF2.PdfFileReader(pdfFileObj)
❶ >>> pdfReader.numPages
19
❷ >>> pageObj = pdfReader.getPage(0)
❸ >>> pageObj.extractText()
'OOFFFFIICCIIAALL  BBOOAARRDD  MMIINNUUTTEESS   Meeting of March 7,
2015        \n    The Board of Elementary and Secondary Education shall
provide leadership and create policies for education that expand opportunities
for children, empower families and communities, and advance Louisiana in an
increasingly competitive global market. BOARD  of ELEMENTARY and  SECONDARY
EDUCATION  '
>>> pdfFileObj.close()
```

First, import the PyPDF2 module. Then open *meetingminutes.pdf* in read binary mode and store it in pdfFileObj. To get a PdfFileReader object that represents this PDF, call PyPDF2.PdfFileReader() and pass it pdfFileObj. Store this PdfFileReader object in pdfReader.

The total number of pages in the document is stored in the numPages attribute of a PdfFileReader object ❶. The example PDF has 19 pages, but let's extract text from only the first page.

To extract text from a page, you need to get a Page object, which represents a single page of a PDF, from a PdfFileReader object. You can get a Page object by calling the getPage() method ❷ on a PdfFileReader object and passing it the page number of the page you're interested in—in our case, 0.

PyPDF2 uses a *zero-based index* for getting pages: The first page is page 0, the second is page 1, and so on. This is always the case, even if pages are numbered differently within the document. For example, say your PDF is a three-page excerpt from a longer report, and its pages are numbered 42, 43, and 44. To get the first page of this document, you would want to call pdfReader.getPage(0), not getPage(42) or getPage(1).

Once you have your Page object, call its extractText() method to return a string of the page's text ❸. The text extraction isn't perfect: The text *Charles E. "Chas" Roemer, President* from the PDF is absent from the string returned by extractText(), and the spacing is sometimes off. Still, this approximation of the PDF text content may be good enough for your program.

### Decrypting PDFs

Some PDF documents have an encryption feature that will keep them from being read until whoever is opening the document provides a password. Enter the following into the interactive shell with the PDF you downloaded, which has been encrypted with the password *rosebud*:

```
>>> import PyPDF2
>>> pdfReader = PyPDF2.PdfFileReader(open('encrypted.pdf', 'rb'))
```

```
❶ >>> pdfReader.isEncrypted
   True
   >>> pdfReader.getPage(0)
❷ Traceback (most recent call last):
     File "<pyshell#173>", line 1, in <module>
       pdfReader.getPage()
     --snip--
     File "C:\Python34\lib\site-packages\PyPDF2\pdf.py", line 1173, in getObject
       raise utils.PdfReadError("file has not been decrypted")
   PyPDF2.utils.PdfReadError: file has not been decrypted
>>> pdfReader = PyPDF2.PdfFileReader(open('encrypted.pdf', 'rb'))
❸ >>> pdfReader.decrypt('rosebud')
   1
   >>> pageObj = pdfReader.getPage(0)
```

All `PdfFileReader` objects have an `isEncrypted` attribute that is `True` if the PDF is encrypted and `False` if it isn't ❶. Any attempt to call a function that reads the file before it has been decrypted with the correct password will result in an error ❷.

**NOTE**    *Due to a bug in PyPDF2 version 1.26.0, calling getPage() on an encrypted PDF before calling decrypt() on it causes future getPage() calls to fail with the following error: IndexError: list index out of range. This is why our example reopened the file with a new PdfFileReader object.*

To read an encrypted PDF, call the `decrypt()` function and pass the password as a string ❸. After you call `decrypt()` with the correct password, you'll see that calling `getPage()` no longer causes an error. If given the wrong password, the `decrypt()` function will return 0 and `getPage()` will continue to fail. Note that the `decrypt()` method decrypts only the `PdfFileReader` object, not the actual PDF file. After your program terminates, the file on your hard drive remains encrypted. Your program will have to call `decrypt()` again the next time it is run.

### Creating PDFs

PyPDF2's counterpart to `PdfFileReader` is `PdfFileWriter`, which can create new PDF files. But PyPDF2 cannot write arbitrary text to a PDF like Python can do with plaintext files. Instead, PyPDF2's PDF-writing capabilities are limited to copying pages from other PDFs, rotating pages, overlaying pages, and encrypting files.

PyPDF2 doesn't allow you to directly edit a PDF. Instead, you have to create a new PDF and then copy content over from an existing document. The examples in this section will follow this general approach:

1. Open one or more existing PDFs (the source PDFs) into `PdfFileReader` objects.

2. Create a new `PdfFileWriter` object.

3. Copy pages from the `PdfFileReader` objects into the `PdfFileWriter` object.

4. Finally, use the `PdfFileWriter` object to write the output PDF.

Creating a `PdfFileWriter` object creates only a value that represents a PDF document in Python. It doesn't create the actual PDF file. For that, you must call the `PdfFileWriter`'s `write()` method.

The `write()` method takes a regular `File` object that has been opened in *write-binary* mode. You can get such a `File` object by calling Python's `open()` function with two arguments: the string of what you want the PDF's filename to be and `'wb'` to indicate the file should be opened in write-binary mode.

If this sounds a little confusing, don't worry—you'll see how this works in the following code examples.

### Copying Pages

You can use PyPDF2 to copy pages from one PDF document to another. This allows you to combine multiple PDF files, cut unwanted pages, or reorder pages.

Download *meetingminutes.pdf* and *meetingminutes2.pdf* from *https://nostarch .com/automatestuff2/* and place the PDFs in the current working directory. Enter the following into the interactive shell:

```
>>> import PyPDF2
>>> pdf1File = open('meetingminutes.pdf', 'rb')
>>> pdf2File = open('meetingminutes2.pdf', 'rb')
❶ >>> pdf1Reader = PyPDF2.PdfFileReader(pdf1File)
❷ >>> pdf2Reader = PyPDF2.PdfFileReader(pdf2File)
❸ >>> pdfWriter = PyPDF2.PdfFileWriter()

>>> for pageNum in range(pdf1Reader.numPages):
❹       pageObj = pdf1Reader.getPage(pageNum)
❺       pdfWriter.addPage(pageObj)

>>> for pageNum in range(pdf2Reader.numPages):
❹       pageObj = pdf2Reader.getPage(pageNum)
❺       pdfWriter.addPage(pageObj)

❻ >>> pdfOutputFile = open('combinedminutes.pdf', 'wb')
>>> pdfWriter.write(pdfOutputFile)
>>> pdfOutputFile.close()
>>> pdf1File.close()
>>> pdf2File.close()
```

Open both PDF files in read binary mode and store the two resulting `File` objects in `pdf1File` and `pdf2File`. Call `PyPDF2.PdfFileReader()` and pass it `pdf1File` to get a `PdfFileReader` object for *meetingminutes.pdf* ❶. Call it again and pass it `pdf2File` to get a `PdfFileReader` object for *meetingminutes2.pdf* ❷. Then create a new `PdfFileWriter` object, which represents a blank PDF document ❸.

Next, copy all the pages from the two source PDFs and add them to the PdfFileWriter object. Get the Page object by calling getPage() on a PdfFileReader object ❹. Then pass that Page object to your PdfFileWriter's addPage() method ❺. These steps are done first for pdf1Reader and then again for pdf2Reader. When you're done copying pages, write a new PDF called *combinedminutes.pdf* by passing a File object to the PdfFileWriter's write() method ❻.

**NOTE**  *PyPDF2 cannot insert pages in the middle of a PdfFileWriter object; the addPage() method will only add pages to the end.*

You have now created a new PDF file that combines the pages from *meetingminutes.pdf* and *meetingminutes2.pdf* into a single document. Remember that the File object passed to PyPDF2.PdfFileReader() needs to be opened in read-binary mode by passing 'rb' as the second argument to open(). Likewise, the File object passed to PyPDF2.PdfFileWriter() needs to be opened in write-binary mode with 'wb'.

### Rotating Pages

The pages of a PDF can also be rotated in 90-degree increments with the rotateClockwise() and rotateCounterClockwise() methods. Pass one of the integers 90, 180, or 270 to these methods. Enter the following into the interactive shell, with the *meetingminutes.pdf* file in the current working directory:

```
>>> import PyPDF2
>>> minutesFile = open('meetingminutes.pdf', 'rb')
>>> pdfReader = PyPDF2.PdfFileReader(minutesFile)
❶ >>> page = pdfReader.getPage(0)
❷ >>> page.rotateClockwise(90)
{'/Contents': [IndirectObject(961, 0), IndirectObject(962, 0),
--snip--
}
>>> pdfWriter = PyPDF2.PdfFileWriter()
>>> pdfWriter.addPage(page)
❸ >>> resultPdfFile = open('rotatedPage.pdf', 'wb')
>>> pdfWriter.write(resultPdfFile)
>>> resultPdfFile.close()
>>> minutesFile.close()
```

Here we use getPage(0) to select the first page of the PDF ❶, and then we call rotateClockwise(90) on that page ❷. We write a new PDF with the rotated page and save it as *rotatedPage.pdf* ❸.

The resulting PDF will have one page, rotated 90 degrees clockwise, as shown in Figure 15-2. The return values from rotateClockwise() and rotateCounterClockwise() contain a lot of information that you can ignore.

*Figure 15-2: The* rotatedPage.pdf *file with the page rotated 90 degrees clockwise*

## Overlaying Pages

PyPDF2 can also overlay the contents of one page over another, which is useful for adding a logo, timestamp, or watermark to a page. With Python, it's easy to add watermarks to multiple files and only to pages your program specifies.

Download *watermark.pdf* from *https://nostarch.com/automatestuff2/* and place the PDF in the current working directory along with *meetingminutes.pdf*. Then enter the following into the interactive shell:

```
>>> import PyPDF2
>>> minutesFile = open('meetingminutes.pdf', 'rb')
❶ >>> pdfReader = PyPDF2.PdfFileReader(minutesFile)
❷ >>> minutesFirstPage = pdfReader.getPage(0)
❸ >>> pdfWatermarkReader = PyPDF2.PdfFileReader(open('watermark.pdf', 'rb'))
❹ >>> minutesFirstPage.mergePage(pdfWatermarkReader.getPage(0))
❺ >>> pdfWriter = PyPDF2.PdfFileWriter()
❻ >>> pdfWriter.addPage(minutesFirstPage)

❼ >>> for pageNum in range(1, pdfReader.numPages):
        pageObj = pdfReader.getPage(pageNum)
        pdfWriter.addPage(pageObj)

>>> resultPdfFile = open('watermarkedCover.pdf', 'wb')
>>> pdfWriter.write(resultPdfFile)
>>> minutesFile.close()
>>> resultPdfFile.close()
```

Here we make a `PdfFileReader` object of *meetingminutes.pdf* ❶. We call getPage(0) to get a `Page` object for the first page and store this object in minutesFirstPage ❷. We then make a `PdfFileReader` object for *watermark .pdf* ❸ and call mergePage() on minutesFirstPage ❹. The argument we pass to mergePage() is a `Page` object for the first page of *watermark.pdf.*

Now that we've called mergePage() on minutesFirstPage, minutesFirstPage represents the watermarked first page. We make a `PdfFileWriter` object ❺ and add the watermarked first page ❻. Then we loop through the rest of the pages in *meetingminutes.pdf* and add them to the `PdfFileWriter` object ❼. Finally, we open a new PDF called *watermarkedCover.pdf* and write the contents of the PdfFileWriter to the new PDF.

Figure 15-3 shows the results. Our new PDF, *watermarkedCover.pdf*, has all the contents of the *meetingminutes.pdf*, and the first page is watermarked.

Figure 15-3: The original PDF (left), the watermark PDF (center), and the merged PDF (right)

### Encrypting PDFs

A `PdfFileWriter` object can also add encryption to a PDF document. Enter the following into the interactive shell:

```
>>> import PyPDF2
>>> pdfFile = open('meetingminutes.pdf', 'rb')
>>> pdfReader = PyPDF2.PdfFileReader(pdfFile)
>>> pdfWriter = PyPDF2.PdfFileWriter()
>>> for pageNum in range(pdfReader.numPages):
        pdfWriter.addPage(pdfReader.getPage(pageNum))

❶ >>> pdfWriter.encrypt('swordfish')
>>> resultPdf = open('encryptedminutes.pdf', 'wb')
>>> pdfWriter.write(resultPdf)
>>> resultPdf.close()
```

Before calling the write() method to save to a file, call the encrypt() method and pass it a password string ❶. PDFs can have a *user password* (allowing you to view the PDF) and an *owner password* (allowing you to set permissions for printing, commenting, extracting text, and other features). The user password and owner password are the first and second arguments to encrypt(), respectively. If only one string argument is passed to encrypt(), it will be used for both passwords.

In this example, we copied the pages of *meetingminutes.pdf* to a PdfFileWriter object. We encrypted the PdfFileWriter with the password *swordfish*, opened a new PDF called *encryptedminutes.pdf*, and wrote the contents of the PdfFileWriter to the new PDF. Before anyone can view *encryptedminutes.pdf*, they'll have to enter this password. You may want to delete the original, unencrypted *meetingminutes.pdf* file after ensuring its copy was correctly encrypted.

# Project: Combining Select Pages from Many PDFs

Say you have the boring job of merging several dozen PDF documents into a single PDF file. Each of them has a cover sheet as the first page, but you don't want the cover sheet repeated in the final result. Even though there are lots of free programs for combining PDFs, many of them simply merge entire files together. Let's write a Python program to customize which pages you want in the combined PDF.

At a high level, here's what the program will do:

1. Find all PDF files in the current working directory.
2. Sort the filenames so the PDFs are added in order.
3. Write each page, excluding the first page, of each PDF to the output file.

In terms of implementation, your code will need to do the following:

1. Call os.listdir() to find all the files in the working directory and remove any non-PDF files.
2. Call Python's sort() list method to alphabetize the filenames.
3. Create a PdfFileWriter object for the output PDF.
4. Loop over each PDF file, creating a PdfFileReader object for it.
5. Loop over each page (except the first) in each PDF file.
6. Add the pages to the output PDF.
7. Write the output PDF to a file named *allminutes.pdf*.

For this project, open a new file editor tab and save it as *combinePdfs.py*.

## Step 1: Find All PDF Files

First, your program needs to get a list of all files with the *.pdf* extension in the current working directory and sort them. Make your code look like the following:

```
#! python3
# combinePdfs.py - Combines all the PDFs in the current working directory into
# into a single PDF.

❶ import PyPDF2, os
```

```
     # Get all the PDF filenames.
     pdfFiles = []
     for filename in os.listdir('.'):
         if filename.endswith('.pdf'):
           ❷ pdfFiles.append(filename)
   ❸ pdfFiles.sort(key = str.lower)

   ❹ pdfWriter = PyPDF2.PdfFileWriter()

     # TODO: Loop through all the PDF files.

     # TODO: Loop through all the pages (except the first) and add them.

     # TODO: Save the resulting PDF to a file.
```

After the shebang line and the descriptive comment about what the program does, this code imports the os and PyPDF2 modules ❶. The os.listdir('.') call will return a list of every file in the current working directory. The code loops over this list and adds only those files with the *.pdf* extension to pdfFiles ❷. Afterward, this list is sorted in alphabetical order with the key = str.lower keyword argument to sort() ❸.

A PdfFileWriter object is created to hold the combined PDF pages ❹. Finally, a few comments outline the rest of the program.

## Step 2: Open Each PDF

Now the program must read each PDF file in pdfFiles. Add the following to your program:

```
#! python3
# combinePdfs.py - Combines all the PDFs in the current working directory into
# a single PDF.

import PyPDF2, os

# Get all the PDF filenames.
pdfFiles = []
--snip--

# Loop through all the PDF files.
for filename in pdfFiles:
    pdfFileObj = open(filename, 'rb')
    pdfReader = PyPDF2.PdfFileReader(pdfFileObj)
    # TODO: Loop through all the pages (except the first) and add them.

# TODO: Save the resulting PDF to a file.
```

For each PDF, the loop opens a filename in read-binary mode by calling open() with 'rb' as the second argument. The open() call returns a File object, which gets passed to PyPDF2.PdfFileReader() to create a PdfFileReader object for that PDF file.

### Step 3: Add Each Page

For each PDF, you'll want to loop over every page except the first. Add this code to your program:

```
#! python3
# combinePdfs.py - Combines all the PDFs in the current working directory into
# a single PDF.

import PyPDF2, os

--snip--

# Loop through all the PDF files.
for filename in pdfFiles:
--snip--
    # Loop through all the pages (except the first) and add them.
❶ for pageNum in range(1, pdfReader.numPages):
        pageObj = pdfReader.getPage(pageNum)
        pdfWriter.addPage(pageObj)

# TODO: Save the resulting PDF to a file.
```

The code inside the for loop copies each Page object individually to the PdfFileWriter object. Remember, you want to skip the first page. Since PyPDF2 considers 0 to be the first page, your loop should start at 1 ❶ and then go up to, but not include, the integer in pdfReader.numPages.

### Step 4: Save the Results

After these nested for loops are done looping, the pdfWriter variable will contain a PdfFileWriter object with the pages for all the PDFs combined. The last step is to write this content to a file on the hard drive. Add this code to your program:

```
#! python3
# combinePdfs.py - Combines all the PDFs in the current working directory into
# a single PDF.

import PyPDF2, os

--snip--

# Loop through all the PDF files.
for filename in pdfFiles:
--snip--
    # Loop through all the pages (except the first) and add them.
    for pageNum in range(1, pdfReader.numPages):
        --snip--

# Save the resulting PDF to a file.
pdfOutput = open('allminutes.pdf', 'wb')
```

```
pdfWriter.write(pdfOutput)
pdfOutput.close()
```

Passing `'wb'` to open() opens the output PDF file, *allminutes.pdf*, in write-binary mode. Then, passing the resulting File object to the write() method creates the actual PDF file. A call to the close() method finishes the program.

### Ideas for Similar Programs

Being able to create PDFs from the pages of other PDFs will let you make programs that can do the following:

- Cut out specific pages from PDFs.
- Reorder pages in a PDF.
- Create a PDF from only those pages that have some specific text, identified by extractText().

# Word Documents

Python can create and modify Word documents, which have the *.docx* file extension, with the docx module. You can install the module by running `pip install --user -U python-docx==0.8.10`. (Appendix A has full details on installing third-party modules.)

**NOTE** *When using pip to first install Python-Docx, be sure to install python-docx, not docx. The package name docx is for a different module that this book does not cover. However, when you are going to import the module from the python-docx package, you'll need to run import docx, not import python-docx.*

If you don't have Word, LibreOffice Writer and OpenOffice Writer are free alternative applications for Windows, macOS, and Linux that can be used to open *.docx* files. You can download them from *https://www.libreoffice .org/* and *https://openoffice.org/*, respectively. The full documentation for Python-Docx is available at *https://python-docx.readthedocs.io/*. Although there is a version of Word for macOS, this chapter will focus on Word for Windows.

Compared to plaintext, *.docx* files have a lot of structure. This structure is represented by three different data types in Python-Docx. At the highest level, a Document object represents the entire document. The Document object contains a list of Paragraph objects for the paragraphs in the document. (A new paragraph begins whenever the user presses ENTER or RETURN while typing in a Word document.) Each of these Paragraph objects contains a list of one or more Run objects. The single-sentence paragraph in Figure 15-4 has four runs.

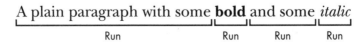

*Figure 15-4: The Run objects identified in a Paragraph object*

The text in a Word document is more than just a string. It has font, size, color, and other styling information associated with it. A *style* in Word is a collection of these attributes. A Run object is a contiguous run of text with the same style. A new Run object is needed whenever the text style changes.

## Reading Word Documents

Let's experiment with the docx module. Download *demo.docx* from *https://nostarch.com/automatestuff2/* and save the document to the working directory. Then enter the following into the interactive shell:

```
>>> import docx
❶ >>> doc = docx.Document('demo.docx')
❷ >>> len(doc.paragraphs)
7
❸ >>> doc.paragraphs[0].text
'Document Title'
❹ >>> doc.paragraphs[1].text
'A plain paragraph with some bold and some italic'
❺ >>> len(doc.paragraphs[1].runs)
4
❻ >>> doc.paragraphs[1].runs[0].text
'A plain paragraph with some '
❼ >>> doc.paragraphs[1].runs[1].text
'bold'
❽ >>> doc.paragraphs[1].runs[2].text
' and some '
❾ >>> doc.paragraphs[1].runs[3].text
'italic'
```

At ❶, we open a *.docx* file in Python, call docx.Document(), and pass the filename *demo.docx*. This will return a Document object, which has a paragraphs attribute that is a list of Paragraph objects. When we call len() on doc.paragraphs, it returns 7, which tells us that there are seven Paragraph objects in this document ❷. Each of these Paragraph objects has a text attribute that contains a string of the text in that paragraph (without the style information). Here, the first text attribute contains 'DocumentTitle' ❸, and the second contains 'A plain paragraph with some bold and some italic' ❹.

Each Paragraph object also has a runs attribute that is a list of Run objects. Run objects also have a text attribute, containing just the text in that particular run. Let's look at the text attributes in the second Paragraph object, 'A plain paragraph with some bold and some italic'. Calling len() on this Paragraph object tells us that there are four Run objects ❺. The first run object contains 'A plain paragraph with some ' ❻. Then, the text changes to a bold style, so 'bold' starts a new Run object ❼. The text returns to an unbolded style after that, which results in a third Run object, ' and some ' ❽. Finally, the fourth and last Run object contains 'italic' in an italic style ❾.

With Python-Docx, your Python programs will now be able to read the text from a *.docx* file and use it just like any other string value.

### Getting the Full Text from a .docx File

If you care only about the text, not the styling information, in the Word document, you can use the getText() function. It accepts a filename of a *.docx* file and returns a single string value of its text. Open a new file editor tab and enter the following code, saving it as *readDocx.py*:

```
#! python3

import docx

def getText(filename):
    doc = docx.Document(filename)
    fullText = []
    for para in doc.paragraphs:
        fullText.append(para.text)
    return '\n'.join(fullText)
```

The getText() function opens the Word document, loops over all the Paragraph objects in the paragraphs list, and then appends their text to the list in fullText. After the loop, the strings in fullText are joined together with newline characters.

The *readDocx.py* program can be imported like any other module. Now if you just need the text from a Word document, you can enter the following:

```
>>> import readDocx
>>> print(readDocx.getText('demo.docx'))
Document Title
A plain paragraph with some bold and some italic
Heading, level 1
Intense quote
first item in unordered list
first item in ordered list
```

You can also adjust getText() to modify the string before returning it. For example, to indent each paragraph, replace the append() call in *readDocx.py* with this:

```
fullText.append('    ' + para.text)
```

To add a double space between paragraphs, change the join() call code to this:

```
return '\n\n'.join(fullText)
```

As you can see, it takes only a few lines of code to write functions that will read a *.docx* file and return a string of its content to your liking.

## Styling Paragraph and Run Objects

In Word for Windows, you can see the styles by pressing CTRL-ALT-SHIFT-S to display the Styles pane, which looks like Figure 15-5. On macOS, you can view the Styles pane by clicking the **View ▸ Styles** menu item.

Figure 15-5: Display the Styles pane by pressing CTRL-ALT-SHIFT-S on Windows.

Word and other word processors use styles to keep the visual presentation of similar types of text consistent and easy to change. For example, perhaps you want to set body paragraphs in 11-point, Times New Roman, left-justified, ragged-right text. You can create a style with these settings and assign it to all body paragraphs. Then, if you later want to change the presentation of all body paragraphs in the document, you can just change the style, and all those paragraphs will be automatically updated.

For Word documents, there are three types of styles: *paragraph styles* can be applied to Paragraph objects, *character styles* can be applied to Run objects, and *linked styles* can be applied to both kinds of objects. You can give both Paragraph and Run objects styles by setting their style attribute to a string. This string should be the name of a style. If style is set to None, then there will be no style associated with the Paragraph or Run object.

The string values for the default Word styles are as follows:

| | | | |
|---|---|---|---|
| 'Normal' | 'Heading 5' | 'List Bullet' | 'List Paragraph' |
| 'Body Text' | 'Heading 6' | 'List Bullet 2' | 'MacroText' |
| 'Body Text 2' | 'Heading 7' | 'List Bullet 3' | 'No Spacing' |
| 'Body Text 3' | 'Heading 8' | 'List Continue' | 'Quote' |
| 'Caption' | 'Heading 9' | 'List Continue 2' | 'Subtitle' |
| 'Heading 1' | 'Intense Quote' | 'List Continue 3' | 'TOC Heading' |
| 'Heading 2' | 'List' | 'List Number ' | 'Title' |
| 'Heading 3' | 'List 2' | 'List Number 2' | |
| 'Heading 4' | 'List 3' | 'List Number 3' | |

When using a linked style for a `Run` object, you will need to add `' Char'` to the end of its name. For example, to set the Quote linked style for a `Paragraph` object, you would use `paragraphObj.style = 'Quote'`, but for a `Run` object, you would use `runObj.style = 'Quote Char'`.

In the current version of Python-Docx (0.8.10), the only styles that can be used are the default Word styles and the styles in the opened *.docx*. New styles cannot be created—though this may change in future versions of Python-Docx.

### Creating Word Documents with Nondefault Styles

If you want to create Word documents that use styles beyond the default ones, you will need to open Word to a blank Word document and create the styles yourself by clicking the **New Style** button at the bottom of the Styles pane (Figure 15-6 shows this on Windows).

This will open the Create New Style from Formatting dialog, where you can enter the new style. Then, go back into the interactive shell and open this blank Word document with `docx.Document()`, using it as the base for your Word document. The name you gave this style will now be available to use with Python-Docx.

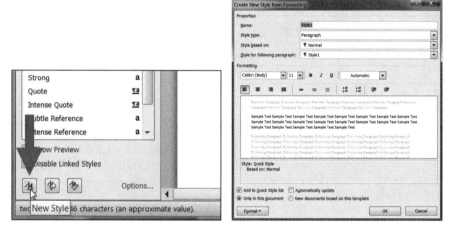

*Figure 15-6: The New Style button (left) and the Create New Style from Formatting dialog (right)*

### Run Attributes

Runs can be further styled using text attributes. Each attribute can be set to one of three values: `True` (the attribute is always enabled, no matter what other styles are applied to the run), `False` (the attribute is always disabled), or `None` (defaults to whatever the run's style is set to).

Table 15-1 lists the text attributes that can be set on `Run` objects.

**Table 15-1:** Run Object text Attributes

| Attribute | Description |
| --- | --- |
| bold | The text appears in bold. |
| italic | The text appears in italic. |
| underline | The text is underlined. |
| strike | The text appears with strikethrough. |
| double_strike | The text appears with double strikethrough. |
| all_caps | The text appears in capital letters. |
| small_caps | The text appears in capital letters, with lowercase letters two points smaller. |
| shadow | The text appears with a shadow. |
| outline | The text appears outlined rather than solid. |
| rtl | The text is written right-to-left. |
| imprint | The text appears pressed into the page. |
| emboss | The text appears raised off the page in relief. |

For example, to change the styles of *demo.docx*, enter the following into the interactive shell:

```
>>> import docx
>>> doc = docx.Document('demo.docx')
>>> doc.paragraphs[0].text
'Document Title'
>>> doc.paragraphs[0].style # The exact id may be different:
_ParagraphStyle('Title') id: 3095631007984
>>> doc.paragraphs[0].style = 'Normal'
>>> doc.paragraphs[1].text
'A plain paragraph with some bold and some italic'
>>> (doc.paragraphs[1].runs[0].text, doc.paragraphs[1].runs[1].text, doc.
paragraphs[1].runs[2].text, doc.paragraphs[1].runs[3].text)
('A plain paragraph with some ', 'bold', ' and some ', 'italic')
>>> doc.paragraphs[1].runs[0].style = 'QuoteChar'
>>> doc.paragraphs[1].runs[1].underline = True
>>> doc.paragraphs[1].runs[3].underline = True
>>> doc.save('restyled.docx')
```

Here, we use the text and style attributes to easily see what's in the paragraphs in our document. We can see that it's simple to divide a paragraph into runs and access each run individually. So we get the first, second, and fourth runs in the second paragraph; style each run; and save the results to a new document.

The words *Document Title* at the top of *restyled.docx* will have the Normal style instead of the Title style, the Run object for the text *A plain paragraph with some* will have the QuoteChar style, and the two Run objects for the words *bold* and *italic* will have their underline attributes set to `True`. Figure 15-7 shows how the styles of paragraphs and runs look in *restyled.docx*.

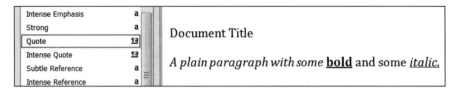

*Figure 15-7: The* restyled.docx *file*

You can find more complete documentation on Python-Docx's use of styles at *https://python-docx.readthedocs.io/en/latest/user/styles.html*.

## Writing Word Documents

Enter the following into the interactive shell:

```
>>> import docx
>>> doc = docx.Document()
>>> doc.add_paragraph('Hello, world!')
<docx.text.Paragraph object at 0x0000000003B56F60>
>>> doc.save('helloworld.docx')
```

To create your own *.docx* file, call `docx.Document()` to return a new, blank Word `Document` object. The `add_paragraph()` document method adds a new paragraph of text to the document and returns a reference to the `Paragraph` object that was added. When you're done adding text, pass a filename string to the `save()` document method to save the `Document` object to a file.

This will create a file named *helloworld.docx* in the current working directory that, when opened, looks like Figure 15-8.

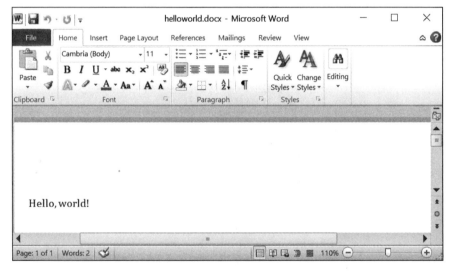

*Figure 15-8: The Word document created using* add_paragraph('Hello, world!')

You can add paragraphs by calling the add_paragraph() method again with the new paragraph's text. Or to add text to the end of an existing paragraph, you can call the paragraph's add_run() method and pass it a string. Enter the following into the interactive shell:

```
>>> import docx
>>> doc = docx.Document()
>>> doc.add_paragraph('Hello world!')
<docx.text.Paragraph object at 0x000000000366AD30>
>>> paraObj1 = doc.add_paragraph('This is a second paragraph.')
>>> paraObj2 = doc.add_paragraph('This is a yet another paragraph.')
>>> paraObj1.add_run(' This text is being added to the second paragraph.')
<docx.text.Run object at 0x0000000003A2C860>
>>> doc.save('multipleParagraphs.docx')
```

The resulting document will look like Figure 15-9. Note that the text *This text is being added to the second paragraph.* was added to the Paragraph object in paraObj1, which was the second paragraph added to doc. The add_paragraph() and add_run() functions return paragraph and Run objects, respectively, to save you the trouble of extracting them as a separate step.

Keep in mind that as of Python-Docx version 0.8.10, new Paragraph objects can be added only to the end of the document, and new Run objects can be added only to the end of a Paragraph object.

The save() method can be called again to save the additional changes you've made.

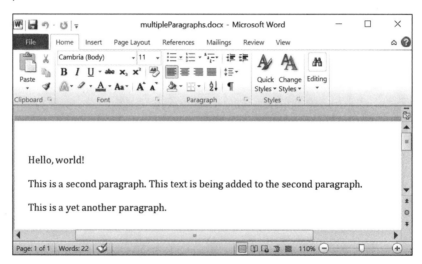

*Figure 15-9: The document with multiple Paragraph and Run objects added*

Both add_paragraph() and add_run() accept an optional second argument that is a string of the Paragraph or Run object's style. Here's an example:

```
>>> doc.add_paragraph('Hello, world!', 'Title')
```

This line adds a paragraph with the text *Hello, world!* in the Title style.

## Adding Headings

Calling add_heading() adds a paragraph with one of the heading styles. Enter the following into the interactive shell:

```
>>> doc = docx.Document()
>>> doc.add_heading('Header 0', 0)
<docx.text.Paragraph object at 0x00000000036CB3C8>
>>> doc.add_heading('Header 1', 1)
<docx.text.Paragraph object at 0x00000000036CB630>
>>> doc.add_heading('Header 2', 2)
<docx.text.Paragraph object at 0x00000000036CB828>
>>> doc.add_heading('Header 3', 3)
<docx.text.Paragraph object at 0x00000000036CB2E8>
>>> doc.add_heading('Header 4', 4)
<docx.text.Paragraph object at 0x00000000036CB3C8>
>>> doc.save('headings.docx')
```

The arguments to add_heading() are a string of the heading text and an integer from 0 to 4. The integer 0 makes the heading the Title style, which is used for the top of the document. Integers 1 to 4 are for various heading levels, with 1 being the main heading and 4 the lowest subheading. The add_heading() function returns a Paragraph object to save you the step of extracting it from the Document object as a separate step.

The resulting *headings.docx* file will look like Figure 15-10.

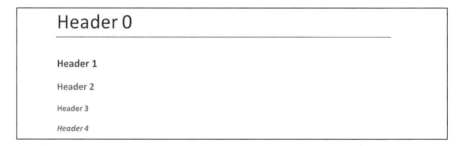

*Figure 15-10: The* headings.docx *document with headings 0 to 4*

## Adding Line and Page Breaks

To add a line break (rather than starting a whole new paragraph), you can call the add_break() method on the Run object you want to have the break appear after. If you want to add a page break instead, you need to pass the value docx.enum.text.WD_BREAK.PAGE as a lone argument to add_break(), as is done in the middle of the following example:

```
>>> doc = docx.Document()
>>> doc.add_paragraph('This is on the first page!')
<docx.text.Paragraph object at 0x0000000003785518>
>>> doc.paragraphs[0].runs[0].add_break(docx.enum.text.WD_BREAK.PAGE)
>>> doc.add_paragraph('This is on the second page!')
<docx.text.Paragraph object at 0x00000000037855F8>
>>> doc.save('twoPage.docx')
```

This creates a two-page Word document with *This is on the first page!* on the first page and *This is on the second page!* on the second. Even though there was still plenty of space on the first page after the text *This is on the first page!*, we forced the next paragraph to begin on a new page by inserting a page break after the first run of the first paragraph ❶.

### Adding Pictures

Document objects have an add_picture() method that will let you add an image to the end of the document. Say you have a file *zophie.png* in the current working directory. You can add *zophie.png* to the end of your document with a width of 1 inch and height of 4 centimeters (Word can use both imperial and metric units) by entering the following:

```
>>> doc.add_picture('zophie.png', width=docx.shared.Inches(1),
height=docx.shared.Cm(4))
<docx.shape.InlineShape object at 0x00000000036C7D30>
```

The first argument is a string of the image's filename. The optional width and height keyword arguments will set the width and height of the image in the document. If left out, the width and height will default to the normal size of the image.

You'll probably prefer to specify an image's height and width in familiar units such as inches and centimeters, so you can use the docx.shared.Inches() and docx.shared.Cm() functions when you're specifying the width and height keyword arguments.

## Creating PDFs from Word Documents

The PyPDF2 module doesn't allow you to create PDF documents directly, but there's a way to generate PDF files with Python if you're on Windows and have Microsoft Word installed. You'll need to install the Pywin32 package by running pip install --user -U pywin32==224. With this and the docx module, you can create Word documents and then convert them to PDFs with the following script.

Open a new file editor tab, enter the following code, and save it as *convertWordToPDF.py*:

```
# This script runs on Windows only, and you must have Word installed.
import win32com.client # install with "pip install pywin32==224"
import docx
wordFilename = 'your_word_document.docx'
pdfFilename = 'your_pdf_filename.pdf'

doc = docx.Document()
# Code to create Word document goes here.
doc.save(wordFilename)

wdFormatPDF = 17 # Word's numeric code for PDFs.
wordObj = win32com.client.Dispatch('Word.Application')
```

```
docObj = wordObj.Documents.Open(wordFilename)
docObj.SaveAs(pdfFilename, FileFormat=wdFormatPDF)
docObj.Close()
wordObj.Quit()
```

To write a program that produces PDFs with your own content, you must use the docx module to create a Word document, then use the Pywin32 package's win32com.client module to convert it to a PDF. Replace the # Code to create Word document goes here. comment with docx function calls to create your own content for the PDF in a Word document.

This may seem like a convoluted way to produce PDFs, but as it turns out, professional software solutions are often just as complicated.

## Summary

Text information isn't just for plaintext files; in fact, it's pretty likely that you deal with PDFs and Word documents much more often. You can use the PyPDF2 module to read and write PDF documents. Unfortunately, reading text from PDF documents might not always result in a perfect translation to a string because of the complicated PDF file format, and some PDFs might not be readable at all. In these cases, you're out of luck unless future updates to PyPDF2 support additional PDF features.

Word documents are more reliable, and you can read them with the python-docx package's docx module. You can manipulate text in Word documents via Paragraph and Run objects. These objects can also be given styles, though they must be from the default set of styles or styles already in the document. You can add new paragraphs, headings, breaks, and pictures to the document, though only to the end.

Many of the limitations that come with working with PDFs and Word documents are because these formats are meant to be nicely displayed for human readers, rather than easy to parse by software. The next chapter takes a look at two other common formats for storing information: JSON and CSV files. These formats are designed to be used by computers, and you'll see that Python can work with these formats much more easily.

## Practice Questions

1. A string value of the PDF filename is *not* passed to the PyPDF2 .PdfFileReader() function. What do you pass to the function instead?

2. What modes do the File objects for PdfFileReader() and PdfFileWriter() need to be opened in?

3. How do you acquire a Page object for page 5 from a PdfFileReader object?

4. What PdfFileReader variable stores the number of pages in the PDF document?

5. If a PdfFileReader object's PDF is encrypted with the password swordfish, what must you do before you can obtain Page objects from it?

6. What methods do you use to rotate a page?

7. What method returns a `Document` object for a file named *demo.docx*?

8. What is the difference between a `Paragraph` object and a `Run` object?

9. How do you obtain a list of `Paragraph` objects for a `Document` object that's stored in a variable named doc?

10. What type of object has `bold`, `underline`, `italic`, `strike`, and `outline` variables?

11. What is the difference between setting the `bold` variable to `True`, `False`, or `None`?

12. How do you create a `Document` object for a new Word document?

13. How do you add a paragraph with the text `'Hello, there!'` to a `Document` object stored in a variable named doc?

14. What integers represent the levels of headings available in Word documents?

# Practice Projects

For practice, write programs that do the following.

## PDF Paranoia

Using the `os.walk()` function from Chapter 10, write a script that will go through every PDF in a folder (and its subfolders) and encrypt the PDFs using a password provided on the command line. Save each encrypted PDF with an *_encrypted.pdf* suffix added to the original filename. Before deleting the original file, have the program attempt to read and decrypt the file to ensure that it was encrypted correctly.

Then, write a program that finds all encrypted PDFs in a folder (and its subfolders) and creates a decrypted copy of the PDF using a provided password. If the password is incorrect, the program should print a message to the user and continue to the next PDF.

## Custom Invitations as Word Documents

Say you have a text file of guest names. This *guests.txt* file has one name per line, as follows:

```
Prof. Plum
Miss Scarlet
Col. Mustard
Al Sweigart
RoboCop
```

Write a program that would generate a Word document with custom invitations that look like Figure 15-11.

Since Python-Docx can use only those styles that already exist in the Word document, you will have to first add these styles to a blank Word file and then open that file with Python-Docx. There should be one invitation per page in the resulting Word document, so call add_break() to add a page break after the last paragraph of each invitation. This way, you will need to open only one Word document to print all of the invitations at once.

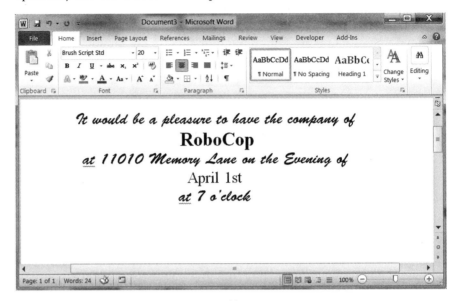

Figure 15-11: The Word document generated by your custom invite script

You can download a sample *guests.txt* file from *https://nostarch.com /automatestuff2/*.

## Brute-Force PDF Password Breaker

Say you have an encrypted PDF that you have forgotten the password to, but you remember it was a single English word. Trying to guess your forgotten password is quite a boring task. Instead you can write a program that will decrypt the PDF by trying every possible English word until it finds one that works. This is called a *brute-force password attack*. Download the text file *dictionary.txt* from *https://nostarch.com/automatestuff2/*. This *dictionary file* contains over 44,000 English words with one word per line.

Using the file-reading skills you learned in Chapter 9, create a list of word strings by reading this file. Then loop over each word in this list, passing it to the decrypt() method. If this method returns the integer 0, the password was wrong and your program should continue to the next password. If decrypt() returns 1, then your program should break out of the loop and print the hacked password. You should try both the uppercase and lowercase form of each word. (On my laptop, going through all 88,000 uppercase and lowercase words from the dictionary file takes a couple of minutes. This is why you shouldn't use a simple English word for your passwords.)

# 16

## WORKING WITH CSV FILES AND JSON DATA

In Chapter 15, you learned how to extract text from PDF and Word documents. These files were in a binary format, which required special Python modules to access their data. CSV and JSON files, on the other hand, are just plaintext files. You can view them in a text editor, such as Mu. But Python also comes with the special csv and json modules, each providing functions to help you work with these file formats.

CSV stands for "comma-separated values," and CSV files are simplified spreadsheets stored as plaintext files. Python's csv module makes it easy to parse CSV files.

JSON (pronounced "JAY-sawn" or "Jason"—it doesn't matter how because either way people will say you're pronouncing it wrong) is a format that stores information as JavaScript source code in plaintext files. (JSON is short for JavaScript Object Notation.) You don't need to know the JavaScript programming language to use JSON files, but the JSON format is useful to know because it's used in many web applications.

# The csv Module

Each line in a CSV file represents a row in the spreadsheet, and commas separate the cells in the row. For example, the spreadsheet *example.xlsx* from *https://nostarch.com/automatestuff2/* would look like this in a CSV file:

```
4/5/2015 13:34,Apples,73
4/5/2015 3:41,Cherries,85
4/6/2015 12:46,Pears,14
4/8/2015 8:59,Oranges,52
4/10/2015 2:07,Apples,152
4/10/2015 18:10,Bananas,23
4/10/2015 2:40,Strawberries,98
```

I will use this file for this chapter's interactive shell examples. You can download *example.csv* from *https://nostarch.com/automatestuff2/* or enter the text into a text editor and save it as *example.csv*.

CSV files are simple, lacking many of the features of an Excel spreadsheet. For example, CSV files:

- Don't have types for their values—everything is a string
- Don't have settings for font size or color
- Don't have multiple worksheets
- Can't specify cell widths and heights
- Can't have merged cells
- Can't have images or charts embedded in them

The advantage of CSV files is simplicity. CSV files are widely supported by many types of programs, can be viewed in text editors (including Mu), and are a straightforward way to represent spreadsheet data. The CSV format is exactly as advertised: it's just a text file of comma-separated values.

Since CSV files are just text files, you might be tempted to read them in as a string and then process that string using the techniques you learned in Chapter 9. For example, since each cell in a CSV file is separated by a comma, maybe you could just call split(',') on each line of text to get the comma-separated values as a list of strings. But not every comma in a CSV file represents the boundary between two cells. CSV files also have their own set of escape characters to allow commas and other characters to be included *as part of the values*. The split() method doesn't handle these escape characters. Because of these potential pitfalls, you should always use the csv module for reading and writing CSV files.

## reader Objects

To read data from a CSV file with the csv module, you need to create a reader object. A reader object lets you iterate over lines in the CSV file. Enter the following into the interactive shell, with *example.csv* in the current working directory:

```
❶ >>> import csv
❷ >>> exampleFile = open('example.csv')
❸ >>> exampleReader = csv.reader(exampleFile)
❹ >>> exampleData = list(exampleReader)
❺ >>> exampleData
  [['4/5/2015 13:34', 'Apples', '73'], ['4/5/2015 3:41', 'Cherries', '85'],
  ['4/6/2015 12:46', 'Pears', '14'], ['4/8/2015 8:59', 'Oranges', '52'],
  ['4/10/2015 2:07', 'Apples', '152'], ['4/10/2015 18:10', 'Bananas', '23'],
  ['4/10/2015 2:40', 'Strawberries', '98']]
```

The csv module comes with Python, so we can import it ❶ without having to install it first.

To read a CSV file with the csv module, first open it using the open() function ❷, just as you would any other text file. But instead of calling the read() or readlines() method on the File object that open() returns, pass it to the csv.reader() function ❸. This will return a reader object for you to use. Note that you don't pass a filename string directly to the csv.reader() function.

The most direct way to access the values in the reader object is to convert it to a plain Python list by passing it to list() ❹. Using list() on this reader object returns a list of lists, which you can store in a variable like exampleData. Entering exampleData in the shell displays the list of lists ❺.

Now that you have the CSV file as a list of lists, you can access the value at a particular row and column with the expression exampleData[row][col], where row is the index of one of the lists in exampleData, and col is the index of the item you want from that list. Enter the following into the interactive shell:

```
>>> exampleData[0][0]
'4/5/2015 13:34'
>>> exampleData[0][1]
'Apples'
>>> exampleData[0][2]
'73'
>>> exampleData[1][1]
'Cherries'
>>> exampleData[6][1]
'Strawberries'
```

As you can see from the output, exampleData[0][0] goes into the first list and gives us the first string, exampleData[0][2] goes into the first list and gives us the third string, and so on.

### Reading Data from reader Objects in a for Loop

For large CSV files, you'll want to use the reader object in a for loop. This avoids loading the entire file into memory at once. For example, enter the following into the interactive shell:

```
>>> import csv
>>> exampleFile = open('example.csv')
>>> exampleReader = csv.reader(exampleFile)
>>> for row in exampleReader:
        print('Row #' + str(exampleReader.line_num) + ' ' + str(row))

Row #1 ['4/5/2015 13:34', 'Apples', '73']
Row #2 ['4/5/2015 3:41', 'Cherries', '85']
Row #3 ['4/6/2015 12:46', 'Pears', '14']
Row #4 ['4/8/2015 8:59', 'Oranges', '52']
Row #5 ['4/10/2015 2:07', 'Apples', '152']
Row #6 ['4/10/2015 18:10', 'Bananas', '23']
Row #7 ['4/10/2015 2:40', 'Strawberries', '98']
```

After you import the csv module and make a reader object from the CSV file, you can loop through the rows in the reader object. Each row is a list of values, with each value representing a cell.

The print() function call prints the number of the current row and the contents of the row. To get the row number, use the reader object's line_num variable, which contains the number of the current line.

The reader object can be looped over only once. To reread the CSV file, you must call csv.reader to create a reader object.

### writer Objects

A writer object lets you write data to a CSV file. To create a writer object, you use the csv.writer() function. Enter the following into the interactive shell:

```
>>> import csv
❶ >>> outputFile = open('output.csv', 'w', newline='')
❷ >>> outputWriter = csv.writer(outputFile)
>>> outputWriter.writerow(['spam', 'eggs', 'bacon', 'ham'])
21
>>> outputWriter.writerow(['Hello, world!', 'eggs', 'bacon', 'ham'])
32
>>> outputWriter.writerow([1, 2, 3.141592, 4])
16
>>> outputFile.close()
```

First, call open() and pass it 'w' to open a file in write mode ❶. This will create the object you can then pass to csv.writer() ❷ to create a writer object.

On Windows, you'll also need to pass a blank string for the open() function's newline keyword argument. For technical reasons beyond the scope of this book, if you forget to set the newline argument, the rows in *output.csv* will be double-spaced, as shown in Figure 16-1.

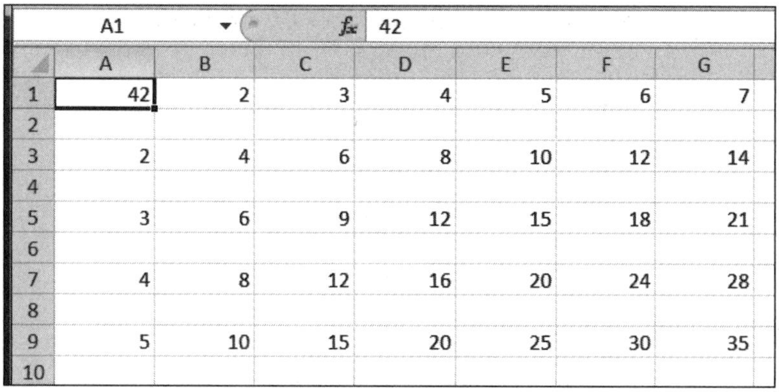

A1 | fx 42

| | A | B | C | D | E | F | G |
|---|---|---|---|---|---|---|---|
| 1 | 42 | 2 | 3 | 4 | 5 | 6 | 7 |
| 2 | | | | | | | |
| 3 | 2 | 4 | 6 | 8 | 10 | 12 | 14 |
| 4 | | | | | | | |
| 5 | 3 | 6 | 9 | 12 | 15 | 18 | 21 |
| 6 | | | | | | | |
| 7 | 4 | 8 | 12 | 16 | 20 | 24 | 28 |
| 8 | | | | | | | |
| 9 | 5 | 10 | 15 | 20 | 25 | 30 | 35 |
| 10 | | | | | | | |

*Figure 16-1: If you forget the `newline=''` keyword argument in `open()`, the CSV file will be double-spaced.*

The `writerow()` method for `writer` objects takes a list argument. Each value in the list is placed in its own cell in the output CSV file. The return value of `writerow()` is the number of characters written to the file for that row (including newline characters).

This code produces an *output.csv* file that looks like this:

```
spam,eggs,bacon,ham
"Hello, world!",eggs,bacon,ham
1,2,3.141592,4
```

Notice how the `writer` object automatically escapes the comma in the value `'Hello, world!'` with double quotes in the CSV file. The `csv` module saves you from having to handle these special cases yourself.

### The delimiter and lineterminator Keyword Arguments

Say you want to separate cells with a tab character instead of a comma and you want the rows to be double-spaced. You could enter something like the following into the interactive shell:

```
>>> import csv
>>> csvFile = open('example.tsv', 'w', newline='')
❶ >>> csvWriter = csv.writer(csvFile, delimiter='\t', lineterminator='\n\n')
>>> csvWriter.writerow(['apples', 'oranges', 'grapes'])
24
>>> csvWriter.writerow(['eggs', 'bacon', 'ham'])
17
>>> csvWriter.writerow(['spam', 'spam', 'spam', 'spam', 'spam', 'spam'])
32
>>> csvFile.close()
```

This changes the delimiter and line terminator characters in your file. The *delimiter* is the character that appears between cells on a row. By default, the delimiter for a CSV file is a comma. The *line terminator* is the character that comes at the end of a row. By default, the line terminator is a newline. You can change characters to different values by using the delimiter and lineterminator keyword arguments with `csv.writer()`.

Passing `delimiter='\t'` and `lineterminator='\n\n'` ❶ changes the character between cells to a tab and the character between rows to two newlines. We then call `writerow()` three times to give us three rows.

This produces a file named *example.tsv* with the following contents:

| | | | | | |
|---|---|---|---|---|---|
| apples | oranges | grapes | | | |
| eggs | bacon | ham | | | |
| spam | spam | spam | spam | spam | spam |

Now that our cells are separated by tabs, we're using the file extension *.tsv*, for tab-separated values.

## DictReader and DictWriter CSV Objects

For CSV files that contain header rows, it's often more convenient to work with the DictReader and DictWriter objects, rather than the reader and writer objects.

The reader and writer objects read and write to CSV file rows by using lists. The DictReader and DictWriter CSV objects perform the same functions but use dictionaries instead, and they use the first row of the CSV file as the keys of these dictionaries.

Go to *https://nostarch.com/automatestuff2/* and download the *exampleWithHeader.csv* file. This file is the same as *example.csv* except it has Timestamp, Fruit, and Quantity as the column headers in the first row.

To read the file, enter the following into the interactive shell:

```
>>> import csv
>>> exampleFile = open('exampleWithHeader.csv')
>>> exampleDictReader = csv.DictReader(exampleFile)
>>> for row in exampleDictReader:
...     print(row['Timestamp'], row['Fruit'], row['Quantity'])
...
4/5/2015 13:34 Apples 73
4/5/2015 3:41 Cherries 85
4/6/2015 12:46 Pears 14
4/8/2015 8:59 Oranges 52
4/10/2015 2:07 Apples 152
4/10/2015 18:10 Bananas 23
4/10/2015 2:40 Strawberries 98
```

Inside the loop, DictReader object sets row to a dictionary object with keys derived from the headers in the first row. (Well, technically, it sets row to an OrderedDict object, which you can use in the same way as a dictionary; the difference between them is beyond the scope of this book.) Using a DictReader object means you don't need additional code to skip the first row's header information, since the DictReader object does this for you.

If you tried to use DictReader objects with *example.csv*, which doesn't have column headers in the first row, the DictReader object would use '4/5/2015 13:34', 'Apples', and '73' as the dictionary keys. To avoid this, you can supply the DictReader() function with a second argument containing made-up header names:

```
>>> import csv
>>> exampleFile = open('example.csv')
>>> exampleDictReader = csv.DictReader(exampleFile, ['time', 'name',
'amount'])
>>> for row in exampleDictReader:
...     print(row['time'], row['name'], row['amount'])
...
4/5/2015 13:34 Apples 73
4/5/2015 3:41 Cherries 85
4/6/2015 12:46 Pears 14
4/8/2015 8:59 Oranges 52
4/10/2015 2:07 Apples 152
4/10/2015 18:10 Bananas 23
4/10/2015 2:40 Strawberries 98
```

Because *example.csv*'s first row doesn't have any text for the heading of each column, we created our own: 'time', 'name', and 'amount'.

DictWriter objects use dictionaries to create CSV files.

```
>>> import csv
>>> outputFile = open('output.csv', 'w', newline='')
>>> outputDictWriter = csv.DictWriter(outputFile, ['Name', 'Pet', 'Phone'])
>>> outputDictWriter.writeheader()
>>> outputDictWriter.writerow({'Name': 'Alice', 'Pet': 'cat', 'Phone': '555-
1234'})
20
>>> outputDictWriter.writerow({'Name': 'Bob', 'Phone': '555-9999'})
15
>>> outputDictWriter.writerow({'Phone': '555-5555', 'Name': 'Carol', 'Pet':
'dog'})
20
>>> outputFile.close()
```

If you want your file to contain a header row, write that row by calling writeheader(). Otherwise, skip calling writeheader() to omit a header row from the file. You then write each row of the CSV file with a writerow() method call, passing a dictionary that uses the headers as keys and contains the data to write to the file.

The *output.csv* file this code creates looks like this:

```
Name,Pet,Phone
Alice,cat,555-1234
Bob,,555-9999
Carol,dog,555-5555
```

Notice that the order of the key-value pairs in the dictionaries you passed to `writerow()` doesn't matter: they're written in the order of the keys given to `DictWriter()`. For example, even though you passed the `Phone` key and value before the `Name` and `Pet` keys and values in the fourth row, the phone number still appeared last in the output.

Notice also that any missing keys, such as `'Pet'` in `{'Name': 'Bob', 'Phone': '555-9999'}`, will simply be empty in the CSV file.

# Project: Removing the Header from CSV Files

Say you have the boring job of removing the first line from several hundred CSV files. Maybe you'll be feeding them into an automated process that requires just the data and not the headers at the top of the columns. You *could* open each file in Excel, delete the first row, and resave the file—but that would take hours. Let's write a program to do it instead.

The program will need to open every file with the *.csv* extension in the current working directory, read in the contents of the CSV file, and rewrite the contents without the first row to a file of the same name. This will replace the old contents of the CSV file with the new, headless contents.

**WARNING** *As always, whenever you write a program that modifies files, be sure to back up the files first, just in case your program does not work the way you expect it to. You don't want to accidentally erase your original files.*

At a high level, the program must do the following:

1. Find all the CSV files in the current working directory.
2. Read in the full contents of each file.
3. Write out the contents, skipping the first line, to a new CSV file.

At the code level, this means the program will need to do the following:

1. Loop over a list of files from `os.listdir()`, skipping the non-CSV files.
2. Create a CSV reader object and read in the contents of the file, using the `line_num` attribute to figure out which line to skip.
3. Create a CSV `writer` object and write out the read-in data to the new file.

For this project, open a new file editor window and save it as *removeCsvHeader.py*.

## Step 1: Loop Through Each CSV File

The first thing your program needs to do is loop over a list of all CSV filenames for the current working directory. Make your *removeCsvHeader.py* look like this:

```
#! python3
# removeCsvHeader.py - Removes the header from all CSV files in the current
```

```
# working directory.

import csv, os

os.makedirs('headerRemoved', exist_ok=True)

# Loop through every file in the current working directory.
for csvFilename in os.listdir('.'):
    if not csvFilename.endswith('.csv'):
        ❶ continue    # skip non-csv files

    print('Removing header from ' + csvFilename + '...')

    # TODO: Read the CSV file in (skipping first row).

    # TODO: Write out the CSV file.
```

The os.makedirs() call will create a headerRemoved folder where all the headless CSV files will be written. A for loop on os.listdir('.') gets you partway there, but it will loop over *all* files in the working directory, so you'll need to add some code at the start of the loop that skips filenames that don't end with .csv. The continue statement ❶ makes the for loop move on to the next filename when it comes across a non-CSV file.

Just so there's *some* output as the program runs, print out a message saying which CSV file the program is working on. Then, add some TODO comments for what the rest of the program should do.

## Step 2: Read in the CSV File

The program doesn't remove the first line from the CSV file. Rather, it creates a new copy of the CSV file without the first line. Since the copy's filename is the same as the original filename, the copy will overwrite the original.

The program will need a way to track whether it is currently looping on the first row. Add the following to *removeCsvHeader.py*.

```
#! python3
# removeCsvHeader.py - Removes the header from all CSV files in the current
# working directory.

--snip--

    # Read the CSV file in (skipping first row).
    csvRows = []
    csvFileObj = open(csvFilename)
    readerObj = csv.reader(csvFileObj)
    for row in readerObj:
        if readerObj.line_num == 1:
            continue    # skip first row
        csvRows.append(row)
    csvFileObj.close()

    # TODO: Write out the CSV file.
```

The reader object's line_num attribute can be used to determine which line in the CSV file it is currently reading. Another for loop will loop over the rows returned from the CSV reader object, and all rows but the first will be appended to csvRows.

As the for loop iterates over each row, the code checks whether readerObj.line_num is set to 1. If so, it executes a continue to move on to the next row without appending it to csvRows. For every row afterward, the condition will be always be False, and the row will be appended to csvRows.

### Step 3: Write Out the CSV File Without the First Row

Now that csvRows contains all rows but the first row, the list needs to be written out to a CSV file in the *headerRemoved* folder. Add the following to *removeCsvHeader.py*:

```
#! python3
# removeCsvHeader.py - Removes the header from all CSV files in the current
# working directory.
--snip--

# Loop through every file in the current working directory.
❶ for csvFilename in os.listdir('.'):
    if not csvFilename.endswith('.csv'):
        continue    # skip non-CSV files

    --snip--

    # Write out the CSV file.
    csvFileObj = open(os.path.join('headerRemoved', csvFilename), 'w',
                newline='')
    csvWriter = csv.writer(csvFileObj)
    for row in csvRows:
        csvWriter.writerow(row)
    csvFileObj.close()
```

The CSV writer object will write the list to a CSV file in headerRemoved using csvFilename (which we also used in the CSV reader). This will overwrite the original file.

Once we create the writer object, we loop over the sublists stored in csvRows and write each sublist to the file.

After the code is executed, the outer for loop ❶ will loop to the next filename from os.listdir('.'). When that loop is finished, the program will be complete.

To test your program, download *removeCsvHeader.zip* from *https://nostarch .com/automatestuff2/* and unzip it to a folder. Run the *removeCsvHeader.py* program in that folder. The output will look like this:

```
Removing header from NAICS_data_1048.csv...
Removing header from NAICS_data_1218.csv...
--snip--
```

```
Removing header from NAICS_data_9834.csv...
Removing header from NAICS_data_9986.csv...
```

This program should print a filename each time it strips the first line from a CSV file.

### Ideas for Similar Programs

The programs that you could write for CSV files are similar to the kinds you could write for Excel files, since they're both spreadsheet files. You could write programs to do the following:

- Compare data between different rows in a CSV file or between multiple CSV files.
- Copy specific data from a CSV file to an Excel file, or vice versa.
- Check for invalid data or formatting mistakes in CSV files and alert the user to these errors.
- Read data from a CSV file as input for your Python programs.

## JSON and APIs

JavaScript Object Notation is a popular way to format data as a single human-readable string. JSON is the native way that JavaScript programs write their data structures and usually resembles what Python's pprint() function would produce. You don't need to know JavaScript in order to work with JSON-formatted data.

Here's an example of data formatted as JSON:

```
{"name": "Zophie", "isCat": true,
 "miceCaught": 0, "napsTaken": 37.5,
 "felineIQ": null}
```

JSON is useful to know, because many websites offer JSON content as a way for programs to interact with the website. This is known as providing an *application programming interface (API)*. Accessing an API is the same as accessing any other web page via a URL. The difference is that the data returned by an API is formatted (with JSON, for example) for machines; APIs aren't easy for people to read.

Many websites make their data available in JSON format. Facebook, Twitter, Yahoo, Google, Tumblr, Wikipedia, Flickr, Data.gov, Reddit, IMDb, Rotten Tomatoes, LinkedIn, and many other popular sites offer APIs for programs to use. Some of these sites require registration, which is almost always free. You'll have to find documentation for what URLs your program needs to request in order to get the data you want, as well as the general format of the JSON data structures that are returned. This documentation should be provided by whatever site is offering the API; if they have a "Developers" page, look for the documentation there.

Using APIs, you could write programs that do the following:

- Scrape raw data from websites. (Accessing APIs is often more convenient than downloading web pages and parsing HTML with Beautiful Soup.)
- Automatically download new posts from one of your social network accounts and post them to another account. For example, you could take your Tumblr posts and post them to Facebook.
- Create a "movie encyclopedia" for your personal movie collection by pulling data from IMDb, Rotten Tomatoes, and Wikipedia and putting it into a single text file on your computer.

You can see some examples of JSON APIs in the resources at *https://nostarch.com/automatestuff2/*.

JSON isn't the only way to format data into a human-readable string. There are many others, including XML (eXtensible Markup Language), TOML (Tom's Obvious, Minimal Language), YML (Yet another Markup Language), INI (Initialization), or even the outdated ASN.1 (Abstract Syntax Notation One) formats, all of which provide a structure for representing data as human-readable text. This book won't cover these, because JSON has quickly become the most widely used alternate format, but there are third-party Python modules that readily handle them.

# The json Module

Python's json module handles all the details of translating between a string with JSON data and Python values for the json.loads() and json.dumps() functions. JSON can't store *every* kind of Python value. It can contain values of only the following data types: strings, integers, floats, Booleans, lists, dictionaries, and NoneType. JSON cannot represent Python-specific objects, such as File objects, CSV reader or writer objects, Regex objects, or Selenium WebElement objects.

## Reading JSON with the loads() Function

To translate a string containing JSON data into a Python value, pass it to the json.loads() function. (The name means "load string," not "loads.") Enter the following into the interactive shell:

```
>>> stringOfJsonData = '{"name": "Zophie", "isCat": true, "miceCaught": 0,
"felineIQ": null}'
>>> import json
>>> jsonDataAsPythonValue = json.loads(stringOfJsonData)
>>> jsonDataAsPythonValue
{'isCat': True, 'miceCaught': 0, 'name': 'Zophie', 'felineIQ': None}
```

After you import the json module, you can call loads() and pass it a string of JSON data. Note that JSON strings always use double quotes. It will return that data as a Python dictionary. Python dictionaries are not ordered, so the key-value pairs may appear in a different order when you print `jsonDataAsPythonValue`.

### Writing JSON with the dumps() Function

The `json.dumps()` function (which means "dump string," not "dumps") will translate a Python value into a string of JSON-formatted data. Enter the following into the interactive shell:

```
>>> pythonValue = {'isCat': True, 'miceCaught': 0, 'name': 'Zophie',
'felineIQ': None}
>>> import json
>>> stringOfJsonData = json.dumps(pythonValue)
>>> stringOfJsonData
'{"isCat": true, "felineIQ": null, "miceCaught": 0, "name": "Zophie" }'
```

The value can only be one of the following basic Python data types: dictionary, list, integer, float, string, Boolean, or `None`.

# Project: Fetching Current Weather Data

Checking the weather seems fairly trivial: Open your web browser, click the address bar, type the URL to a weather website (or search for one and then click the link), wait for the page to load, look past all the ads, and so on.

Actually, there are a lot of boring steps you could skip if you had a program that downloaded the weather forecast for the next few days and printed it as plaintext. This program uses the `requests` module from Chapter 12 to download data from the web.

Overall, the program does the following:

1. Reads the requested location from the command line
2. Downloads JSON weather data from OpenWeatherMap.org
3. Converts the string of JSON data to a Python data structure
4. Prints the weather for today and the next two days

So the code will need to do the following:

1. Join strings in `sys.argv` to get the location.
2. Call `requests.get()` to download the weather data.
3. Call `json.loads()` to convert the JSON data to a Python data structure.
4. Print the weather forecast.

For this project, open a new file editor window and save it as *getOpen Weather.py*. Then visit *https://openweathermap.org/api/* in your browser and sign up for a free account to obtain an *API key*, also called an app ID, which for the OpenWeatherMap service is a string code that looks something like `'30144aba38018987d84710d0e319281e'`. You don't need to pay for this service unless you plan on making more than 60 API calls per minute. Keep the API key secret; anyone who knows it can write scripts that use your account's usage quota.

## Step 1: Get Location from the Command Line Argument

The input for this program will come from the command line. Make *getOpenWeather.py* look like this:

```
#! python3
# getOpenWeather.py - Prints the weather for a location from the command line.

APPID = 'YOUR_APPID_HERE'

import json, requests, sys

# Compute location from command line arguments.
if len(sys.argv) < 2:
    print('Usage: getOpenWeather.py city_name, 2-letter_country_code')
    sys.exit()
location = ' '.join(sys.argv[1:])

# TODO: Download the JSON data from OpenWeatherMap.org's API.

# TODO: Load JSON data into a Python variable.
```

In Python, command line arguments are stored in the sys.argv list. The APPID variable should be set to the API key for your account. Without this key, your requests to the weather service will fail. After the #! shebang line and import statements, the program will check that there is more than one command line argument. (Recall that sys.argv will always have at least one element, sys.argv[0], which contains the Python script's filename.) If there is only one element in the list, then the user didn't provide a location on the command line, and a "usage" message will be provided to the user before the program ends.

The OpenWeatherMap service requires that the query be formatted as the city name, a comma, and a two-letter country code (like "US" for the United States). You can find a list of these codes at *https://en.wikipedia.org /wiki/ISO_3166-1_alpha-2*. Our script displays the weather for the first city listed in the retrieved JSON text. Unfortunately, cities that share a name, like Portland, Oregon, and Portland, Maine, will both be included, though the JSON text will include longitude and latitude information to differentiate between the cities.

Command line arguments are split on spaces. The command line argument San Francisco, US would make sys.argv hold ['getOpenWeather.py', 'San', 'Francisco,', 'US']. Therefore, call the join() method to join all the strings except for the first in sys.argv. Store this joined string in a variable named location.

### Step 2: Download the JSON Data

*OpenWeatherMap.org* provides real-time weather information in JSON format. First you must sign up for a free API key on the site. (This key is used to limit how frequently you make requests on their server, to keep their bandwidth costs down.) Your program simply has to download the page at *https://api .openweathermap.org/data/2.5/forecast/daily?q=<Location>&cnt=3&APPID=<API key>*, where *<Location>* is the name of the city whose weather you want and *<API key>* is your personal API key. Add the following to *getOpenWeather.py*.

```python
#! python3
# getOpenWeather.py - Prints the weather for a location from the command line.

--snip--

# Download the JSON data from OpenWeatherMap.org's API.
url ='https://api.openweathermap.org/data/2.5/forecast/daily?q=%s&cnt=3&APPID=%s ' % (location,
APPID)
response = requests.get(url)
response.raise_for_status()

# Uncomment to see the raw JSON text:
#print(response.text)

# TODO: Load JSON data into a Python variable.
```

We have location from our command line arguments. To make the URL we want to access, we use the %s placeholder and insert whatever string is stored in location into that spot in the URL string. We store the result in url and pass url to requests.get(). The requests.get() call returns a Response object, which you can check for errors by calling raise_for_status(). If no exception is raised, the downloaded text will be in response.text.

### Step 3: Load JSON Data and Print Weather

The response.text member variable holds a large string of JSON-formatted data. To convert this to a Python value, call the json.loads() function. The JSON data will look something like this:

```
{'city': {'coord': {'lat': 37.7771, 'lon': -122.42},
          'country': 'United States of America',
          'id': '5391959',
          'name': 'San Francisco',
          'population': 0},
 'cnt': 3,
 'cod': '200',
 'list': [{'clouds': 0,
           'deg': 233,
           'dt': 1402344000,
           'humidity': 58,
           'pressure': 1012.23,
```

```
                'speed': 1.96,
                'temp': {'day': 302.29,
                         'eve': 296.46,
                         'max': 302.29,
                         'min': 289.77,
                         'morn': 294.59,
                         'night': 289.77},
                'weather': [{'description': 'sky is clear',
                             'icon': '01d',
--snip--
```

You can see this data by passing weatherData to pprint.pprint(). You may want to check *https://openweathermap.org/* for more documentation on what these fields mean. For example, the online documentation will tell you that the 302.29 after 'day' is the daytime temperature in Kelvin, not Celsius or Fahrenheit.

The weather descriptions you want are after 'main' and 'description'. To neatly print them out, add the following to *getOpenWeather.py*.

```
! python3
# getOpenWeather.py - Prints the weather for a location from the command line.

--snip--

# Load JSON data into a Python variable.
weatherData = json.loads(response.text)

# Print weather descriptions.
❶ w = weatherData['list']
print('Current weather in %s:' % (location))
print(w[0]['weather'][0]['main'], '-', w[0]['weather'][0]['description'])
print()
print('Tomorrow:')
print(w[1]['weather'][0]['main'], '-', w[1]['weather'][0]['description'])
print()
print('Day after tomorrow:')
print(w[2]['weather'][0]['main'], '-', w[2]['weather'][0]['description'])
```

Notice how the code stores weatherData['list'] in the variable w to save you some typing ❶. You use w[0], w[1], and w[2] to retrieve the dictionaries for today, tomorrow, and the day after tomorrow's weather, respectively. Each of these dictionaries has a 'weather' key, which contains a list value. You're interested in the first list item, a nested dictionary with several more keys, at index 0. Here, we print the values stored in the 'main' and 'description' keys, separated by a hyphen.

When this program is run with the command line argument getOpen Weather.py San Francisco, CA, the output looks something like this:

```
Current weather in San Francisco, CA:
Clear - sky is clear
```

```
Tomorrow:
Clouds - few clouds

Day after tomorrow:
Clear - sky is clear
```

(The weather is one of the reasons I like living in San Francisco!)

### Ideas for Similar Programs

Accessing weather data can form the basis for many types of programs. You can create similar programs to do the following:

- Collect weather forecasts for several campsites or hiking trails to see which one will have the best weather.
- Schedule a program to regularly check the weather and send you a frost alert if you need to move your plants indoors. (Chapter 17 covers scheduling, and Chapter 18 explains how to send email.)
- Pull weather data from multiple sites to show all at once, or calculate and show the average of the multiple weather predictions.

## Summary

CSV and JSON are common plaintext formats for storing data. They are easy for programs to parse while still being human readable, so they are often used for simple spreadsheets or web app data. The csv and json modules greatly simplify the process of reading and writing to CSV and JSON files.

The last few chapters have taught you how to use Python to parse information from a wide variety of file formats. One common task is taking data from a variety of formats and parsing it for the particular information you need. These tasks are often specific to the point that commercial software is not optimally helpful. By writing your own scripts, you can make the computer handle large amounts of data presented in these formats.

In Chapter 18, you'll break away from data formats and learn how to make your programs communicate with you by sending emails and text messages.

## Practice Questions

1. What are some features Excel spreadsheets have that CSV spreadsheets don't?

2. What do you pass to csv.reader() and csv.writer() to create reader and writer objects?

3. What modes do File objects for reader and writer objects need to be opened in?

4. What method takes a list argument and writes it to a CSV file?

5. What do the delimiter and lineterminator keyword arguments do?

6. What function takes a string of JSON data and returns a Python data structure?

7. What function takes a Python data structure and returns a string of JSON data?

# Practice Project

For practice, write a program that does the following.

## Excel-to-CSV Converter

Excel can save a spreadsheet to a CSV file with a few mouse clicks, but if you had to convert hundreds of Excel files to CSVs, it would take hours of clicking. Using the openpyxl module from Chapter 12, write a program that reads all the Excel files in the current working directory and outputs them as CSV files.

A single Excel file might contain multiple sheets; you'll have to create one CSV file per *sheet*. The filenames of the CSV files should be *<excel filename>_<sheet title>.csv*, where *<excel filename>* is the filename of the Excel file without the file extension (for example, 'spam_data', not 'spam_data.xlsx') and *<sheet title>* is the string from the Worksheet object's title variable.

This program will involve many nested for loops. The skeleton of the program will look something like this:

```
for excelFile in os.listdir('.'):
    # Skip non-xlsx files, load the workbook object.
    for sheetName in wb.get_sheet_names():
        # Loop through every sheet in the workbook.
        sheet = wb.get_sheet_by_name(sheetName)

        # Create the CSV filename from the Excel filename and sheet title.
        # Create the csv.writer object for this CSV file.

        # Loop through every row in the sheet.
        for rowNum in range(1, sheet.max_row + 1):
            rowData = []    # append each cell to this list
            # Loop through each cell in the row.
            for colNum in range(1, sheet.max_column + 1):
                # Append each cell's data to rowData.

            # Write the rowData list to the CSV file.

        csvFile.close()
```

Download the ZIP file *excelSpreadsheets.zip* from *https://nostarch.com /automatestuff2/* and unzip the spreadsheets into the same directory as your program. You can use these as the files to test the program on.

# 17

## KEEPING TIME, SCHEDULING TASKS, AND LAUNCHING PROGRAMS

Running programs while you're sitting at your computer is fine, but it's also useful to have programs run without your direct supervision. Your computer's clock can schedule programs to run code at some specified time and date or at regular intervals. For example, your program could scrape a website every hour to check for changes or do a CPU-intensive task at 4 AM while you sleep. Python's time and datetime modules provide these functions.

You can also write programs that launch other programs on a schedule by using the subprocess and threading modules. Often, the fastest way to program is to take advantage of applications that other people have already written.

# The time Module

Your computer's system clock is set to a specific date, time, and time zone. The built-in time module allows your Python programs to read the system clock for the current time. The time.time() and time.sleep() functions are the most useful in the time module.

## The time.time() Function

The *Unix epoch* is a time reference commonly used in programming: 12 AM on January 1, 1970, Coordinated Universal Time (UTC). The time.time() function returns the number of seconds since that moment as a float value. (Recall that a float is just a number with a decimal point.) This number is called an *epoch timestamp*. For example, enter the following into the interactive shell:

```
>>> import time
>>> time.time()
1543813875.3518236
```

Here I'm calling time.time() on December 2, 2018, at 9:11 PM Pacific Standard Time. The return value is how many seconds have passed between the Unix epoch and the moment time.time() was called.

Epoch timestamps can be used to *profile* code, that is, measure how long a piece of code takes to run. If you call time.time() at the beginning of the code block you want to measure and again at the end, you can subtract the first timestamp from the second to find the elapsed time between those two calls. For example, open a new file editor tab and enter the following program:

```
  import time
❶ def calcProd():
      # Calculate the product of the first 100,000 numbers.
      product = 1
      for i in range(1, 100000):
          product = product * i
      return product

❷ startTime = time.time()
  prod = calcProd()
❸ endTime = time.time()
❹ print('The result is %s digits long.' % (len(str(prod))))
❺ print('Took %s seconds to calculate.' % (endTime - startTime))
```

At ❶, we define a function calcProd() to loop through the integers from 1 to 99,999 and return their product. At ❷, we call time.time() and store it in startTime. Right after calling calcProd(), we call time.time() again and store it in endTime ❸. We end by printing the length of the product returned by calcProd() ❹ and how long it took to run calcProd() ❺.

Save this program as *calcProd.py* and run it. The output will look something like this:

```
The result is 456569 digits long.
Took 2.844162940979004 seconds to calculate.
```

**NOTE** *Another way to profile your code is to use the* `cProfile.run()` *function, which provides a much more informative level of detail than the simple* `time.time()` *technique. The* `cProfile.run()` *function is explained at* https://docs.python.org/3/library /profile.html.

The return value from `time.time()` is useful, but not human-readable. The `time.ctime()` function returns a string description of the current time. You can also optionally pass the number of seconds since the Unix epoch, as returned by `time.time()`, to get a string value of that time. Enter the following into the interactive shell:

```
>>> import time
>>> time.ctime()
'Mon Jun 15 14:00:38 2020'
>>> thisMoment = time.time()
>>> time.ctime(thisMoment)
'Mon Jun 15 14:00:45 2020'
```

## The time.sleep() Function

If you need to pause your program for a while, call the `time.sleep()` function and pass it the number of seconds you want your program to stay paused. Enter the following into the interactive shell:

```
>>> import time
>>> for i in range(3):
❶     print('Tick')
❷     time.sleep(1)
❸     print('Tock')
❹     time.sleep(1)

Tick
Tock
Tick
Tock
Tick
Tock
❺ >>> time.sleep(5)
```

The `for` loop will print Tick ❶, pause for 1 second ❷, print Tock ❸, pause for 1 second ❹, print Tick, pause, and so on until Tick and Tock have each been printed three times.

The time.sleep() function will *block*—that is, it will not return and release your program to execute other code—until after the number of seconds you passed to time.sleep() has elapsed. For example, if you enter time.sleep(5) ❺, you'll see that the next prompt (>>>) doesn't appear until 5 seconds have passed.

## Rounding Numbers

When working with times, you'll often encounter float values with many digits after the decimal. To make these values easier to work with, you can shorten them with Python's built-in round() function, which rounds a float to the precision you specify. Just pass in the number you want to round, plus an optional second argument representing how many digits after the decimal point you want to round it to. If you omit the second argument, round() rounds your number to the nearest whole integer. Enter the following into the interactive shell:

```
>>> import time
>>> now = time.time()
>>> now
1543814036.6147408
>>> round(now, 2)
1543814036.61
>>> round(now, 4)
1543814036.6147
>>> round(now)
1543814037
```

After importing time and storing time.time() in now, we call round(now, 2) to round now to two digits after the decimal, round(now, 4) to round to four digits after the decimal, and round(now) to round to the nearest integer.

## Project: Super Stopwatch

Say you want to track how much time you spend on boring tasks you haven't automated yet. You don't have a physical stopwatch, and it's surprisingly difficult to find a free stopwatch app for your laptop or smartphone that isn't covered in ads and doesn't send a copy of your browser history to marketers. (It says it can do this in the license agreement you agreed to. You did read the license agreement, didn't you?) You can write a simple stopwatch program yourself in Python.

At a high level, here's what your program will do:

1. Track the amount of time elapsed between presses of the ENTER key, with each key press starting a new "lap" on the timer.
2. Print the lap number, total time, and lap time.

This means your code will need to do the following:

1. Find the current time by calling `time.time()` and store it as a timestamp at the start of the program, as well as at the start of each lap.
2. Keep a lap counter and increment it every time the user presses ENTER.
3. Calculate the elapsed time by subtracting timestamps.
4. Handle the `KeyboardInterrupt` exception so the user can press CTRL-C to quit.

Open a new file editor tab and save it as *stopwatch.py*.

## Step 1: Set Up the Program to Track Times

The stopwatch program will need to use the current time, so you'll want to import the time module. Your program should also print some brief instructions to the user before calling `input()`, so the timer can begin after the user presses ENTER. Then the code will start tracking lap times.

Enter the following code into the file editor, writing a TODO comment as a placeholder for the rest of the code:

```
#! python3
# stopwatch.py - A simple stopwatch program.

import time

# Display the program's instructions.
print('Press ENTER to begin. Afterward, press ENTER to "click" the stopwatch.
Press Ctrl-C to quit.')
input()                    # press Enter to begin
print('Started.')
startTime = time.time()    # get the first lap's start time
lastTime = startTime
lapNum = 1

# TODO: Start tracking the lap times.
```

Now that we've written the code to display the instructions, start the first lap, note the time, and set our lap count to 1.

## Step 2: Track and Print Lap Times

Now let's write the code to start each new lap, calculate how long the previous lap took, and calculate the total time elapsed since starting the stopwatch. We'll display the lap time and total time and increase the lap count for each new lap. Add the following code to your program:

```
#! python3
# stopwatch.py - A simple stopwatch program.

import time

--snip--
```

```
           # Start tracking the lap times.
❶ try:
    ❷ while True:
        input()
        ❸ lapTime = round(time.time() - lastTime, 2)
        ❹ totalTime = round(time.time() - startTime, 2)
        ❺ print('Lap #%s: %s (%s)' % (lapNum, totalTime, lapTime), end='')
        lapNum += 1
        lastTime = time.time() # reset the last lap time
❻ except KeyboardInterrupt:
    # Handle the Ctrl-C exception to keep its error message from displaying.
    print('\nDone.')
```

If the user presses CTRL-C to stop the stopwatch, the KeyboardInterrupt exception will be raised, and the program will crash if its execution is not a try statement. To prevent crashing, we wrap this part of the program in a try statement ❶. We'll handle the exception in the except clause ❻, so when CTRL-C is pressed and the exception is raised, the program execution moves to the except clause to print Done, instead of the KeyboardInterrupt error message. Until this happens, the execution is inside an infinite loop ❷ that calls input() and waits until the user presses ENTER to end a lap. When a lap ends, we calculate how long the lap took by subtracting the start time of the lap, lastTime, from the current time, time.time() ❸. We calculate the total time elapsed by subtracting the overall start time of the stopwatch, startTime, from the current time ❹.

Since the results of these time calculations will have many digits after the decimal point (such as 4.766272783279419), we use the round() function to round the float value to two digits at ❸ and ❹.

At ❺, we print the lap number, total time elapsed, and the lap time. Since the user pressing ENTER for the input() call will print a newline to the screen, pass end='' to the print() function to avoid double-spacing the output. After printing the lap information, we get ready for the next lap by adding 1 to the count lapNum and setting lastTime to the current time, which is the start time of the next lap.

## Ideas for Similar Programs

Time tracking opens up several possibilities for your programs. Although you can download apps to do some of these things, the benefit of writing programs yourself is that they will be free and not bloated with ads and useless features. You could write similar programs to do the following:

- Create a simple timesheet app that records when you type a person's name and uses the current time to clock them in or out.

- Add a feature to your program to display the elapsed time since a process started, such as a download that uses the requests module. (See Chapter 12.)

- Intermittently check how long a program has been running and offer the user a chance to cancel tasks that are taking too long.

# The datetime Module

The time module is useful for getting a Unix epoch timestamp to work with. But if you want to display a date in a more convenient format, or do arithmetic with dates (for example, figuring out what date was 205 days ago or what date is 123 days from now), you should use the datetime module.

The datetime module has its own datetime data type. datetime values represent a specific moment in time. Enter the following into the interactive shell:

```
>>> import datetime
❶ >>> datetime.datetime.now()
❷ datetime.datetime(2019, 2, 27, 11, 10, 49, 55, 53)
❸ >>> dt = datetime.datetime(2019, 10, 21, 16, 29, 0)
❹ >>> dt.year, dt.month, dt.day
(2019, 10, 21)
❺ >>> dt.hour, dt.minute, dt.second
(16, 29, 0)
```

Calling datetime.datetime.now() ❶ returns a datetime object ❷ for the current date and time, according to your computer's clock. This object includes the year, month, day, hour, minute, second, and microsecond of the current moment. You can also retrieve a datetime object for a specific moment by using the datetime.datetime() function ❸, passing it integers representing the year, month, day, hour, and second of the moment you want. These integers will be stored in the datetime object's year, month, day ❹, hour, minute, and second ❺ attributes.

A Unix epoch timestamp can be converted to a datetime object with the datetime.datetime.fromtimestamp() function. The date and time of the datetime object will be converted for the local time zone. Enter the following into the interactive shell:

```
>>> import datetime, time
>>> datetime.datetime.fromtimestamp(1000000)
datetime.datetime(1970, 1, 12, 5, 46, 40)
>>> datetime.datetime.fromtimestamp(time.time())
datetime.datetime(2019, 10, 21, 16, 30, 0, 604980)
```

Calling datetime.datetime.fromtimestamp() and passing it 1000000 returns a datetime object for the moment 1,000,000 seconds after the Unix epoch. Passing time.time(), the Unix epoch timestamp for the current moment, returns a datetime object for the current moment. So the expressions datetime .datetime.now() and datetime.datetime.fromtimestamp(time.time()) do the same thing; they both give you a datetime object for the present moment.

You can compare datetime objects with each other using comparison operators to find out which one precedes the other. The later datetime object is the "greater" value. Enter the following into the interactive shell:

```
❶ >>> halloween2019 = datetime.datetime(2019, 10, 31, 0, 0, 0)
❷ >>> newyears2020 = datetime.datetime(2020, 1, 1, 0, 0, 0)
>>> oct31_2019 = datetime.datetime(2019, 10, 31, 0, 0, 0)
```

❸ `>>> halloween2019 == oct31_2019`
`True`
❹ `>>> halloween2019 > newyears2020`
`False`
❺ `>>> newyears2020 > halloween2019`
`True`
`>>> newyears2020 != oct31_2019`
`True`

Make a datetime object for the first moment (midnight) of October 31, 2019, and store it in `halloween2019` ❶. Make a datetime object for the first moment of January 1, 2020, and store it in `newyears2020` ❷. Then make another object for midnight on October 31, 2019, and store it in `oct31_2019`. Comparing `halloween2019` and `oct31_2019` shows that they're equal ❸. Comparing `newyears2020` and `halloween2019` shows that `newyears2020` is greater (later) than `halloween2019` ❹ ❺.

### The timedelta Data Type

The datetime module also provides a `timedelta` data type, which represents a *duration* of time rather than a *moment* in time. Enter the following into the interactive shell:

❶ `>>> delta = datetime.timedelta(days=11, hours=10, minutes=9, seconds=8)`
❷ `>>> delta.days, delta.seconds, delta.microseconds`
`(11, 36548, 0)`
`>>> delta.total_seconds()`
`986948.0`
`>>> str(delta)`
`'11 days, 10:09:08'`

To create a `timedelta` object, use the `datetime.timedelta()` function. The `datetime.timedelta()` function takes keyword arguments `weeks`, `days`, `hours`, `minutes`, `seconds`, `milliseconds`, and `microseconds`. There is no `month` or `year` keyword argument, because "a month" or "a year" is a variable amount of time depending on the particular month or year. A `timedelta` object has the total duration represented in days, seconds, and microseconds. These numbers are stored in the `days`, `seconds`, and `microseconds` attributes, respectively. The `total_seconds()` method will return the duration in number of seconds alone. Passing a `timedelta` object to `str()` will return a nicely formatted, human-readable string representation of the object.

In this example, we pass keyword arguments to `datetime.delta()` to specify a duration of 11 days, 10 hours, 9 minutes, and 8 seconds, and store the returned `timedelta` object in `delta` ❶. This `timedelta` object's `days` attributes stores `11`, and its `seconds` attribute stores `36548` (10 hours, 9 minutes, and 8 seconds, expressed in seconds) ❷. Calling `total_seconds()` tells us that 11 days, 10 hours, 9 minutes, and 8 seconds is 986,948 seconds. Finally, passing the `timedelta` object to `str()` returns a string that plainly describes the duration.

The arithmetic operators can be used to perform *date arithmetic* on datetime values. For example, to calculate the date 1,000 days from now, enter the following into the interactive shell:

```
>>> dt = datetime.datetime.now()
>>> dt
datetime.datetime(2018, 12, 2, 18, 38, 50, 636181)
>>> thousandDays = datetime.timedelta(days=1000)
>>> dt + thousandDays
datetime.datetime(2021, 8, 28, 18, 38, 50, 636181)
```

First, make a datetime object for the current moment and store it in dt. Then make a timedelta object for a duration of 1,000 days and store it in thousandDays. Add dt and thousandDays together to get a datetime object for the date 1,000 days from now. Python will do the date arithmetic to figure out that 1,000 days after December 2, 2018, will be August 18, 2021. This is useful because when you calculate 1,000 days from a given date, you have to remember how many days are in each month and factor in leap years and other tricky details. The datetime module handles all of this for you.

timedelta objects can be added or subtracted with datetime objects or other timedelta objects using the + and - operators. A timedelta object can be multiplied or divided by integer or float values with the * and / operators. Enter the following into the interactive shell:

```
❶ >>> oct21st = datetime.datetime(2019, 10, 21, 16, 29, 0)
❷ >>> aboutThirtyYears = datetime.timedelta(days=365 * 30)
>>> oct21st
datetime.datetime(2019, 10, 21, 16, 29)
>>> oct21st - aboutThirtyYears
datetime.datetime(1989, 10, 28, 16, 29)
>>> oct21st - (2 * aboutThirtyYears)
datetime.datetime(1959, 11, 5, 16, 29)
```

Here we make a datetime object for October 21, 2019, ❶ and a timedelta object for a duration of about 30 years (we're assuming 365 days for each of those years) ❷. Subtracting aboutThirtyYears from oct21st gives us a datetime object for the date 30 years before October 21, 2019. Subtracting 2 * aboutThirtyYears from oct21st returns a datetime object for the date 60 years before October 21, 2019.

## Pausing Until a Specific Date

The time.sleep() method lets you pause a program for a certain number of seconds. By using a while loop, you can pause your programs until a specific date. For example, the following code will continue to loop until Halloween 2016:

```
import datetime
import time
halloween2016 = datetime.datetime(2016, 10, 31, 0, 0, 0)
while datetime.datetime.now() < halloween2016:
    time.sleep(1)
```

The `time.sleep(1)` call will pause your Python program so that the computer doesn't waste CPU processing cycles simply checking the time over and over. Rather, the `while` loop will just check the condition once per second and continue with the rest of the program after Halloween 2016 (or whenever you program it to stop).

## Converting datetime Objects into Strings

Epoch timestamps and `datetime` objects aren't very friendly to the human eye. Use the `strftime()` method to display a datetime object as a string. (The *f* in the name of the `strftime()` function stands for *format*.)

The `strftime()` method uses directives similar to Python's string formatting. Table 17-1 has a full list of `strftime()` directives.

**Table 17-1:** `strftime()` Directives

| `strftime()` directive | Meaning |
|---|---|
| %Y | Year with century, as in '2014' |
| %y | Year without century, '00' to '99' (1970 to 2069) |
| %m | Month as a decimal number, '01' to '12' |
| %B | Full month name, as in 'November' |
| %b | Abbreviated month name, as in 'Nov' |
| %d | Day of the month, '01' to '31' |
| %j | Day of the year, '001' to '366' |
| %w | Day of the week, '0' (Sunday) to '6' (Saturday) |
| %A | Full weekday name, as in 'Monday' |
| %a | Abbreviated weekday name, as in 'Mon' |
| %H | Hour (24-hour clock), '00' to '23' |
| %I | Hour (12-hour clock), '01' to '12' |
| %M | Minute, '00' to '59' |
| %S | Second, '00' to '59' |
| %p | 'AM' or 'PM' |
| %% | Literal '%' character |

Pass `strftime()` a custom format string containing formatting directives (along with any desired slashes, colons, and so on), and `strftime()` will return the datetime object's information as a formatted string. Enter the following into the interactive shell:

```
>>> oct21st = datetime.datetime(2019, 10, 21, 16, 29, 0)
>>> oct21st.strftime('%Y/%m/%d %H:%M:%S')
'2019/10/21 16:29:00'
>>> oct21st.strftime('%I:%M %p')
'04:29 PM'
>>> oct21st.strftime("%B of '%y")
"October of '19"
```

Here we have a datetime object for October 21, 2019, at 4:29 PM, stored in oct21st. Passing strftime() the custom format string '%Y/%m/%d %H:%M:%S' returns a string containing 2019, 10, and 21 separated by slashes and 16, 29, and 00 separated by colons. Passing '%I:%M% p' returns '04:29 PM', and passing "%B of '%y" returns "October of '19". Note that strftime() doesn't begin with datetime.datetime.

### Converting Strings into datetime Objects

If you have a string of date information, such as '2019/10/21 16:29:00' or 'October 21, 2019', and need to convert it to a datetime object, use the datetime.datetime.strptime() function. The strptime() function is the inverse of the strftime() method. A custom format string using the same directives as strftime() must be passed so that strptime() knows how to parse and understand the string. (The *p* in the name of the strptime() function stands for *parse*.)

Enter the following into the interactive shell:

```
❶ >>> datetime.datetime.strptime('October 21, 2019', '%B %d, %Y')
datetime.datetime(2019, 10, 21, 0, 0)
>>> datetime.datetime.strptime('2019/10/21 16:29:00', '%Y/%m/%d %H:%M:%S')
datetime.datetime(2019, 10, 21, 16, 29)
>>> datetime.datetime.strptime("October of '19", "%B of '%y")
datetime.datetime(2019, 10, 1, 0, 0)
>>> datetime.datetime.strptime("November of '63", "%B of '%y")
datetime.datetime(2063, 11, 1, 0, 0)
```

To get a datetime object from the string 'October 21, 2019', pass that string as the first argument to strptime() and the custom format string that corresponds to 'October 21, 2019' as the second argument ❶. The string with the date information must match the custom format string exactly, or Python will raise a ValueError exception.

# Review of Python's Time Functions

Dates and times in Python can involve quite a few different data types and functions. Here's a review of the three different types of values used to represent time:

- A Unix epoch timestamp (used by the time module) is a float or integer value of the number of seconds since 12 AM on January 1, 1970, UTC.
- A datetime object (of the datetime module) has integers stored in the attributes year, month, day, hour, minute, and second.
- A timedelta object (of the datetime module) represents a time duration, rather than a specific moment.

Here's a review of time functions and their parameters and return values:

`time.time()`   This function returns an epoch timestamp float value of the current moment.

`time.sleep(`*`seconds`*`)`   This function stops the program for the number of seconds specified by the *seconds* argument.

`datetime.datetime(`*`year, month, day, hour, minute, second`*`)`   This function returns a datetime object of the moment specified by the arguments. If *hour*, *minute*, or *second* arguments are not provided, they default to 0.

`datetime.datetime.now()`   This function returns a datetime object of the current moment.

`datetime.datetime.fromtimestamp(`*`epoch`*`)`   This function returns a datetime object of the moment represented by the *epoch* timestamp argument.

`datetime.timedelta(`*`weeks, days, hours, minutes, seconds, milliseconds, microseconds`*`)`   This function returns a timedelta object representing a duration of time. The function's keyword arguments are all optional and do not include *month* or *year*.

`total_seconds()`   This method for timedelta objects returns the number of seconds the timedelta object represents.

`strftime(`*`format`*`)`   This method returns a string of the time represented by the datetime object in a custom format that's based on the *format* string. See Table 17-1 for the format details.

`datetime.datetime.strptime(`*`time_string, format`*`)`   This function returns a datetime object of the moment specified by *time_string*, parsed using the *format* string argument. See Table 17-1 for the format details.

# Multithreading

To introduce the concept of multithreading, let's look at an example situation. Say you want to schedule some code to run after a delay or at a specific time. You could add code like the following at the start of your program:

```
import time, datetime

startTime = datetime.datetime(2029, 10, 31, 0, 0, 0)
while datetime.datetime.now() < startTime:
    time.sleep(1)

print('Program now starting on Halloween 2029')
--snip--
```

This code designates a start time of October 31, 2029, and keeps calling `time.sleep(1)` until the start time arrives. Your program cannot do anything while waiting for the loop of `time.sleep()` calls to finish; it just sits around until Halloween 2029. This is because Python programs by default have a single *thread* of execution.

To understand what a thread of execution is, remember the Chapter 2 discussion of flow control, when you imagined the execution of a program as placing your finger on a line of code in your program and moving to the next line or wherever it was sent by a flow control statement. A *single-threaded* program has only one finger. But a *multithreaded* program has multiple fingers. Each finger still moves to the next line of code as defined by the flow control statements, but the fingers can be at different places in the program, executing different lines of code at the same time. (All of the programs in this book so far have been single threaded.)

Rather than having all of your code wait until the time.sleep() function finishes, you can execute the delayed or scheduled code in a separate thread using Python's threading module. The separate thread will pause for the time.sleep calls. Meanwhile, your program can do other work in the original thread.

To make a separate thread, you first need to make a Thread object by calling the threading.Thread() function. Enter the following code in a new file and save it as *threadDemo.py*:

```
import threading, time
print('Start of program.')

❶ def takeANap():
      time.sleep(5)
      print('Wake up!')

❷ threadObj = threading.Thread(target=takeANap)
❸ threadObj.start()

print('End of program.')
```

At ❶, we define a function that we want to use in a new thread. To create a Thread object, we call threading.Thread() and pass it the keyword argument target=takeANap ❷. This means the function we want to call in the new thread is takeANap(). Notice that the keyword argument is target=takeANap, not target =takeANap(). This is because you want to pass the takeANap() function itself as the argument, not call takeANap() and pass its return value.

After we store the Thread object created by threading.Thread() in threadObj, we call threadObj.start() ❸ to create the new thread and start executing the target function in the new thread. When this program is run, the output will look like this:

```
Start of program.
End of program.
Wake up!
```

This can be a bit confusing. If print('End of program.') is the last line of the program, you might think that it should be the last thing printed. The reason Wake up! comes after it is that when threadObj.start() is called, the target function for threadObj is run in a new thread of execution. Think of it as a second finger appearing at the start of the takeANap() function. The main thread continues to print('End of program.'). Meanwhile, the new thread that has been executing the time.sleep(5) call, pauses for 5 seconds. After it wakes from its 5-second nap, it prints 'Wake up!' and then returns from the takeANap() function. Chronologically, 'Wake up!' is the last thing printed by the program.

Normally a program terminates when the last line of code in the file has run (or the sys.exit() function is called). But *threadDemo.py* has two threads. The first is the original thread that began at the start of the program and ends after print('End of program.'). The second thread is created when threadObj.start() is called, begins at the start of the takeANap() function, and ends after takeANap() returns.

A Python program will not terminate until all its threads have terminated. When you ran *threadDemo.py*, even though the original thread had terminated, the second thread was still executing the time.sleep(5) call.

## *Passing Arguments to the Thread's Target Function*

If the target function you want to run in the new thread takes arguments, you can pass the target function's arguments to threading.Thread(). For example, say you wanted to run this print() call in its own thread:

```
>>> print('Cats', 'Dogs', 'Frogs', sep=' & ')
Cats & Dogs & Frogs
```

This print() call has three regular arguments, 'Cats', 'Dogs', and 'Frogs', and one keyword argument, sep=' & '. The regular arguments can be passed as a list to the args keyword argument in threading.Thread(). The keyword argument can be specified as a dictionary to the kwargs keyword argument in threading.Thread().

Enter the following into the interactive shell:

```
>>> import threading
>>> threadObj = threading.Thread(target=print, args=['Cats', 'Dogs', 'Frogs'],
kwargs={'sep': ' & '})
>>> threadObj.start()
Cats & Dogs & Frogs
```

To make sure the arguments 'Cats', 'Dogs', and 'Frogs' get passed to print() in the new thread, we pass args=['Cats', 'Dogs', 'Frogs'] to threading .Thread(). To make sure the keyword argument sep=' & ' gets passed to print() in the new thread, we pass kwargs={'sep': '& '} to threading.Thread().

The threadObj.start() call will create a new thread to call the print() function, and it will pass 'Cats', 'Dogs', and 'Frogs' as arguments and ' & ' for the sep keyword argument.

This is an incorrect way to create the new thread that calls print():

```
threadObj = threading.Thread(target=print('Cats', 'Dogs', 'Frogs', sep=' & '))
```

What this ends up doing is calling the print() function and passing its return value (print()'s return value is always None) as the target keyword argument. It *doesn't* pass the print() function itself. When passing arguments to a function in a new thread, use the threading.Thread() function's args and kwargs keyword arguments.

## Concurrency Issues

You can easily create several new threads and have them all running at the same time. But multiple threads can also cause problems called *concurrency issues*. These issues happen when threads read and write variables at the same time, causing the threads to trip over each other. Concurrency issues can be hard to reproduce consistently, making them hard to debug.

Multithreaded programming is its own wide subject and beyond the scope of this book. What you have to keep in mind is this: to avoid concurrency issues, never let multiple threads read or write the same variables. When you create a new Thread object, make sure its target function uses only local variables in that function. This will avoid hard-to-debug concurrency issues in your programs.

**NOTE** *A beginner's tutorial on multithreaded programming is available at* https://nostarch.com/automatestuff2/.

# Project: Multithreaded XKCD Downloader

In Chapter 12, you wrote a program that downloaded all of the XKCD comic strips from the XKCD website. This was a single-threaded program: it downloaded one comic at a time. Much of the program's running time was spent establishing the network connection to begin the download and writing the downloaded images to the hard drive. If you have a broadband internet connection, your single-threaded program wasn't fully utilizing the available bandwidth.

A multithreaded program that has some threads downloading comics while others are establishing connections and writing the comic image files to disk uses your internet connection more efficiently and downloads the collection of comics more quickly. Open a new file editor tab and save it as *threadedDownloadXkcd.py*. You will modify this program to add multithreading. The completely modified source code is available to download from *https://nostarch.com/automatestuff2/*.

## Step 1: Modify the Program to Use a Function

This program will mostly be the same downloading code from Chapter 12, so I'll skip the explanation for the requests and Beautiful Soup code. The main changes you need to make are importing the threading module and making a downloadXkcd() function, which takes starting and ending comic numbers as parameters.

For example, calling downloadXkcd(140, 280) would loop over the downloading code to download the comics at *https://xkcd.com/140/, https://xkcd.com/141/, https://xkcd.com/142/,* and so on, up to *https://xkcd.com/279/.* Each thread that you create will call downloadXkcd() and pass a different range of comics to download.

Add the following code to your *threadedDownloadXkcd.py* program:

```python
#! python3
# threadedDownloadXkcd.py - Downloads XKCD comics using multiple threads.

import requests, os, bs4, threading
❶ os.makedirs('xkcd', exist_ok=True)    # store comics in ./xkcd

❷ def downloadXkcd(startComic, endComic):
    ❸ for urlNumber in range(startComic, endComic):
        # Download the page.
        print('Downloading page https://xkcd.com/%s...' % (urlNumber))
        ❹ res = requests.get('https://xkcd.com/%s' % (urlNumber))
        res.raise_for_status()

        ❺ soup = bs4.BeautifulSoup(res.text, 'html.parser')

        # Find the URL of the comic image.
        ❻ comicElem = soup.select('#comic img')
        if comicElem == []:
            print('Could not find comic image.')
        else:
            ❼ comicUrl = comicElem[0].get('src')
            # Download the image.
            print('Downloading image %s...' % (comicUrl))
            ❽ res = requests.get('https:' + comicUrl)
            res.raise_for_status()

            # Save the image to ./xkcd.
            imageFile = open(os.path.join('xkcd', os.path.basename(comicUrl)),
'wb')
            for chunk in res.iter_content(100000):
                imageFile.write(chunk)
            imageFile.close()

# TODO: Create and start the Thread objects.
# TODO: Wait for all threads to end.
```

After importing the modules we need, we make a directory to store comics in ❶ and start defining downloadxkcd() ❷. We loop through all the numbers in the specified range ❸ and download each page ❹. We use Beautiful Soup to look through the HTML of each page ❺ and find the comic image ❻. If no comic image is found on a page, we print a message. Otherwise, we get the URL of the image ❼ and download the image ❽. Finally, we save the image to the directory we created.

### Step 2: Create and Start Threads

Now that we've defined `downloadXkcd()`, we'll create the multiple threads that each call `downloadXkcd()` to download different ranges of comics from the XKCD website. Add the following code to *threadedDownloadXkcd.py* after the `downloadXkcd()` function definition:

```python3
#! python3
# threadedDownloadXkcd.py - Downloads XKCD comics using multiple threads.

--snip--

# Create and start the Thread objects.
downloadThreads = []                # a list of all the Thread objects
for i in range(0, 140, 10):    # loops 14 times, creates 14 threads
    start = i
    end = i + 9
    if start == 0:
        start = 1 # There is no comic 0, so set it to 1.
    downloadThread = threading.Thread(target=downloadXkcd, args=(start, end))
    downloadThreads.append(downloadThread)
    downloadThread.start()
```

First we make an empy list `downloadThreads`; the list will help us keep track of the many `Thread` objects we'll create. Then we start our `for` loop. Each time through the loop, we create a `Thread` object with `threading.Thread()`, append the `Thread` object to the list, and call `start()` to start running `downloadXkcd()` in the new thread. Since the `for` loop sets the `i` variable from 0 to 140 at steps of 10, `i` will be set to 0 on the first iteration, 10 on the second iteration, 20 on the third, and so on. Since we pass `args=(start, end)` to `threading.Thread()`, the two arguments passed to `downloadXkcd()` will be 1 and 9 on the first iteration, 10 and 19 on the second iteration, 20 and 29 on the third, and so on.

As the `Thread` object's `start()` method is called and the new thread begins to run the code inside `downloadXkcd()`, the main thread will continue to the next iteration of the `for` loop and create the next thread.

### Step 3: Wait for All Threads to End

The main thread moves on as normal while the other threads we create download comics. But say there's some code you don't want to run in the main thread until all the threads have completed. Calling a `Thread` object's `join()` method will block until that thread has finished. By using a `for` loop to iterate over all the `Thread` objects in the `downloadThreads` list, the main thread can call the `join()` method on each of the other threads. Add the following to the bottom of your program:

```python3
#! python3
# threadedDownloadXkcd.py - Downloads XKCD comics using multiple threads.
```

```
--snip--

# Wait for all threads to end.
for downloadThread in downloadThreads:
    downloadThread.join()
print('Done.')
```

The 'Done.' string will not be printed until all of the join() calls have returned. If a Thread object has already completed when its join() method is called, then the method will simply return immediately. If you wanted to extend this program with code that runs only after all of the comics downloaded, you could replace the print('Done.') line with your new code.

## Launching Other Programs from Python

Your Python program can start other programs on your computer with the Popen() function in the built-in subprocess module. (The *P* in the name of the Popen() function stands for *process*.) If you have multiple instances of an application open, each of those instances is a separate process of the same program. For example, if you open multiple windows of your web browser at the same time, each of those windows is a different process of the web browser program. See Figure 17-1 for an example of multiple calculator processes open at once.

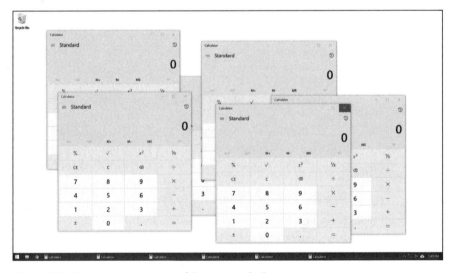

*Figure 17-1: Six running processes of the same calculator program*

Every process can have multiple threads. Unlike threads, a process cannot directly read and write another process's variables. If you think of a multithreaded program as having multiple fingers following source code, then having multiple processes of the same program open is like having a friend with a separate copy of the program's source code. You are both independently executing the same program.

If you want to start an external program from your Python script, pass the program's filename to subprocess.Popen(). (On Windows, right-click the application's **Start** menu item and select **Properties** to view the application's filename. On macOS, CTRL-click the application and select **Show Package Contents** to find the path to the executable file.) The Popen() function will then immediately return. Keep in mind that the launched program is not run in the same thread as your Python program.

On a Windows computer, enter the following into the interactive shell:

```
>>> import subprocess
>>> subprocess.Popen('C:\\Windows\\System32\\calc.exe')
<subprocess.Popen object at 0x0000000003055A58>
```

On Ubuntu Linux, you would enter the following:

```
>>> import subprocess
>>> subprocess.Popen('/snap/bin/gnome-calculator')
<subprocess.Popen object at 0x7f2bcf93b20>
```

On macOS, the process is slightly different. See "Opening Files with Default Applications" on page 409.

The return value is a Popen object, which has two useful methods: poll() and wait().

You can think of the poll() method as asking your driver "Are we there yet?" over and over until you arrive. The poll() method will return None if the process is still running at the time poll() is called. If the program has terminated, it will return the process's integer *exit code*. An exit code is used to indicate whether the process terminated without errors (an exit code of 0) or whether an error caused the process to terminate (a nonzero exit code—generally 1, but it may vary depending on the program).

The wait() method is like waiting until the driver has arrived at your destination. The wait() method will block until the launched process has terminated. This is helpful if you want your program to pause until the user finishes with the other program. The return value of wait() is the process's integer exit code.

On Windows, enter the following into the interactive shell. Note that the wait() call will block until you quit the launched MS Paint program.

```
>>> import subprocess
❶ >>> paintProc = subprocess.Popen('c:\\Windows\\System32\\mspaint.exe')
❷ >>> paintProc.poll() == None
True
❸ >>> paintProc.wait() # Doesn't return until MS Paint closes.
0
>>> paintProc.poll()
0
```

Here we open an MS Paint process ❶. While it's still running, we check whether poll() returns None ❷. It should, as the process is still running. Then we close the MS Paint program and call wait() on the terminated process ❸. Now wait() and poll()return 0, indicating that the process terminated without errors.

*Unlike* mspaint.exe, *if you run* calc.exe *on Windows 10 using* subprocess.Popen(), *you'll notice that* wait() *instantly returns even though the calculator app is still running. This is because* calc.exe *launches the calculator app and then instantly closes itself. Windows' calculator program is a "Trusted Microsoft Store app," and its specifics are beyond the scope of this book. Suffice it to say, programs can run in many application- and operating system–specific ways.*

### Passing Command Line Arguments to the Popen() Function

You can pass command line arguments to processes you create with Popen(). To do so, you pass a list as the sole argument to Popen(). The first string in this list will be the executable filename of the program you want to launch; all the subsequent strings will be the command line arguments to pass to the program when it starts. In effect, this list will be the value of sys.argv for the launched program.

Most applications with a graphical user interface (GUI) don't use command line arguments as extensively as command line–based or terminal-based programs do. But most GUI applications will accept a single argument for a file that the applications will immediately open when they start. For example, if you're using Windows, create a simple text file called *C:\Users \Al\hello.txt* and then enter the following into the interactive shell:

```
>>> subprocess.Popen(['C:\\Windows\\notepad.exe', 'C:\\Users\Al\\hello.txt'])
<subprocess.Popen object at 0x00000000032DCEB8>
```

This will not only launch the Notepad application but also have it immediately open the *C:\Users\Al\hello.txt* file.

### Task Scheduler, launchd, and cron

If you are computer savvy, you may know about Task Scheduler on Windows, launchd on macOS, or the cron scheduler on Linux. These well-documented and reliable tools all allow you to schedule applications to launch at specific times. If you'd like to learn more about them, you can find links to tutorials at *https://nostarch.com/automatestuff2/*.

Using your operating system's built-in scheduler saves you from writing your own clock-checking code to schedule your programs. However, use the time.sleep() function if you just need your program to pause briefly. Or instead of using the operating system's scheduler, your code can loop until a certain date and time, calling time.sleep(1) each time through the loop.

## Opening Websites with Python

The webbrowser.open() function can launch a web browser from your program to a specific website, rather than opening the browser application with subprocess.Popen(). See "Project: *mapIt.py* with the webbrowser Module" on page 268 for more details.

## Running Other Python Scripts

You can launch a Python script from Python just like any other application. Simply pass the *python.exe* executable to Popen() and the filename of the *.py* script you want to run as its argument. For example, the following would run the *hello.py* script from Chapter 1:

```
>>> subprocess.Popen(['C:\\Users\\<YOUR USERNAME>\\AppData\\Local\\Programs\\
Python\\Python38\\python.exe', 'hello.py'])
<subprocess.Popen object at 0x000000000331CF28>
```

Pass Popen() a list containing a string of the Python executable's path and a string of the script's filename. If the script you're launching needs command line arguments, add them to the list after the script's filename. The location of the Python executable on Windows is *C:\Users\<YOUR USERNAME>\AppData\Local\Programs\Python\Python38\python.exe*. On macOS, it is */Library/Frameworks/Python.framework/Versions/3.8/bin/python3*. On Linux, it is */usr/bin/python3.8*.

Unlike importing the Python program as a module, when your Python program launches another Python program, the two are run in separate processes and will not be able to share each other's variables.

## Opening Files with Default Applications

Double-clicking a *.txt* file on your computer will automatically launch the application associated with the *.txt* file extension. Your computer will have several of these file extension associations set up already. Python can also open files this way with Popen().

Each operating system has a program that performs the equivalent of double-clicking a document file to open it. On Windows, this is the start program. On macOS, this is the open program. On Ubuntu Linux, this is the see program. Enter the following into the interactive shell, passing 'start', 'open', or 'see' to Popen() depending on your system:

```
>>> fileObj = open('hello.txt', 'w')
>>> fileObj.write('Hello, world!')
12
>>> fileObj.close()
>>> import subprocess
>>> subprocess.Popen(['start', 'hello.txt'], shell=True)
```

Here we write Hello, world! to a new *hello.txt* file. Then we call Popen(), passing it a list containing the program name (in this example, 'start' for Windows) and the filename. We also pass the shell=True keyword argument, which is needed only on Windows. The operating system knows all of the file associations and can figure out that it should launch, say, *Notepad.exe* to handle the *hello.txt* file.

On macOS, the open program is used for opening both document files and programs. Enter the following into the interactive shell if you have a Mac:

```
>>> subprocess.Popen(['open', '/Applications/Calculator.app/'])
<subprocess.Popen object at 0x10202ff98>
```

The Calculator app should open.

## Project: Simple Countdown Program

Just like it's hard to find a simple stopwatch application, it can be hard to find a simple countdown application. Let's write a countdown program that plays an alarm at the end of the countdown.

At a high level, here's what your program will do:

1. Count down from 60.
2. Play a sound file (*alarm.wav*) when the countdown reaches zero.

This means your code will need to do the following:

1. Pause for 1 second in between displaying each number in the countdown by calling time.sleep().
2. Call subprocess.Popen() to open the sound file with the default application.

Open a new file editor tab and save it as *countdown.py*.

### Step 1: Count Down

This program will require the time module for the time.sleep() function and the subprocess module for the subprocess.Popen() function. Enter the following code and save the file as *countdown.py*:

```
#! python3
# countdown.py - A simple countdown script.

import time, subprocess

❶ timeLeft = 60
   while timeLeft > 0:
❷     print(timeLeft, end='')
❸     time.sleep(1)
```

```
❹ timeLeft = timeLeft - 1

# TODO: At the end of the countdown, play a sound file.
```

After importing `time` and `subprocess`, make a variable called `timeLeft` to hold the number of seconds left in the countdown ❶. It can start at 60—or you can change the value here to whatever you need, or even have it get set from a command line argument.

In a `while` loop, you display the remaining count ❷, pause for 1 second ❸, and then decrement the `timeLeft` variable ❹ before the loop starts over again. The loop will keep looping as long as `timeLeft` is greater than 0. After that, the countdown will be over.

## Step 2: Play the Sound File

While there are third-party modules to play sound files of various formats, the quick and easy way is to just launch whatever application the user already uses to play sound files. The operating system will figure out from the *.wav* file extension which application it should launch to play the file. This *.wav* file could easily be some other sound file format, such as *.mp3* or *.ogg*.

You can use any sound file that is on your computer to play at the end of the countdown, or you can download *alarm.wav* from *https://nostarch.com/automatestuff2/*.

Add the following to your code:

```
#! python3
# countdown.py - A simple countdown script.

import time, subprocess

--snip--

# At the end of the countdown, play a sound file.
subprocess.Popen(['start', 'alarm.wav'], shell=True)
```

After the `while` loop finishes, *alarm.wav* (or the sound file you choose) will play to notify the user that the countdown is over. On Windows, be sure to include `'start'` in the list you pass to `Popen()` and pass the keyword argument `shell=True`. On macOS, pass `'open'` instead of `'start'` and remove `shell=True`.

Instead of playing a sound file, you could save a text file somewhere with a message like *Break time is over!* and use `Popen()` to open it at the end of the countdown. This will effectively create a pop-up window with a message. Or you could use the `webbrowser.open()` function to open a specific website at the end of the countdown. Unlike some free countdown application you'd find online, your own countdown program's alarm can be anything you want!

### Ideas for Similar Programs

A countdown is a simple delay before continuing the program's execution. This can also be used for other applications and features, such as the following:

- Use `time.sleep()` to give the user a chance to press CTRL-C to cancel an action, such as deleting files. Your program can print a "Press CTRL-C to cancel" message and then handle any `KeyboardInterrupt` exceptions with `try` and `except` statements.

- For a long-term countdown, you can use `timedelta` objects to measure the number of days, hours, minutes, and seconds until some point (a birthday? an anniversary?) in the future.

## Summary

The Unix epoch (January 1, 1970, at midnight, UTC) is a standard reference time for many programming languages, including Python. While the `time.time()` function module returns an epoch timestamp (that is, a float value of the number of seconds since the Unix epoch), the `datetime` module is better for performing date arithmetic and formatting or parsing strings with date information.

The `time.sleep()` function will block (that is, not return) for a certain number of seconds. It can be used to add pauses to your program. But if you want to schedule your programs to start at a certain time, the instructions at *https://nostarch.com/automatestuff2/* can tell you how to use the scheduler already provided by your operating system.

The `threading` module is used to create multiple threads, which is useful when you need to download multiple files or do other tasks simultaneously. But make sure the thread reads and writes only local variables, or you might run into concurrency issues.

Finally, your Python programs can launch other applications with the `subprocess.Popen()` function. Command line arguments can be passed to the `Popen()` call to open specific documents with the application. Alternatively, you can use the `start`, `open`, or `see` program with `Popen()` to use your computer's file associations to automatically figure out which application to use to open a document. By using the other applications on your computer, your Python programs can leverage their capabilities for your automation needs.

## Practice Questions

1. What is the Unix epoch?
2. What function returns the number of seconds since the Unix epoch?
3. How can you pause your program for exactly 5 seconds?

4. What does the round() function return?

5. What is the difference between a datetime object and a timedelta object?

6. Using the datetime module, what day of the week was January 7, 2019?

7. Say you have a function named spam(). How can you call this function and run the code inside it in a separate thread?

8. What should you do to avoid concurrency issues with multiple threads?

## Practice Projects

For practice, write programs that do the following.

### Prettified Stopwatch

Expand the stopwatch project from this chapter so that it uses the rjust() and ljust() string methods to "prettify" the output. (These methods were covered in Chapter 6.) Instead of output such as this:

```
Lap #1: 3.56 (3.56)
Lap #2: 8.63 (5.07)
Lap #3: 17.68 (9.05)
Lap #4: 19.11 (1.43)
```

. . . the output will look like this:

```
Lap # 1:   3.56 (  3.56)
Lap # 2:   8.63 (  5.07)
Lap # 3:  17.68 (  9.05)
Lap # 4:  19.11 (  1.43)
```

Note that you will need string versions of the lapNum, lapTime, and totalTime integer and float variables in order to call the string methods on them.

Next, use the pyperclip module introduced in Chapter 6 to copy the text output to the clipboard so the user can quickly paste the output to a text file or email.

### Scheduled Web Comic Downloader

Write a program that checks the websites of several web comics and automatically downloads the images if the comic was updated since the program's last visit. Your operating system's scheduler (Scheduled Tasks on Windows, launchd on macOS, and cron on Linux) can run your Python program once a day. The Python program itself can download the comic and then copy it to your desktop so that it is easy to find. This will free you from having to check the website yourself to see whether it has updated. (A list of web comics is available at *https://nostarch.com/automatestuff2/.*)

# 18

## SENDING EMAIL AND TEXT MESSAGES

Checking and replying to email is a huge time sink. Of course, you can't just write a program to handle all your email for you, since each message requires its own response. But you can still automate plenty of email-related tasks once you know how to write programs that can send and receive email.

For example, maybe you have a spreadsheet full of customer records and want to send each customer a different form letter depending on their age and location details. Commercial software might not be able to do this for you; fortunately, you can write your own program to send these emails, saving yourself a lot of time copying and pasting form emails.

You can also write programs to send emails and SMS texts to notify you of things even while you're away from your computer. If you're automating a task that takes a couple of hours to do, you don't want to go back to your computer every few minutes to check on the program's status. Instead, the program can just text your phone when it's done—freeing you to focus on more important things while you're away from your computer.

This chapter features the EZGmail module, a simple way to send and read emails from Gmail accounts, as well as a Python module for using the standard SMTP and IMAP email protocols.

**WARNING**    *I highly recommend you set up a separate email account for any scripts that send or receive emails. This will prevent bugs in your programs from affecting your personal email account (by deleting emails or accidentally spamming your contacts, for example). It's a good idea to first do a dry run by commenting out the code that actually sends or deletes emails and replacing it with a temporary print() call. This way you can test your program before running it for real.*

## Sending and Receiving Email with the Gmail API

Gmail owns close to a third of the email client market share, and most likely you have at least one Gmail email address. Because of additional security and anti-spam measures, it is easier to control a Gmail account through the *EZGmail module* than through smtplib and imapclient, discussed later in this chapter. EZGmail is a module I wrote that works on top of the official Gmail API and provides functions that make it easy to use Gmail from Python. You can find full details on EZGmail at *https://github.com/asweigart /ezgmail/*. EZGmail is not produced by or affiliated with Google; find the official Gmail API documentation at *https://developers.google.com/gmail/api /v1/reference/*.

To install EZGmail, run pip install --user --upgrade ezgmail on Windows (or use pip3 on macOS and Linux). The --upgrade option will ensure that you install the latest version of the package, which is necessary for interacting with a constantly changing online service like the Gmail API.

### Enabling the Gmail API

Before you write code, you must first sign up for a Gmail email account at *https://gmail.com/*. Then, go to *https://developers.google.com/gmail/api /quickstart/python/*, click the **Enable the Gmail API** button on that page, and fill out the form that appears.

After you've filled out the form, the page will present a link to the *credentials.json* file, which you'll need to download and place in the same folder as your *.py* file. The *credentials.json* file contains the Client ID and Client Secret information, which you should treat the same as your Gmail password and not share with anyone else.

Then, in the interactive shell, enter the following code:

```
>>> import ezgmail, os
>>> os.chdir(r'C:\path\to\credentials_json_file')
>>> ezgmail.init()
```

Make sure you set your current working directory to the same folder that *credentials.json* is in and that you're connected to the internet. The ezgmail.init() function will open your browser to a Google sign-in page.

Enter your Gmail address and password. The page may warn you "This app isn't verified," but this is fine; click **Advanced** and then **Go to Quickstart** (**unsafe**). (If you write Python scripts for others and don't want this warning appearing for them, you'll need to learn about Google's app verification process, which is beyond the scope of this book.) When the next page prompts you with "Quickstart wants to access your Google Account," click **Allow** and then close the browser.

A *token.json* file will be generated to give your Python scripts access to the Gmail account you entered. The browser will only open to the login page if it can't find an existing *token.json* file. With *credentials.json* and *token .json*, your Python scripts can send and read emails from your Gmail account without requiring you to include your Gmail password in your source code.

### Sending Mail from a Gmail Account

Once you have a *token.json* file, the EZGmail module should be able to send email with a single function call:

```
>>> import ezgmail
>>> ezgmail.send('recipient@example.com', 'Subject line', 'Body of the email')
```

If you want to attach files to your email, you can provide an extra list argument to the send() function:

```
>>> ezgmail.send('recipient@example.com', 'Subject line', 'Body of the email',
['attachment1.jpg', 'attachment2.mp3'])
```

Note that as part of its security and anti-spam features, Gmail might not send repeated emails with the exact same text (since these are likely spam) or emails that contain *.exe* or *.zip* file attachments (since they are likely viruses).

You can also supply the optional keyword arguments cc and bcc to send carbon copies and blind carbon copies:

```
>>> import ezgmail
>>> ezgmail.send('recipient@example.com', 'Subject line', 'Body of the email',
cc='friend@example.com', bcc='otherfriend@example.com,someoneelse@example.com')
```

If you need to remember which Gmail address the *token.json* file is configured for, you can examine ezgmail.EMAIL_ADDRESS. Note that this variable is populated only after ezgmail.init() or any other EZGmail function is called:

```
>>> import ezgmail
>>> ezgmail.init()
>>> ezgmail.EMAIL_ADDRESS
'example@gmail.com'
```

Be sure to treat the *token.json* file the same as your password. If someone else obtains this file, they can access your Gmail account (though they won't be able to change your Gmail password). To revoke previously issued *token .json* files, go to *https://security.google.com/settings/security/permissions?pli=1/* and revoke access to the Quickstart app. You will need to run `ezgmail.init()` and go through the login process again to obtain a new *token.json* file.

## Reading Mail from a Gmail Account

Gmail organizes emails that are replies to each other into conversation threads. When you log in to Gmail in your web browser or through an app, you're really looking at email threads rather than individual emails (even if the thread has only one email in it).

EZGmail has `GmailThread` and `GmailMessage` objects to represent conversation threads and individual emails, respectively. A `GmailThread` object has a `messages` attribute that holds a list of `GmailMessage` objects. The `unread()` function returns a list of `GmailThread` objects for all unread emails, which can then be passed to `ezgmail.summary()` to print a summary of the conversation threads in that list:

```
>>> import ezgmail
>>> unreadThreads = ezgmail.unread() # List of GmailThread objects.
>>> ezgmail.summary(unreadThreads)
Al, Jon - Do you want to watch RoboCop this weekend? - Dec 09
Jon - Thanks for stopping me from buying Bitcoin. - Dec 09
```

The `summary()` function is handy for displaying a quick summary of the email threads, but to access specific messages (and parts of messages), you'll want to examine the `messages` attribute of the `GmailThread` object. The `messages` attribute contains a list of the `GmailMessage` objects that make up the thread, and these have `subject`, `body`, `timestamp`, `sender`, and `recipient` attributes that describe the email:

```
>>> len(unreadThreads)
2
>>> str(unreadThreads[0])
"<GmailThread len=2 snippet= Do you want to watch RoboCop this weekend?'>"
>>> len(unreadThreads[0].messages)
2
>>> str(unreadThreads[0].messages[0])
"<GmailMessage from='Al Sweigart <al@inventwithpython.com>' to='Jon Doe
<example@gmail.com>' timestamp=datetime.datetime(2018, 12, 9, 13, 28, 48)
subject='RoboCop' snippet='Do you want to watch RoboCop this weekend?'>"
>>> unreadThreads[0].messages[0].subject
'RoboCop'
>>> unreadThreads[0].messages[0].body
'Do you want to watch RoboCop this weekend?\r\n'
>>> unreadThreads[0].messages[0].timestamp
datetime.datetime(2018, 12, 9, 13, 28, 48)
>>> unreadThreads[0].messages[0].sender
'Al Sweigart <al@inventwithpython.com>'
>>> unreadThreads[0].messages[0].recipient
'Jon Doe <example@gmail.com>'
```

Similar to the `ezgmail.unread()` function, the `ezgmail.recent()` function will return the 25 most recent threads in your Gmail account. You can pass an optional `maxResults` keyword argument to change this limit:

```
>>> recentThreads = ezgmail.recent()
>>> len(recentThreads)
25
>>> recentThreads = ezgmail.recent(maxResults=100)
>>> len(recentThreads)
46
```

## Searching Mail from a Gmail Account

In addition to using `ezgmail.unread()` and `ezgmail.recent()`, you can search for specific emails, the same way you would if you entered queries into the *https://gmail.com/* search box, by calling `ezgmail.search()`:

```
>>> resultThreads = ezgmail.search('RoboCop')
>>> len(resultThreads)
1
>>> ezgmail.summary(resultThreads)
Al, Jon - Do you want to watch RoboCop this weekend? - Dec 09
```

The previous `search()` call should yield the same results as if you had entered "RoboCop" into the search box, as in Figure 18-1.

*Figure 18-1: Searching for "RoboCop" emails at the Gmail website*

Like `unread()` and `recent()`, the `search()` function returns a list of `GmailThread` objects. You can also pass any of the special search operators that you can enter into the search box to the `search()` function, such as the following:

`'label:UNREAD'`   For unread emails

`'from:al@inventwithpython.com'`   For emails from *al@inventwithpython.com*

`'subject:hello'`   For emails with "hello" in the subject

`'has:attachment'`   For emails with file attachments

You can view a full list of search operators at *https://support.google.com/mail/answer/7190?hl=en/*.

## Downloading Attachments from a Gmail Account

The `GmailMessage` objects have an attachments attribute that is a list of filenames for the message's attached files. You can pass any of these names to a GmailMessage object's downloadAttachment() method to download the files. You can also download all of them at once with downloadAllAttachments(). By default, EZGmail saves attachments to the current working directory, but you can pass an additional downloadFolder keyword argument to downloadAttachment() and downloadAllAttachments() as well. For example:

```
>>> import ezgmail
>>> threads = ezgmail.search('vacation photos')
>>> threads[0].messages[0].attachments
['tulips.jpg', 'canal.jpg', 'bicycles.jpg']
>>> threads[0].messages[0].downloadAttachment('tulips.jpg')
>>> threads[0].messages[0].downloadAllAttachments(downloadFolder='vacat
ion2019')
['tulips.jpg', 'canal.jpg', 'bicycles.jpg']
```

If a file already exists with the attachment's filename, the downloaded attachment will automatically overwrite it.

EZGmail contains additional features, and you can find the full documentation at *https://github.com/asweigart/ezgmail/*.

## SMTP

Much as HTTP is the protocol used by computers to send web pages across the internet, *Simple Mail Transfer Protocol (SMTP)* is the protocol used for sending email. SMTP dictates how email messages should be formatted, encrypted, and relayed between mail servers and all the other details that your computer handles after you click Send. You don't need to know these technical details, though, because Python's smtplib module simplifies them into a few functions.

SMTP just deals with sending emails to others. A different protocol, called IMAP, deals with retrieving emails sent to you and is described in "IMAP" on page 424.

In addition to SMTP and IMAP, most web-based email providers today have other security measures in place to protect against spam, phishing, and other malicious email usage. These measures prevent Python scripts from logging in to an email account with the smtplib and imapclient modules. However, many of these services have APIs and specific Python modules that allow scripts to access them. This chapter covers Gmail's module. For others, you'll need to consult their online documentation.

## Sending Email

You may be familiar with sending emails from Outlook or Thunderbird or through a website such as Gmail or Yahoo Mail. Unfortunately, Python doesn't offer you a nice graphical user interface like those services. Instead, you call functions to perform each major step of SMTP, as shown in the following interactive shell example.

*Don't enter this example in the interactive shell; it won't work, because* smtp.example
.com, bob@example.com, MY_SECRET_PASSWORD, *and* alice@example.com
*are just placeholders. This code is just an overview of the process of sending email
with Python.*

```
>>> import smtplib
>>> smtpObj = smtplib.SMTP('smtp.example.com', 587)
>>> smtpObj.ehlo()
(250, b'mx.example.com at your service, [216.172.148.131]\nSIZE 35882577\
n8BITMIME\nSTARTTLS\nENHANCEDSTATUSCODES\nCHUNKING')
>>> smtpObj.starttls()
(220, b'2.0.0 Ready to start TLS')
>>> smtpObj.login('bob@example.com', 'MY_SECRET_PASSWORD')
(235, b'2.7.0 Accepted')
>>> smtpObj.sendmail('bob@example.com', 'alice@example.com', 'Subject: So
long.\nDear Alice, so long and thanks for all the fish. Sincerely, Bob')
{}
>>> smtpObj.quit()
(221, b'2.0.0 closing connection ko10sm23097611pbd.52 - gsmtp')
```

In the following sections, we'll go through each step, replacing the
placeholders with your information to connect and log in to an SMTP
server, send an email, and disconnect from the server.

## Connecting to an SMTP Server

If you've ever set up Thunderbird, Outlook, or another program to connect to your email account, you may be familiar with configuring the SMTP
server and port. These settings will be different for each email provider, but
a web search for *<your provider> smtp settings* should turn up the server and
port to use.

The domain name for the SMTP server will usually be the name of
your email provider's domain name, with *smtp.* in front of it. For example,
Verizon's SMTP server is at *smtp.verizon.net*. Table 18-1 lists some common
email providers and their SMTP servers. (The port is an integer value and will
almost always be 587. It's used by the command encryption standard, TLS.)

**Table 18-1:** Email Providers and Their SMTP Servers

| Provider | SMTP server domain name |
|---|---|
| Gmail* | *smtp.gmail.com* |
| Outlook.com/Hotmail.com* | *smtp-mail.outlook.com* |
| Yahoo Mail* | *smtp.mail.yahoo.com* |
| AT&T | *smpt.mail.att.net* (port 465) |
| Comcast | *smtp.comcast.net* |
| Verizon | *smtp.verizon.net* (port 465) |

*Additional security measures prevent Python from being able to log in to
these servers with the smtplib module. The EZGmail module can bypass this
difficulty for Gmail accounts.

Once you have the domain name and port information for your email provider, create an SMTP object by calling smptlib.SMTP(), passing the domain name as a string argument, and passing the port as an integer argument. The SMTP object represents a connection to an SMTP mail server and has methods for sending emails. For example, the following call creates an SMTP object for connecting to an imaginary email server:

```
>>> smtpObj = smtplib.SMTP('smtp.example.com', 587)
>>> type(smtpObj)
<class 'smtplib.SMTP'>
```

Entering type(smtpObj) shows you that there's an SMTP object stored in smtpObj. You'll need this SMTP object in order to call the methods that log you in and send emails. If the smptlib.SMTP() call is not successful, your SMTP server might not support TLS on port 587. In this case, you will need to create an SMTP object using smtplib.SMTP_SSL() and port 465 instead.

```
>>> smtpObj = smtplib.SMTP_SSL('smtp.example.com', 465)
```

**NOTE** *If you are not connected to the internet, Python will raise a socket.gaierror: [Errno 11004] getaddrinfo failed or similar exception.*

For your programs, the differences between TLS and SSL aren't important. You only need to know which encryption standard your SMTP server uses so you know how to connect to it. In all of the interactive shell examples that follow, the smtpObj variable will contain an SMTP object returned by the smtplib.SMTP() or smtplib.SMTP_SSL() function.

## Sending the SMTP "Hello" Message

Once you have the SMTP object, call its oddly named ehlo() method to "say hello" to the SMTP email server. This greeting is the first step in SMTP and is important for establishing a connection to the server. You don't need to know the specifics of these protocols. Just be sure to call the ehlo() method first thing after getting the SMTP object or else the later method calls will result in errors. The following is an example of an ehlo() call and its return value:

```
>>> smtpObj.ehlo()
(250, b'mx.example.com at your service, [216.172.148.131]\nSIZE 35882577\
n8BITMIME\nSTARTTLS\nENHANCEDSTATUSCODES\nCHUNKING')
```

If the first item in the returned tuple is the integer 250 (the code for "success" in SMTP), then the greeting succeeded.

## Starting TLS Encryption

If you are connecting to port 587 on the SMTP server (that is, you're using TLS encryption), you'll need to call the starttls() method next. This required step enables encryption for your connection. If you are connecting to port 465 (using SSL), then encryption is already set up, and you should skip this step.

Here's an example of the starttls() method call:

```
>>> smtpObj.starttls()
(220, b'2.0.0 Ready to start TLS')
```

The starttls() method puts your SMTP connection in TLS mode. The 220 in the return value tells you that the server is ready.

### Logging In to the SMTP Server

Once your encrypted connection to the SMTP server is set up, you can log in with your username (usually your email address) and email password by calling the login() method.

```
>>> smtpObj.login('my_email_address@example.com', 'MY_SECRET_PASSWORD')
(235, b'2.7.0 Accepted')
```

Pass a string of your email address as the first argument and a string of your password as the second argument. The 235 in the return value means authentication was successful. Python raises an smtplib.SMTP AuthenticationError exception for incorrect passwords.

**WARNING** *Be careful about putting passwords in your source code. If anyone ever copies your program, they'll have access to your email account! It's a good idea to call input() and have the user type in the password. It may be inconvenient to have to enter a password each time you run your program, but this approach prevents you from leaving your password in an unencrypted file on your computer where a hacker or laptop thief could easily get it.*

### Sending an Email

Once you are logged in to your email provider's SMTP server, you can call the sendmail() method to actually send the email. The sendmail() method call looks like this:

```
>>> smtpObj.sendmail('my_email_address@example.com', 'recipient@example.com',
'Subject: So long.\nDear Alice, so long and thanks for all the fish.
Sincerely, Bob')
{}
```

The sendmail() method requires three arguments:

- Your email address as a string (for the email's "from" address)
- The recipient's email address as a string, or a list of strings for multiple recipients (for the "to" address)
- The email body as a string

The start of the email body string *must* begin with 'Subject: \n' for the subject line of the email. The '\n' newline character separates the subject line from the main body of the email.

The return value from sendmail() is a dictionary. There will be one key-value pair in the dictionary for each recipient for whom email delivery *failed*. An empty dictionary means all recipients were *successfully* sent the email.

### Disconnecting from the SMTP Server

Be sure to call the quit() method when you are done sending emails. This will disconnect your program from the SMTP server.

```
>>> smtpObj.quit()
(221, b'2.0.0 closing connection ko10sm23097611pbd.52 - gsmtp')
```

The 221 in the return value means the session is ending.

To review all the steps for connecting and logging in to the server, sending email, and disconnecting, see "Sending Email" on page 420.

## IMAP

Just as SMTP is the protocol for sending email, the *Internet Message Access Protocol (IMAP)* specifies how to communicate with an email provider's server to retrieve emails sent to your email address. Python comes with an imaplib module, but in fact the third-party imapclient module is easier to use. This chapter provides an introduction to using IMAPClient; the full documentation is at *https://imapclient.readthedocs.io/*.

The imapclient module downloads emails from an IMAP server in a rather complicated format. Most likely, you'll want to convert them from this format into simple string values. The pyzmail module does the hard job of parsing these email messages for you. You can find the complete documentation for PyzMail at *https://www.magiksys.net/pyzmail/*.

Install imapclient and pyzmail from a Terminal window with pip install --user -U imapclient==2.1.0 and pip install --user -U pyzmail36== 1.0.4 on Windows (or using pip3 on macOS and Linux). Appendix A has steps on how to install third-party modules.

## Retrieving and Deleting Emails with IMAP

Finding and retrieving an email in Python is a multistep process that requires both the imapclient and pyzmail third-party modules. Just to give you an overview, here's a full example of logging in to an IMAP server, searching for emails, fetching them, and then extracting the text of the email messages from them.

```
>>> import imapclient
>>> imapObj = imapclient.IMAPClient('imap.example.com', ssl=True)
>>> imapObj.login('my_email_address@example.com', 'MY_SECRET_PASSWORD')
```

```
'my_email_address@example.com Jane Doe authenticated (Success)'
>>> imapObj.select_folder('INBOX', readonly=True)
>>> UIDs = imapObj.search(['SINCE 05-Jul-2019'])
>>> UIDs
[40032, 40033, 40034, 40035, 40036, 40037, 40038, 40039, 40040, 40041]
>>> rawMessages = imapObj.fetch([40041], ['BODY[]', 'FLAGS'])
>>> import pyzmail
>>> message = pyzmail.PyzMessage.factory(rawMessages[40041][b'BODY[]'])
>>> message.get_subject()
'Hello!'
>>> message.get_addresses('from')
[('Edward Snowden', 'esnowden@nsa.gov')]
>>> message.get_addresses('to')
[('Jane Doe', 'jdoe@example.com')]
>>> message.get_addresses('cc')
[]
>>> message.get_addresses('bcc')
[]
>>> message.text_part != None
True
>>> message.text_part.get_payload().decode(message.text_part.charset)
'Follow the money.\r\n\r\n-Ed\r\n'
>>> message.html_part != None
True
>>> message.html_part.get_payload().decode(message.html_part.charset)
'<div dir="ltr"><div>So long, and thanks for all the fish!<br><br></div>-
Al<br></div>\r\n'
>>> imapObj.logout()
```

You don't have to memorize these steps. After we go through each step in detail, you can come back to this overview to refresh your memory.

### Connecting to an IMAP Server

Just like you needed an SMTP object to connect to an SMTP server and send email, you need an IMAPClient object to connect to an IMAP server and receive email. First you'll need the domain name of your email provider's IMAP server. This will be different from the SMTP server's domain name. Table 18-2 lists the IMAP servers for several popular email providers.

**Table 18-2:** Email Providers and Their IMAP Servers

| Provider | IMAP server domain name |
| --- | --- |
| Gmail* | imap.gmail.com |
| Outlook.com/Hotmail.com* | imap-mail.outlook.com |
| Yahoo Mail* | imap.mail.yahoo.com |
| AT&T | imap.mail.att.net |
| Comcast | imap.comcast.net |
| Verizon | incoming.verizon.net |

*Additional security measures prevent Python from being able to log in to these servers with the imapclient module.

Once you have the domain name of the IMAP server, call the `imapclient`
`.IMAPClient()` function to create an `IMAPClient` object. Most email providers
require SSL encryption, so pass the `ssl=True` keyword argument. Enter the
following into the interactive shell (using your provider's domain name):

```
>>> import imapclient
>>> imapObj = imapclient.IMAPClient('imap.example.com', ssl=True)
```

In all of the interactive shell examples in the following sections, the
`imapObj` variable contains an `IMAPClient` object returned from the `imapclient`
`.IMAPClient()` function. In this context, a *client* is the object that connects
to the server.

### Logging In to the IMAP Server

Once you have an `IMAPClient` object, call its `login()` method, passing in the
username (this is usually your email address) and password as strings.

```
>>> imapObj.login('my_email_address@example.com', 'MY_SECRET_PASSWORD')
'my_email_address@example.com Jane Doe authenticated (Success)'
```

**WARNING** *Remember to never write a password directly into your code! Instead, design your program to accept the password returned from input().*

If the IMAP server rejects this username/password combination,
Python raises an `imaplib.error` exception.

### Searching for Email

Once you're logged on, actually retrieving an email that you're interested
in is a two-step process. First, you must select a folder you want to search
through. Then, you must call the `IMAPClient` object's `search()` method, passing in a string of IMAP search keywords.

#### Selecting a Folder

Almost every account has an `INBOX` folder by default, but you can also get a
list of folders by calling the `IMAPClient` object's `list_folders()` method. This
returns a list of tuples. Each tuple contains information about a single
folder. Continue the interactive shell example by entering the following:

```
>>> import pprint
>>> pprint.pprint(imapObj.list_folders())
[(('\\HasNoChildren',), '/', 'Drafts'),
```

```
(('\\HasNoChildren',), '/', 'Filler'),
(('\\HasNoChildren',), '/', 'INBOX'),
(('\\HasNoChildren',), '/', 'Sent'),
--snip--
(('\\HasNoChildren', '\\Flagged'), '/', 'Starred'),
(('\\HasNoChildren', '\\Trash'), '/', 'Trash')]
```

The three values in each of the tuples—for example, (('\\ HasNoChildren',), '/', 'INBOX')—are as follows:

- A tuple of the folder's flags. (Exactly what these flags represent is beyond the scope of this book, and you can safely ignore this field.)
- The delimiter used in the name string to separate parent folders and subfolders.
- The full name of the folder.

To select a folder to search through, pass the folder's name as a string into the IMAPClient object's select_folder() method.

```
>>> imapObj.select_folder('INBOX', readonly=True)
```

You can ignore select_folder()'s return value. If the selected folder does not exist, Python raises an imaplib.error exception.

The readonly=True keyword argument prevents you from accidentally making changes or deletions to any of the emails in this folder during the subsequent method calls. Unless you *want* to delete emails, it's a good idea to always set readonly to True.

### Performing the Search

With a folder selected, you can now search for emails with the IMAPClient object's search() method. The argument to search() is a list of strings, each formatted to the IMAP's search keys. Table 18-3 describes the various search keys.

Note that some IMAP servers may have slightly different implementations for how they handle their flags and search keys. It may require some experimentation in the interactive shell to see exactly how they behave.

You can pass multiple IMAP search key strings in the list argument to the search() method. The messages returned are the ones that match *all* the search keys. If you want to match *any* of the search keys, use the OR search key. For the NOT and OR search keys, one and two complete search keys follow the NOT and OR, respectively.

**Table 18-3:** IMAP Search Keys

| Search key | Meaning |
| --- | --- |
| `'ALL'` | Returns all messages in the folder. You may run into imaplib size limits if you request all the messages in a large folder. See "Size Limits" on page 429. |
| `'BEFORE date'`, `'ON date'`, `'SINCE date'` | These three search keys return, respectively, messages that were received by the IMAP server before, on, or after the given date. The date must be formatted like `05-Jul-2019`. Also, while `'SINCE 05-Jul-2019'` will match messages on and after July 5, `'BEFORE 05-Jul-2019'` will match only messages before July 5 but not on July 5 itself. |
| `'SUBJECT string'`, `'BODY string'`, `'TEXT string'` | Returns messages where *string* is found in the subject, body, or either, respectively. If *string* has spaces in it, then enclose it with double quotes: `'TEXT "search with spaces"'`. |
| `'FROM string'`, `'TO string'`, `'CC string'`, `'BCC string'` | Returns all messages where *string* is found in the "from" email address, "to" addresses, "cc" (carbon copy) addresses, or "bcc" (blind carbon copy) addresses, respectively. If there are multiple email addresses in *string*, then separate them with spaces and enclose them all with double quotes: `'CC "firstcc@example.com secondcc@example.com"'`. |
| `'SEEN'`, `'UNSEEN'` | Returns all messages with and without the \\*Seen* flag, respectively. An email obtains the \\*Seen* flag if it has been accessed with a `fetch()` method call (described later) or if it is clicked when you're checking your email in an email program or web browser. It's more common to say the email has been "read" rather than "seen," but they mean the same thing. |
| `'ANSWERED'`, `'UNANSWERED'` | Returns all messages with and without the \\*Answered* flag, respectively. A message obtains the \\*Answered* flag when it is replied to. |
| `'DELETED'`, `'UNDELETED'` | Returns all messages with and without the \\*Deleted* flag, respectively. Email messages deleted with the `delete_messages()` method are given the \\*Deleted* flag but are not permanently deleted until the `expunge()` method is called (see "Deleting Emails" on page 432). Note that some email providers automatically expunge emails. |
| `'DRAFT'`, `'UNDRAFT'` | Returns all messages with and without the \\*Draft* flag, respectively. Draft messages are usually kept in a separate `Drafts` folder rather than in the `INBOX` folder. |
| `'FLAGGED'`, `'UNFLAGGED'` | Returns all messages with and without the \\*Flagged* flag, respectively. This flag is usually used to mark email messages as "Important" or "Urgent." |
| `'LARGER N'`, `'SMALLER N'` | Returns all messages larger or smaller than *N* bytes, respectively. |
| `'NOT search-key'` | Returns the messages that *search-key* would *not* have returned. |
| `'OR search-key1 search-key2'` | Returns the messages that match *either* the first or second *search-key*. |

Here are some example search() method calls along with their meanings:

`imapObj.search(['ALL'])`   Returns every message in the currently selected folder.

`imapObj.search(['ON 05-Jul-2019'])`   Returns every message sent on July 5, 2019.

`imapObj.search(['SINCE 01-Jan-2019', 'BEFORE 01-Feb-2019', 'UNSEEN'])` Returns every message sent in January 2019 that is unread. (Note that this means *on and after* January 1 and *up to but not including* February 1.)

`imapObj.search(['SINCE 01-Jan-2019', 'FROM alice@example.com'])`   Returns every message from *alice@example.com* sent since the start of 2019.

`imapObj.search(['SINCE 01-Jan-2019', 'NOT FROM alice@example.com'])` Returns every message sent from everyone except *alice@example.com* since the start of 2019.

`imapObj.search(['OR FROM alice@example.com FROM bob@example.com'])` Returns every message ever sent from *alice@example.com* or *bob@example.com*.

`imapObj.search(['FROM alice@example.com', 'FROM bob@example.com'])` Trick example! This search never returns any messages, because messages must match *all* search keywords. Since there can be only one "from" address, it is impossible for a message to be from both *alice@example.com* and *bob@example.com*.

The search() method doesn't return the emails themselves but rather unique IDs (UIDs) for the emails, as integer values. You can then pass these UIDs to the fetch() method to obtain the email content.

Continue the interactive shell example by entering the following:

```
>>> UIDs = imapObj.search(['SINCE 05-Jul-2019'])
>>> UIDs
[40032, 40033, 40034, 40035, 40036, 40037, 40038, 40039, 40040, 40041]
```

Here, the list of message IDs (for messages received July 5 onward) returned by search() is stored in UIDs. The list of UIDs returned on your computer will be different from the ones shown here; they are unique to a particular email account. When you later pass UIDs to other function calls, use the UID values you received, not the ones printed in this book's examples.

### Size Limits

If your search matches a large number of email messages, Python might raise an exception that says imaplib.error: got more than 10000 bytes. When this happens, you will have to disconnect and reconnect to the IMAP server and try again.

This limit is in place to prevent your Python programs from eating up too much memory. Unfortunately, the default size limit is often too small. You can change this limit from 10,000 bytes to 10,000,000 bytes by running this code:

```
>>> import imaplib
>>> imaplib._MAXLINE = 10000000
```

This should prevent this error message from coming up again. You may want to make these two lines part of every IMAP program you write.

### Fetching an Email and Marking It as Read

Once you have a list of UIDs, you can call the IMAPClient object's fetch() method to get the actual email content.

The list of UIDs will be fetch()'s first argument. The second argument should be the list ['BODY[]'], which tells fetch() to download all the body content for the emails specified in your UID list.

Let's continue our interactive shell example.

```
>>> rawMessages = imapObj.fetch(UIDs, ['BODY[]'])
>>> import pprint
>>> pprint.pprint(rawMessages)
{40040: {'BODY[]': 'Delivered-To: my_email_address@example.com\r\n'
                   'Received: by 10.76.71.167 with SMTP id '
--snip--
                   '\r\n'
                   '-------=_Part_6000970_707736290.1404819487066--\r\n',
          'SEQ': 5430}}
```

Import pprint and pass the return value from fetch(), stored in the variable rawMessages, to pprint.pprint() to "pretty print" it, and you'll see that this return value is a nested dictionary of messages with UIDs as the keys. Each message is stored as a dictionary with two keys: 'BODY[]' and 'SEQ'. The 'BODY[]' key maps to the actual body of the email. The 'SEQ' key is for a *sequence number*, which has a similar role to the UID. You can safely ignore it.

As you can see, the message content in the 'BODY[]' key is pretty unintelligible. It's in a format called RFC 822, which is designed for IMAP servers to read. But you don't need to understand the RFC 822 format; later in this chapter, the pyzmail module will make sense of it for you.

When you selected a folder to search through, you called select_folder() with the readonly=True keyword argument. Doing this prevents you from accidentally deleting an email—but it also means that emails will not get marked as read if you fetch them with the fetch() method. If you *do* want emails to be marked as read when you fetch them, you'll need to pass readonly=False to select_folder(). If the selected folder is already in read-only mode, you can reselect the current folder with another call to select_folder(), this time with the readonly=False keyword argument:

```
>>> imapObj.select_folder('INBOX', readonly=False)
```

### Getting Email Addresses from a Raw Message

The raw messages returned from the fetch() method still aren't very useful to people who just want to read their email. The pyzmail module parses these raw messages and returns them as PyzMessage objects, which make the subject, body, "To" field, "From" field, and other sections of the email easily accessible to your Python code.

Continue the interactive shell example with the following (using UIDs from your own email account, not the ones shown here):

```
>>> import pyzmail
>>> message = pyzmail.PyzMessage.factory(rawMessages[40041][b'BODY[]'])
```

First, import pyzmail. Then, to create a PyzMessage object of an email, call the pyzmail.PyzMessage.factory() function and pass it the 'BODY[]' section of the raw message. (Note that the b prefix means this is a bytes value, not a string value. The difference isn't too important; just remember to include the b prefix in your code.) Store the result in message. Now message contains a PyzMessage object, which has several methods that make it easy to get the email's subject line, as well as all sender and recipient addresses. The get_subject() method returns the subject as a simple string value. The get_addresses() method returns a list of addresses for the field you pass it. For example, the method calls might look like this:

```
>>> message.get_subject()
'Hello!'
>>> message.get_addresses('from')
[('Edward Snowden', 'esnowden@nsa.gov')]
>>> message.get_addresses('to')
[('Jane Doe', 'my_email_address@example.com')]
>>> message.get_addresses('cc')
[]
>>> message.get_addresses('bcc')
[]
```

Notice that the argument for get_addresses() is 'from', 'to', 'cc', or 'bcc'. The return value of get_addresses() is a list of tuples. Each tuple contains two strings: the first is the name associated with the email address, and the second is the email address itself. If there are no addresses in the requested field, get_addresses() returns a blank list. Here, the 'cc' carbon copy and 'bcc' blind carbon copy fields both contained no addresses and so returned empty lists.

### Getting the Body from a Raw Message

Emails can be sent as plaintext, HTML, or both. Plaintext emails contain only text, while HTML emails can have colors, fonts, images, and other features that make the email message look like a small web page. If an email is only plaintext, its PyzMessage object will have its html_part attributes set to None. Likewise, if an email is only HTML, its PyzMessage object will have its text_part attribute set to None.

Otherwise, the text_part or html_part value will have a get_payload() method that returns the email's body as a value of the *bytes* data type. (The bytes data type is beyond the scope of this book.) But this *still* isn't a string value that we can use. Ugh! The last step is to call the decode() method on the bytes value returned by get_payload(). The decode() method takes one argument: the message's character encoding, stored in the text_part.charset or html_part.charset attribute. This, finally, will return the string of the email's body.

Continue the interactive shell example by entering the following:

```
❶ >>> message.text_part != None
   True
   >>> message.text_part.get_payload().decode(message.text_part.charset)
❷ 'So long, and thanks for all the fish!\r\n\r\n-Al\r\n'
❸ >>> message.html_part != None
   True
❹ >>> message.html_part.get_payload().decode(message.html_part.charset)
   '<div dir="ltr"><div>So long, and thanks for all the fish!<br><br></div>-Al
   <br></div>\r\n'
```

The email we're working with has both plaintext and HTML content, so the PyzMessage object stored in message has text_part and html_part attributes not equal to None ❶ ❸. Calling get_payload() on the message's text_part and then calling decode() on the bytes value returns a string of the text version of the email ❷. Using get_payload() and decode() with the message's html_part returns a string of the HTML version of the email ❹.

### Deleting Emails

To delete emails, pass a list of message UIDs to the IMAPClient object's delete_messages() method. This marks the emails with the *\Deleted* flag. Calling the expunge() method permanently deletes all emails with the */Deleted* flag in the currently selected folder. Consider the following interactive shell example:

```
❶ >>> imapObj.select_folder('INBOX', readonly=False)
❷ >>> UIDs = imapObj.search(['ON 09-Jul-2019'])
   >>> UIDs
   [40066]
   >>> imapObj.delete_messages(UIDs)
❸ {40066: ('\\Seen', '\\Deleted')}
   >>> imapObj.expunge()
   ('Success', [(5452, 'EXISTS')])
```

Here we select the inbox by calling select_folder() on the IMAPClient object and passing 'INBOX' as the first argument; we also pass the keyword argument readonly=False so that we can delete emails ❶. We search the inbox for messages received on a specific date and store the returned message IDs in UIDs ❷. Calling delete_message() and passing it UIDs returns a dictionary; each key-value pair is a message ID and a tuple of the message's flags, which should now include *\Deleted* ❸. Calling `expunge()` then permanently deletes messages with the *\Deleted* flag and returns a success message if there were no problems expunging the emails. Note that some email providers automatically expunge emails deleted with `delete_messages()` instead of waiting for an expunge command from the IMAP client.

### Disconnecting from the IMAP Server

When your program has finished retrieving or deleting emails, simply call the IMAPClient's `logout()` method to disconnect from the IMAP server.

```
>>> imapObj.logout()
```

If your program runs for several minutes or more, the IMAP server may *time out,* or automatically disconnect. In this case, the next method call your program makes on the IMAPClient object should raise an exception like the following:

```
imaplib.abort: socket error: [WinError 10054] An existing connection was
forcibly closed by the remote host
```

In this event, your program will have to call `imapclient.IMAPClient()` to connect again.

Whew! That's it. There were a lot of hoops to jump through, but you now have a way to get your Python programs to log in to an email account and fetch emails. You can always consult the overview in "Retrieving and Deleting Emails with IMAP" on page 424 whenever you need to remember all of the steps.

# Project: Sending Member Dues Reminder Emails

Say you have been "volunteered" to track member dues for the Mandatory Volunteerism Club. This is a truly boring job, involving maintaining a spreadsheet of everyone who has paid each month and emailing reminders to those who haven't. Instead of going through the spreadsheet yourself and copying and pasting the same email to everyone who is behind on dues, let's—you guessed it—write a script that does this for you.

At a high level, here's what your program will do:

1. Read data from an Excel spreadsheet.
2. Find all members who have not paid dues for the latest month.
3. Find their email addresses and send them personalized reminders.

This means your code will need to do the following:

1. Open and read the cells of an Excel document with the `openpyxl` module. (See Chapter 13 for working with Excel files.)
2. Create a dictionary of members who are behind on their dues.

3. Log in to an SMTP server by calling `smtplib.SMTP()`, `ehlo()`, `starttls()`, and `login()`.

4. For all members behind on their dues, send a personalized reminder email by calling the `sendmail()` method.

Open a new file editor tab and save it as *sendDuesReminders.py*.

## Step 1: Open the Excel File

Let's say the Excel spreadsheet you use to track membership dues payments looks like Figure 18-2 and is in a file named *duesRecords.xlsx*. You can download this file from *https://nostarch.com/automatestuff2/*.

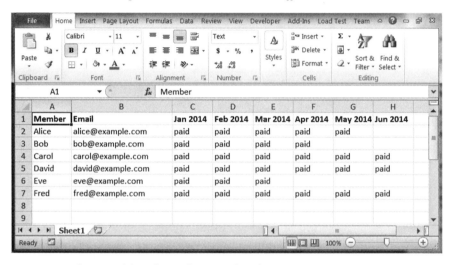

*Figure 18-2: The spreadsheet for tracking member dues payments*

This spreadsheet has every member's name and email address. Each month has a column tracking members' payment statuses. The cell for each member is marked with the text *paid* once they have paid their dues.

The program will have to open *duesRecords.xlsx* and figure out the column for the latest month by reading the `sheet.max_column` attribute. (You can consult Chapter 13 for more information on accessing cells in Excel spreadsheet files with the `openpyxl` module.) Enter the following code into the file editor tab:

```
#! python3
# sendDuesReminders.py - Sends emails based on payment status in spreadsheet.

import openpyxl, smtplib, sys

# Open the spreadsheet and get the latest dues status.
❶ wb = openpyxl.load_workbook('duesRecords.xlsx')
❷ sheet = wb.get_sheet_by_name('Sheet1')
```

```
❸ lastCol = sheet.max_column
❹ latestMonth = sheet.cell(row=1, column=lastCol).value

    # TODO: Check each member's payment status.

    # TODO: Log in to email account.

    # TODO: Send out reminder emails.
```

After importing the openpyxl, smtplib, and sys modules, we open our *duesRecords.xlsx* file and store the resulting Workbook object in wb ❶. Then we get Sheet 1 and store the resulting Worksheet object in sheet ❷. Now that we have a Worksheet object, we can access rows, columns, and cells. We store the highest column in lastCol ❸, and we then use row number 1 and lastCol to access the cell that should hold the most recent month. We get the value in this cell and store it in latestMonth ❹.

## Step 2: Find All Unpaid Members

Once you've determined the column number of the latest month (stored in lastCol), you can loop through all rows after the first row (which has the column headers) to see which members have the text *paid* in the cell for that month's dues. If the member hasn't paid, you can grab the member's name and email address from columns 1 and 2, respectively. This information will go into the unpaidMembers dictionary, which will track all members who haven't paid in the most recent month. Add the following code to *sendDuesReminder.py*.

```
#! python3
# sendDuesReminders.py - Sends emails based on payment status in spreadsheet.

--snip--

# Check each member's payment status.
unpaidMembers = {}
❶ for r in range(2, sheet.max_row + 1):
❷     payment = sheet.cell(row=r, column=lastCol).value
       if payment != 'paid':
❸          name = sheet.cell(row=r, column=1).value
❹          email = sheet.cell(row=r, column=2).value
❺          unpaidMembers[name] = email
```

This code sets up an empty dictionary unpaidMembers and then loops through all the rows after the first ❶. For each row, the value in the most recent column is stored in payment ❷. If payment is not equal to 'paid', then the value of the first column is stored in name ❸, the value of the second column is stored in email ❹, and name and email are added to unpaidMembers ❺.

## Step 3: Send Customized Email Reminders

Once you have a list of all unpaid members, it's time to send them email reminders. Add the following code to your program, except with your real email address and provider information:

```python
#! python3
# sendDuesReminders.py - Sends emails based on payment status in spreadsheet.

--snip--

# Log in to email account.
smtpObj = smtplib.SMTP('smtp.example.com', 587)
smtpObj.ehlo()
smtpObj.starttls()
smtpObj.login('my_email_address@example.com', sys.argv[1])
```

Create an SMTP object by calling smtplib.SMTP() and passing it the domain name and port for your provider. Call ehlo() and starttls(), and then call login() and pass it your email address and sys.argv[1], which will store your password string. You'll enter the password as a command line argument each time you run the program, to avoid saving your password in your source code.

Once your program has logged in to your email account, it should go through the unpaidMembers dictionary and send a personalized email to each member's email address. Add the following to *sendDuesReminders.py*:

```python
#! python3
# sendDuesReminders.py - Sends emails based on payment status in spreadsheet.

--snip--

# Send out reminder emails.
for name, email in unpaidMembers.items():
  ❶ body = "Subject: %s dues unpaid.\nDear %s,\nRecords show that you have not
    paid dues for %s. Please make this payment as soon as possible. Thank you!'" %
    (latestMonth, name, latestMonth)
  ❷ print('Sending email to %s...' % email)
  ❸ sendmailStatus = smtpObj.sendmail('my_email_address@example.com', email,
body)

  ❹ if sendmailStatus != {}:
        print('There was a problem sending email to %s: %s' % (email,
        sendmailStatus))
smtpObj.quit()
```

This code loops through the names and emails in unpaidMembers. For each member who hasn't paid, we customize a message with the latest month and the member's name, and store the message in body ❶. We print output saying that we're sending an email to this member's email address ❷. Then we call sendmail(), passing it the from address and the customized message ❸. We store the return value in sendmailStatus.

Remember that the `sendmail()` method will return a nonempty dictionary value if the SMTP server reported an error sending that particular email. The last part of the `for` loop at ❹ checks if the returned dictionary is nonempty and, if it is, prints the recipient's email address and the returned dictionary.

After the program is done sending all the emails, the `quit()` method is called to disconnect from the SMTP server.

When you run the program, the output will look something like this:

```
Sending email to alice@example.com...
Sending email to bob@example.com...
Sending email to eve@example.com...
```

The recipients will receive an email about their missed payments that looks just like an email you would have sent manually.

## Sending Text Messages with SMS Email Gateways

People are more likely to be near their smartphones than their computers, so text messages are often a more immediate and reliable way of sending notifications than email. Also, text messages are usually shorter, making it more likely that a person will get around to reading them.

The easiest, though not most reliable, way to send text messages is by using an SMS (short message service) email gateway, an email server that a cell phone provider set up to receive text via email and then forward to the recipient as a text message.

You can write a program to send these emails using the `ezgmail` or `smtplib` modules. The phone number and phone company's email server make up the recipient email address. The subject and body of the email will be the body of the text message. For example, to send a text to the phone number 415-555-1234, which is owned by a Verizon customer, you would send an email to *4155551234@vtext.com*.

You can find the SMS email gateway for a cell phone provider by doing a web search for "sms email gateway *provider name*," but Table 18-4 lists the gateways for several popular providers. Many providers have separate email servers for SMS , which limits messages to 160 characters, and MMS (multimedia messaging service), which has no character limit. If you wanted to send a photo, you would have to use the MMS gateway and attach the file to the email.

If you don't know the recipient's cell phone provider, you can try using a *carrier lookup* site, which should provide a phone number's carrier. The best way to find these sites is by searching the web for "find cell phone provider for number." Many of these sites will let you look up numbers for free (though will charge you if you need to look up hundreds or thousands of phone numbers through their API).

**Table 18-4:** SMS Email Gateways for Cell Phone Providers

| Cell phone provider | SMS gateway | MMS gateway |
| --- | --- | --- |
| AT&T | number@txt.att.net | number@mms.att.net |
| Boost Mobile | number@sms.myboostmobile.com | Same as SMS |
| Cricket | number@sms.cricketwireless.net | number@mms.cricketwireless.net |
| Google Fi | number@msg.fi.google.com | Same as SMS |
| Metro PCS | number@mymetropcs.com | Same as SMS |
| Republic Wireless | number@text.republicwireless.com | Same as SMS |
| Sprint | number@messaging.sprintpcs.com | number@pm.sprint.com |
| T-Mobile | number@tmomail.net | Same as SMS |
| U.S. Cellular | number@email.uscc.net | number@mms.uscc.net |
| Verizon | number@vtext.com | number@vzwpix.com |
| Virgin Mobile | number@vmobl.com | number@vmpix.com |
| XFinity Mobile | number@vtext.com | number@mypixmessages.com |

While SMS email gateways are free and simple to use, there are several major disadvantages to them:

- You have no guarantee that the text will arrive promptly, or at all.
- You have no way of knowing if the text failed to arrive.
- The text recipient has no way of replying.
- SMS gateways may block you if you send too many emails, and there's no way to find out how many is "too many."
- Just because the SMS gateway delivers a text message today doesn't mean it will work tomorrow.

Sending texts via an SMS gateway is ideal when you need to send the occasional, nonurgent message. If you need more reliable service, use a non-email SMS gateway service, as described next.

## Sending Text Messages with Twilio

In this section, you'll learn how to sign up for the free Twilio service and use its Python module to send text messages. Twilio is an *SMS gateway service*, which means it allows you to send text messages from your programs via the internet. Although the free trial account comes with a limited amount of credit and the texts will be prefixed with the words *Sent from a Twilio trial account*, this trial service is probably adequate for your personal programs.

But Twilio isn't the only SMS gateway service. If you prefer not to use Twilio, you can find alternative services by searching online for "free sms" "gateway," "python sms api," or even "twilio alternatives."

Before signing up for a Twilio account, install the `twilio` module with `pip install --user --upgrade twilio` on Windows (or use `pip3` on macOS and Linux). Appendix A has more details about installing third-party modules.

*This section is specific to the United States. Twilio does offer SMS texting services for countries other than the United States; see* https://twilio.com/ *for more information. The* `twilio` *module and its functions will work the same outside the United States.*

## Signing Up for a Twilio Account

Go to *https://twilio.com/* and fill out the sign-up form. Once you've signed up for a new account, you'll need to verify a mobile phone number that you want to send texts to. Go to the Verified Caller IDs page and add a phone number you have access to. Twilio will text a code to this number that you must enter to verify the number. (This verification is necessary to prevent people from using the service to spam random phone numbers with text messages.) You will now be able to send texts to this phone number using the `twilio` module.

Twilio provides your trial account with a phone number to use as the sender of text messages. You will need two more pieces of information: your account SID and the auth (authentication) token. You can find this information on the Dashboard page when you are logged in to your Twilio account. These values act as your Twilio username and password when logging in from a Python program.

## Sending Text Messages

Once you've installed the `twilio` module, signed up for a Twilio account, verified your phone number, registered a Twilio phone number, and obtained your account SID and auth token, you will finally be ready to send yourself text messages from your Python scripts.

Compared to all the registration steps, the actual Python code is fairly simple. With your computer connected to the internet, enter the following into the interactive shell, replacing the `accountSID`, `authToken`, `myTwilioNumber`, and `myCellPhone` variable values with your real information:

```
❶ >>> from twilio.rest import Client
   >>> accountSID = 'ACxxxxxxxxxxxxxxxxxxxxxxxxxxxxxxxxx'
   >>> authToken  = 'xxxxxxxxxxxxxxxxxxxxxxxxxxxxxxxxxx'
❷ >>> twilioCli = Client(accountSID, authToken)
   >>> myTwilioNumber = '+14955551234'
   >>> myCellPhone = '+14955558888'
❸ >>> message = twilioCli.messages.create(body='Mr. Watson - Come here - I want
   to see you.', from_=myTwilioNumber, to=myCellPhone)
```

A few moments after typing the last line, you should receive a text message that reads, *Sent from your Twilio trial account - Mr. Watson - Come here - I want to see you.*

Because of the way the twilio module is set up, you need to import it using from twilio.rest import Client, not just import twilio ❶. Store your account SID in accountSID and your auth token in authToken and then call Client() and pass it accountSID and authToken. The call to Client() returns a Client object ❷. This object has a messages attribute, which in turn has a create() method you can use to send text messages. This is the method that will instruct Twilio's servers to send your text message. After storing your Twilio number and cell phone number in myTwilioNumber and myCellPhone, respectively, call create() and pass it keyword arguments specifying the body of the text message, the sender's number (myTwilioNumber), and the recipient's number (myCellPhone) ❸.

The Message object returned from the create() method will have information about the text message that was sent. Continue the interactive shell example by entering the following:

```
>>> message.to
'+14955558888'
>>> message.from_
'+14955551234'
>>> message.body
'Mr. Watson - Come here - I want to see you.'
```

The to, from_, and body attributes should hold your cell phone number, Twilio number, and message, respectively. Note that the sending phone number is in the from_ attribute—with an underscore at the end—not from. This is because from is a keyword in Python (you've seen it used in the from *modulename* import * form of import statement, for example), so it cannot be used as an attribute name. Continue the interactive shell example with the following:

```
>>> message.status
'queued'
>>> message.date_created
datetime.datetime(2019, 7, 8, 1, 36, 18)
>>> message.date_sent == None
True
```

The status attribute should give you a string. The date_created and date_sent attributes should give you a datetime object if the message has been created and sent. It may seem odd that the status attribute is set to 'queued' and the date_sent attribute is set to None when you've already received the text message. This is because you captured the Message object in the message variable *before* the text was actually sent. You will need to refetch the Message object in order to see its most up-to-date status and date_sent. Every Twilio message has a unique string ID (SID) that can be used to fetch the latest update of the Message object. Continue the interactive shell example by entering the following:

```
>>> message.sid
'SM09520de7639ba3af137c6fcb7c5f4b51'
```

```
❶ >>> updatedMessage = twilioCli.messages.get(message.sid)
>>> updatedMessage.status
'delivered'
>>> updatedMessage.date_sent
datetime.datetime(2019, 7, 8, 1, 36, 18)
```

Entering message.sid shows you this message's long SID. By passing this SID to the Twilio client's get() method ❶, you can retrieve a new Message object with the most up-to-date information. In this new Message object, the status and date_sent attributes are correct.

The status attribute will be set to one of the following string values: 'queued', 'sending', 'sent', 'delivered', 'undelivered', or 'failed'. These statuses are self-explanatory, but for more precise details, take a look at the resources at *https://nostarch.com/automatestuff2/*.

---

**RECEIVING TEXT MESSAGES WITH PYTHON**

Unfortunately, receiving text messages with Twilio is a bit more complicated than sending them. Twilio requires that you have a website running its own web application. That's beyond the scope of these pages, but you can find more details in this book's online resources (*https://nostarch.com/automatestuff2/*).

---

## Project: "Just Text Me" Module

The person you'll most often text from your programs is probably you. Texting is a great way to send yourself notifications when you're away from your computer. If you've automated a boring task with a program that takes a couple of hours to run, you could have it notify you with a text when it's finished. Or you may have a regularly scheduled program running that sometimes needs to contact you, such as a weather-checking program that texts you a reminder to pack an umbrella.

As a simple example, here's a small Python program with a textmyself() function that sends a message passed to it as a string argument. Open a new file editor tab and enter the following code, replacing the account SID, auth token, and phone numbers with your own information. Save it as *textMyself.py*.

---

```
#! python3
# textMyself.py - Defines the textmyself() function that texts a message
# passed to it as a string.

# Preset values:
accountSID = 'ACxxxxxxxxxxxxxxxxxxxxxxxxxxxxxxxxxx'
authToken  = 'xxxxxxxxxxxxxxxxxxxxxxxxxxxxxxxxxx'
myNumber = '+15559998888'
twilioNumber = '+15552225678'
```

```
from twilio.rest import Client
```

```
❶ def textmyself(message):
❷     twilioCli = Client(accountSID, authToken)
❸     twilioCli.messages.create(body=message, from_=twilioNumber, to=myNumber)
```

This program stores an account SID, auth token, sending number, and receiving number. It then defined textmyself() to take on argument ❶, make a Client object ❷, and call create() with the message you passed ❸.

If you want to make the textmyself() function available to your other programs, simply place the *textMyself.py* file in the same folder as your Python script. Whenever you want one of your programs to text you, just add the following:

```
import textmyself
textmyself.textmyself('The boring task is finished.')
```

You need to sign up for Twilio and write the texting code only once. After that, it's just two lines of code to send a text from any of your other programs.

## Summary

We communicate with each other on the internet and over cell phone networks in dozens of different ways, but email and texting predominate. Your programs can communicate through these channels, which gives them powerful new notification features. You can even write programs running on different computers that communicate with one another directly via email, with one program sending emails with SMTP and the other retrieving them with IMAP.

Python's smtplib provides functions for using the SMTP to send emails through your email provider's SMTP server. Likewise, the third-party imapclient and pyzmail modules let you access IMAP servers and retrieve emails sent to you. Although IMAP is a bit more involved than SMTP, it's also quite powerful and allows you to search for particular emails, download them, and parse them to extract the subject and body as string values.

As a security and spam precaution, some popular email services like Gmail don't allow you to use the standard SMTP and IMAP protocols to access their services. The EZGmail module acts as a convenient wrapper for the Gmail API, letting your Python scripts access your Gmail account. I highly recommend that you set up a separate Gmail account for your scripts to use so that potential bugs in your program don't cause problems for your personal Gmail account.

Texting is a bit different from email, since, unlike email, more than just an internet connection is needed to send SMS texts. Fortunately, services such as Twilio provide modules to allow you to send text messages from your programs. Once you go through an initial setup process, you'll be able to send texts with just a couple lines of code.

With these modules in your skill set, you'll be able to program the specific conditions under which your programs should send notifications or reminders. Now your programs will have reach far beyond the computer they're running on!

## Practice Questions

1. What is the protocol for sending email? For checking and receiving email?
2. What four `smtplib` functions/methods must you call to log in to an SMTP server?
3. What two `imapclient` functions/methods must you call to log in to an IMAP server?
4. What kind of argument do you pass to `imapObj.search()`?
5. What do you do if your code gets an error message that says `got more than 10000 bytes`?
6. The `imapclient` module handles connecting to an IMAP server and finding emails. What is one module that handles reading the emails that `imapclient` collects?
7. When using the Gmail API, what are the *credentials.json* and *token.json* files?
8. In the Gmail API, what's the difference between "thread" and "message" objects?
9. Using `ezgmail.search()`, how can you find emails that have file attachments?
10. What three pieces of information do you need from Twilio before you can send text messages?

## Practice Projects

For practice, write programs that do the following.

### Random Chore Assignment Emailer

Write a program that takes a list of people's email addresses and a list of chores that need to be done and randomly assigns chores to people. Email each person their assigned chores. If you're feeling ambitious, keep a record of each person's previously assigned chores so that you can make sure the program avoids assigning anyone the same chore they did last time. For another possible feature, schedule the program to run once a week automatically.

Here's a hint: if you pass a list to the `random.choice()` function, it will return a randomly selected item from the list. Part of your code could look like this:

```
chores = ['dishes', 'bathroom', 'vacuum', 'walk dog']
randomChore = random.choice(chores)
chores.remove(randomChore)    # this chore is now taken, so remove it
```

### Umbrella Reminder

Chapter 12 showed you how to use the requests module to scrape data from *https://weather.gov/*. Write a program that runs just before you wake up in the morning and checks whether it's raining that day. If so, have the program text you a reminder to pack an umbrella before leaving the house.

### Auto Unsubscriber

Write a program that scans through your email account, finds all the unsubscribe links in all your emails, and automatically opens them in a browser. This program will have to log in to your email provider's IMAP server and download all of your emails. You can use Beautiful Soup (covered in Chapter 12) to check for any instance where the word *unsubscribe* occurs within an HTML link tag.

Once you have a list of these URLs, you can use webbrowser.open() to automatically open all of these links in a browser.

You'll still have to manually go through and complete any additional steps to unsubscribe yourself from these lists. In most cases, this involves clicking a link to confirm.

But this script saves you from having to go through all of your emails looking for unsubscribe links. You can then pass this script along to your friends so they can run it on their email accounts. (Just make sure your email password isn't hardcoded in the source code!)

### Controlling Your Computer Through Email

Write a program that checks an email account every 15 minutes for any instructions you email it and executes those instructions automatically. For example, BitTorrent is a peer-to-peer downloading system. Using free BitTorrent software such as qBittorrent, you can download large media files on your home computer. If you email the program a (completely legal, not at all piratical) BitTorrent link, the program will eventually check its email, find this message, extract the link, and then launch qBittorrent to start downloading the file. This way, you can have your home computer begin downloads while you're away, and the (completely legal, not at all piratical) download can be finished by the time you return home.

Chapter 17 covers how to launch programs on your computer using the subprocess.Popen() function. For example, the following call would launch the qBittorrent program, along with a torrent file:

```
qbProcess = subprocess.Popen(['C:\\Program Files (x86)\\qBittorrent\\
qbittorrent.exe', 'shakespeare_complete_works.torrent'])
```

Of course, you'll want the program to make sure the emails come from you. In particular, you might want to require that the emails contain a password, since it is fairly trivial for hackers to fake a "from" address in emails. The program should delete the emails it finds so that it doesn't repeat instructions every time it checks the email account. As an extra feature, have the program email or text you a confirmation every time it executes a command. Since you won't be sitting in front of the computer that is running the program, it's a good idea to use the logging functions (see Chapter 11) to write a text file log that you can check if errors come up.

qBittorrent (as well as other BitTorrent applications) has a feature where it can quit automatically after the download completes. Chapter 17 explains how you can determine when a launched application has quit with the `wait()` method for `Popen` objects. The `wait()` method call will block until qBittorrent has stopped, and then your program can email or text you a notification that the download has completed.

There are a lot of possible features you could add to this project. If you get stuck, you can download an example implementation of this program from *https://nostarch.com/automatestuff2/*.

# 19

## MANIPULATING IMAGES

If you have a digital camera or even if you just upload photos from your phone to Facebook, you probably cross paths with digital image files all the time. You may know how to use basic graphics software, such as Microsoft Paint or Paintbrush, or even more advanced applications such as Adobe Photoshop. But if you need to edit a massive number of images, editing them by hand can be a lengthy, boring job.

Enter Python. Pillow is a third-party Python module for interacting with image files. The module has several functions that make it easy to crop, resize, and edit the content of an image. With the power to manipulate images the same way you would with software such as Microsoft Paint or Adobe Photoshop, Python can automatically edit hundreds or thousands of images with ease. You can install Pillow by running `pip install --user -U pillow==6.0.0`. Appendix A has more details on installing modules.

# Computer Image Fundamentals

In order to manipulate an image, you need to understand the basics of how computers deal with colors and coordinates in images and how you can work with colors and coordinates in Pillow. But before you continue, install the pillow module. See Appendix A for help installing third-party modules.

## Colors and RGBA Values

Computer programs often represent a color in an image as an *RGBA value*. An RGBA value is a group of numbers that specify the amount of red, green, blue, and *alpha* (or transparency) in a color. Each of these component values is an integer from 0 (none at all) to 255 (the maximum). These RGBA values are assigned to individual *pixels*; a pixel is the smallest dot of a single color the computer screen can show (as you can imagine, there are millions of pixels on a screen). A pixel's RGB setting tells it precisely what shade of color it should display. Images also have an alpha value to create RGBA values. If an image is displayed on the screen over a background image or desktop wallpaper, the alpha value determines how much of the background you can "see through" the image's pixel.

In Pillow, RGBA values are represented by a tuple of four integer values. For example, the color red is represented by (255, 0, 0, 255). This color has the maximum amount of red, no green or blue, and the maximum alpha value, meaning it is fully opaque. Green is represented by (0, 255, 0, 255), and blue is (0, 0, 255, 255). White, the combination of all colors, is (255, 255, 255, 255), while black, which has no color at all, is (0, 0, 0, 255).

If a color has an alpha value of 0, it is invisible, and it doesn't really matter what the RGB values are. After all, invisible red looks the same as invisible black.

Pillow uses the standard color names that HTML uses. Table 19-1 lists a selection of standard color names and their values.

**Table 19-1:** Standard Color Names and Their RGBA Values

| Name | RGBA value | Name | RGBA value |
|------|------------|------|------------|
| White | (255, 255, 255, 255) | Red | (255, 0, 0, 255) |
| Green | (0, 128, 0, 255) | Blue | (0, 0, 255, 255) |
| Gray | (128, 128, 128, 255) | Yellow | (255, 255, 0, 255) |
| Black | (0, 0, 0, 255) | Purple | (128, 0, 128, 255) |

Pillow offers the ImageColor.getcolor() function so you don't have to memorize RGBA values for the colors you want to use. This function takes a color name string as its first argument, and the string 'RGBA' as its second argument, and it returns an RGBA tuple.

To see how this function works, enter the following into the interactive shell:

```
❶ >>> from PIL import ImageColor
❷ >>> ImageColor.getcolor('red', 'RGBA')
(255, 0, 0, 255)
❸ >>> ImageColor.getcolor('RED', 'RGBA')
(255, 0, 0, 255)
>>> ImageColor.getcolor('Black', 'RGBA')
(0, 0, 0, 255)
>>> ImageColor.getcolor('chocolate', 'RGBA')
(210, 105, 30, 255)
>>> ImageColor.getcolor('CornflowerBlue', 'RGBA')
(100, 149, 237, 255)
```

First, you need to import the `ImageColor` module from PIL ❶ (not from Pillow; you'll see why in a moment). The color name string you pass to `ImageColor.getcolor()` is case-insensitive, so passing `'red'` ❷ and passing `'RED'` ❸ give you the same RGBA tuple. You can also pass more unusual color names, like `'chocolate'` and `'Cornflower Blue'`.

Pillow supports a huge number of color names, from `'aliceblue'` to `'whitesmoke'`. You can find the full list of more than 100 standard color names in the resources at *https://nostarch.com/automatestuff2/*.

## Coordinates and Box Tuples

Image pixels are addressed with x- and y-coordinates, which respectively specify a pixel's horizontal and vertical locations in an image. The *origin* is the pixel at the top-left corner of the image and is specified with the notation (0, 0). The first zero represents the x-coordinate, which starts at zero at the origin and increases going from left to right. The second zero represents the y-coordinate, which starts at zero at the origin and increases going down the image. This bears repeating: y-coordinates increase going downward, which is the opposite of how you may remember y-coordinates being used in math class. Figure 19-1 demonstrates how this coordinate system works.

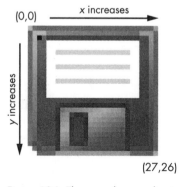

Figure 19-1: The x- and y-coordinates of a 28×27 image of some sort of ancient data storage device

Many of Pillow's functions and methods take a *box tuple* argument. This means Pillow is expecting a tuple of four integer coordinates that represent a rectangular region in an image. The four integers are, in order, as follows:

**Left**   The x-coordinate of the leftmost edge of the box.

**Top**   The y-coordinate of the top edge of the box.

**Right**   The x-coordinate of one pixel to the right of the rightmost edge of the box. This integer must be greater than the left integer.

**Bottom**   The y-coordinate of one pixel lower than the bottom edge of the box. This integer must be greater than the top integer.

Note that the box includes the left and top coordinates and goes up to but does not include the right and bottom coordinates. For example, the box tuple (3, 1, 9, 6) represents all the pixels in the black box in Figure 19-2.

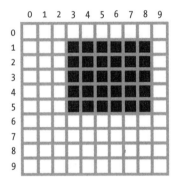

Figure 19-2: The area represented by the box tuple (3, 1, 9, 6)

## Manipulating Images with Pillow

Now that you know how colors and coordinates work in Pillow, let's use Pillow to manipulate an image. Figure 19-3 is the image that will be used for all the interactive shell examples in this chapter. You can download it from *https://nostarch.com/automatestuff2/*.

Once you have the image file *zophie.png* in your current working directory, you'll be ready to load the image of Zophie into Python, like so:

```
>>> from PIL import Image
>>> catIm = Image.open('zophie.png')
```

To load the image, import the Image module from Pillow and call Image .open(), passing it the image's filename. You can then store the loaded image in a variable like CatIm. Pillow's module name is PIL to make it backward compatible with an older module called Python Imaging Library; this is why you must run from PIL import Image instead of from Pillow import Image. Because of the way Pillow's creators set up the pillow module, you must use the import statement from PIL import Image, rather than simply import PIL.

Figure 19-3: My cat, Zophie. The camera adds 10 pounds (which is a lot for a cat).

If the image file isn't in the current working directory, change the working directory to the folder that contains the image file by calling the os.chdir() function.

```
>>> import os
>>> os.chdir('C:\\folder_with_image_file')
```

The Image.open() function returns a value of the Image object data type, which is how Pillow represents an image as a Python value. You can load an Image object from an image file (of any format) by passing the Image.open() function a string of the filename. Any changes you make to the Image object can be saved to an image file (also of any format) with the save() method. All the rotations, resizing, cropping, drawing, and other image manipulations will be done through method calls on this Image object.

To shorten the examples in this chapter, I'll assume you've imported Pillow's Image module and that you have the Zophie image stored in a variable named catIm. Be sure that the *zophie.png* file is in the current working directory so that the Image.open() function can find it. Otherwise, you will also have to specify the full absolute path in the string argument to Image.open().

## Working with the Image Data Type

An Image object has several useful attributes that give you basic information about the image file it was loaded from: its width and height, the filename, and the graphics format (such as JPEG, GIF, or PNG).

For example, enter the following into the interactive shell:

```
>>> from PIL import Image
>>> catIm = Image.open('zophie.png')
```

```
    >>> catIm.size
❶  (816, 1088)
❷  >>> width, height = catIm.size
❸  >>> width
    816
❹  >>> height
    1088
    >>> catIm.filename
    'zophie.png'
    >>> catIm.format
    'PNG'
    >>> catIm.format_description
    'Portable network graphics'
❺  >>> catIm.save('zophie.jpg')
```

After making an Image object from *zophie.png* and storing the Image object in catIm, we can see that the object's size attribute contains a tuple of the image's width and height in pixels ❶. We can assign the values in the tuple to width and height variables ❷ in order to access with width ❸ and height ❹ individually. The filename attribute describes the original file's name. The format and format_description attributes are strings that describe the image format of the original file (with format_description being a bit more verbose).

Finally, calling the save() method and passing it 'zophie.jpg' saves a new image with the filename *zophie.jpg* to your hard drive ❺. Pillow sees that the file extension is *.jpg* and automatically saves the image using the JPEG image format. Now you should have two images, *zophie.png* and *zophie .jpg*, on your hard drive. While these files are based on the same image, they are not identical because of their different formats.

Pillow also provides the Image.new() function, which returns an Image object—much like Image.open(), except the image represented by Image .new()'s object will be blank. The arguments to Image.new() are as follows:

- The string 'RGBA', which sets the color mode to RGBA. (There are other modes that this book doesn't go into.)
- The size, as a two-integer tuple of the new image's width and height.
- The background color that the image should start with, as a four-integer tuple of an RGBA value. You can use the return value of the ImageColor .getcolor() function for this argument. Alternatively, Image.new() also supports just passing the string of the standard color name.

For example, enter the following into the interactive shell:

```
    >>> from PIL import Image
❶  >>> im = Image.new('RGBA', (100, 200), 'purple')
    >>> im.save('purpleImage.png')
❷  >>> im2 = Image.new('RGBA', (20, 20))
    >>> im2.save('transparentImage.png')
```

Here we create an Image object for an image that's 100 pixels wide and 200 pixels tall, with a purple background ❶. This image is then saved to the file *purpleImage.png*. We call Image.new() again to create another Image object, this time passing (20, 20) for the dimensions and nothing for the background color ❷. Invisible black, (0, 0, 0, 0), is the default color used if no color argument is specified, so the second image has a transparent background; we save this 20×20 transparent square in *transparentImage.png*.

## Cropping Images

*Cropping* an image means selecting a rectangular region inside an image and removing everything outside the rectangle. The crop() method on Image objects takes a box tuple and returns an Image object representing the cropped image. The cropping does not happen in place—that is, the original Image object is left untouched, and the crop() method returns a new Image object. Remember that a boxed tuple—in this case, the cropped section—includes the left column and top row of pixels but only goes up to and does *not* include the right column and bottom row of pixels.

Enter the following into the interactive shell:

```
>>> from PIL import Image
>>> catIm = Image.open('zophie.png')
>>> croppedIm = catIm.crop((335, 345, 565, 560))
>>> croppedIm.save('cropped.png')
```

This makes a new Image object for the cropped image, stores the object in croppedIm, and then calls save() on croppedIm to save the cropped image in *cropped.png*. The new file *cropped.png* will be created from the original image, like in Figure 19-4.

Figure 19-4: The new image will be just the cropped section of the original image.

## Copying and Pasting Images onto Other Images

The copy() method will return a new Image object with the same image as the Image object it was called on. This is useful if you need to make changes to an image but also want to keep an untouched version of the original. For example, enter the following into the interactive shell:

```
>>> from PIL import Image
>>> catIm = Image.open('zophie.png')
>>> catCopyIm = catIm.copy()
```

The catIm and catCopyIm variables contain two separate Image objects, which both have the same image on them. Now that you have an Image object stored in catCopyIm, you can modify catCopyIm as you like and save it to a new filename, leaving *zophie.png* untouched. For example, let's try modifying catCopyIm with the paste() method.

The paste() method is called on an Image object and pastes another image on top of it. Let's continue the shell example by pasting a smaller image onto catCopyIm.

```
>>> faceIm = catIm.crop((335, 345, 565, 560))
>>> faceIm.size
(230, 215)
>>> catCopyIm.paste(faceIm, (0, 0))
>>> catCopyIm.paste(faceIm, (400, 500))
>>> catCopyIm.save('pasted.png')
```

First we pass crop() a box tuple for the rectangular area in *zophie.png* that contains Zophie's face. This creates an Image object representing a 230×215 crop, which we store in faceIm. Now we can paste faceIm onto catCopyIm. The paste() method takes two arguments: a "source" Image object and a tuple of the x- and y-coordinates where you want to paste the top-left corner of the source Image object onto the main Image object. Here we call paste() twice on catCopyIm, passing (0, 0) the first time and (400, 500) the second time. This pastes faceIm onto catCopyIm twice: once with the top-left corner of faceIm at (0, 0) on catCopyIm, and once with the top-left corner of faceIm at (400, 500). Finally, we save the modified catCopyIm to *pasted.png*. The *pasted.png* image looks like Figure 19-5.

**NOTE**   *Despite their names, the copy() and paste() methods in Pillow do not use your computer's clipboard.*

Note that the paste() method modifies its Image object *in place*; it does not return an Image object with the pasted image. If you want to call paste() but also keep an untouched version of the original image around, you'll need to first copy the image and then call paste() on that copy.

Figure 19-5: Zophie the cat, with her face pasted twice

Say you want to tile Zophie's head across the entire image, as in Figure 19-6. You can achieve this effect with just a couple for loops. Continue the interactive shell example by entering the following:

```
>>> catImWidth, catImHeight = catIm.size
>>> faceImWidth, faceImHeight = faceIm.size
❶ >>> catCopyTwo = catIm.copy()
❷ >>> for left in range(0, catImWidth, faceImWidth):
        ❸ for top in range(0, catImHeight, faceImHeight):
            print(left, top)
            catCopyTwo.paste(faceIm, (left, top))

0 0
0 215
0 430
0 645
0 860
0 1075
230 0
230 215
--snip--
690 860
690 1075
>>> catCopyTwo.save('tiled.png')
```

*Figure 19-6: Nested for loops used with paste() to duplicate the cat's face (a dupli-cat, if you will)*

Here we store the width of height of catIm in catImWidth and catImHeight. At ❶ we make a copy of catIm and store it in catCopyTwo. Now that we have a copy that we can paste onto, we start looping to paste faceIm onto catCopyTwo. The outer for loop's left variable starts at 0 and increases by faceImWidth(230) ❷. The inner for loop's top variable start at 0 and increases by faceImHeight(215) ❸. These nested for loops produce values for left and top to paste a grid of faceIm images over the catCopyTwo Image object, as in Figure 19-6. To see our nested loops working, we print left and top. After the pasting is complete, we save the modified catCopyTwo to *tiled.png*.

### Resizing an Image

The resize() method is called on an Image object and returns a new Image object of the specified width and height. It accepts a two-integer tuple argument, representing the new width and height of the returned image. Enter the following into the interactive shell:

```
>>> from PIL import Image
>>> catIm = Image.open('zophie.png')
❶ >>> width, height = catIm.size
❷ >>> quartersizedIm = catIm.resize((int(width / 2), int(height / 2)))
>>> quartersizedIm.save('quartersized.png')
❸ >>> svelteIm = catIm.resize((width, height + 300))
>>> svelteIm.save('svelte.png')
```

Here we assign the two values in the catIm.size tuple to the variables width and height ❶. Using width and height instead of catIm.size[0] and catIm.size[1] makes the rest of the code more readable.

The first resize() call passes int(width / 2) for the new width and int(height / 2) for the new height ❷, so the Image object returned from resize() will be half the length and width of the original image, or one-quarter of the original image size overall. The resize() method accepts only integers in its tuple argument, which is why you needed to wrap both divisions by 2 in an int() call.

This resizing keeps the same proportions for the width and height. But the new width and height passed to resize() do not have to be proportional to the original image. The svelteIm variable contains an Image object that has the original width but a height that is 300 pixels taller ❸, giving Zophie a more slender look.

Note that the resize() method does not edit the Image object in place but instead returns a new Image object.

### Rotating and Flipping Images

Images can be rotated with the rotate() method, which returns a new Image object of the rotated image and leaves the original Image object unchanged. The argument to rotate() is a single integer or float representing the number of degrees to rotate the image counterclockwise. Enter the following into the interactive shell:

```
>>> from PIL import Image
>>> catIm = Image.open('zophie.png')
>>> catIm.rotate(90).save('rotated90.png')
>>> catIm.rotate(180).save('rotated180.png')
>>> catIm.rotate(270).save('rotated270.png')
```

Note how you can *chain* method calls by calling save() directly on the Image object returned from rotate(). The first rotate() and save() call makes a new Image object representing the image rotated counterclockwise by 90 degrees and saves the rotated image to *rotated90.png*. The second and third calls do the same, but with 180 degrees and 270 degrees. The results look like Figure 19-7.

Figure 19-7: The original image (left) and the image rotated counterclockwise by 90, 180, and 270 degrees

Notice that the width and height of the image change when the image is rotated 90 or 270 degrees. If you rotate an image by some other amount, the original dimensions of the image are maintained. On Windows, a black background is used to fill in any gaps made by the rotation, like in Figure 19-8. On macOS, transparent pixels are used for the gaps instead.

The rotate() method has an optional expand keyword argument that can be set to True to enlarge the dimensions of the image to fit the entire rotated new image. For example, enter the following into the interactive shell:

```
>>> catIm.rotate(6).save('rotated6.png')
>>> catIm.rotate(6, expand=True).save('rotated6_expanded.png')
```

The first call rotates the image 6 degrees and saves it to *rotate6.png* (see the image on the left of Figure 19-8). The second call rotates the image 6 degrees with expand set to True and saves it to *rotate6_expanded.png* (see the image on the right of Figure 19-8).

Figure 19-8: The image rotated 6 degrees normally (left) and with expand=True (right)

You can also get a "mirror flip" of an image with the transpose() method. You must pass either Image.FLIP_LEFT_RIGHT or Image.FLIP_TOP_BOTTOM to the transpose() method. Enter the following into the interactive shell:

```
>>> catIm.transpose(Image.FLIP_LEFT_RIGHT).save('horizontal_flip.png')
>>> catIm.transpose(Image.FLIP_TOP_BOTTOM).save('vertical_flip.png')
```

Like rotate(), transpose() creates a new Image object. Here we pass Image .FLIP_LEFT_RIGHT to flip the image horizontally and then save the result to *horizontal_flip.png*. To flip the image vertically, we pass Image.FLIP_TOP_BOTTOM and save to *vertical_flip.png*. The results look like Figure 19-9.

Figure 19-9: The original image (left), horizontal flip (center), and vertical flip (right)

## Changing Individual Pixels

The color of an individual pixel can be retrieved or set with the getpixel() and putpixel() methods. These methods both take a tuple representing the x- and y-coordinates of the pixel. The putpixel() method also takes an additional tuple argument for the color of the pixel. This color argument is a four-integer RGBA tuple or a three-integer RGB tuple. Enter the following into the interactive shell:

```
>>> from PIL import Image
❶ >>> im = Image.new('RGBA', (100, 100))
❷ >>> im.getpixel((0, 0))
(0, 0, 0, 0)
❸ >>> for x in range(100):
        for y in range(50):
            ❹ im.putpixel((x, y), (210, 210, 210))

>>> from PIL import ImageColor
❺ >>> for x in range(100):
        for y in range(50, 100):
            ❻ im.putpixel((x, y), ImageColor.getcolor('darkgray', 'RGBA'))
>>> im.getpixel((0, 0))
(210, 210, 210, 255)
>>> im.getpixel((0, 50))
(169, 169, 169, 255)
>>> im.save('putPixel.png')
```

At ❶ we make a new image that is a 100×100 transparent square. Calling getpixel() on some coordinates in this image returns (0, 0, 0, 0) because the image is transparent ❷. To color pixels in this image, we can use nested for loops to go through all the pixels in the top half of the image ❸ and color each pixel using putpixel() ❹. Here we pass putpixel() the RGB tuple (210, 210, 210), a light gray.

Say we want to color the bottom half of the image dark gray but don't know the RGB tuple for dark gray. The putpixel() method doesn't accept a standard color name like 'darkgray', so you have to use ImageColor.getcolor() to get a color tuple from 'darkgray'. Loop through the pixels in the bottom half of the image ❺ and pass putpixel() the return value of ImageColor.getcolor() ❻, and you should now have an image that is light gray in its top half and dark gray in the bottom half, as shown in Figure 19-10. You can call getpixel() on some coordinates to confirm that the color at any given pixel is what you expect. Finally, save the image to *putPixel.png*.

Figure 19-10: The
putPixel.png *image*

Of course, drawing one pixel at a time onto an image isn't very convenient. If you need to draw shapes, use the ImageDraw functions explained later in this chapter.

## Project: Adding a Logo

Say you have the boring job of resizing thousands of images and adding a small logo watermark to the corner of each. Doing this with a basic graphics program such as Paintbrush or Paint would take forever. A fancier graphics application such as Photoshop can do batch processing, but that software costs hundreds of dollars. Let's write a script to do it instead.

Say that Figure 19-11 is the logo you want to add to the bottom-right corner of each image: a black cat icon with a white border, with the rest of the image transparent.

Figure 19-11: The logo to be
added to the image

At a high level, here's what the program should do:

1. Load the logo image.
2. Loop over all *.png* and *.jpg* files in the working directory.
3. Check whether the image is wider or taller than 300 pixels.
4. If so, reduce the width or height (whichever is larger) to 300 pixels and scale down the other dimension proportionally.
5. Paste the logo image into the corner.
6. Save the altered images to another folder.

This means the code will need to do the following:

1. Open the *catlogo.png* file as an `Image` object.
2. Loop over the strings returned from `os.listdir('.')`.
3. Get the width and height of the image from the size attribute.
4. Calculate the new width and height of the resized image.
5. Call the `resize()` method to resize the image.
6. Call the `paste()` method to paste the logo.
7. Call the `save()` method to save the changes, using the original filename.

## Step 1: Open the Logo Image

For this project, open a new file editor tab, enter the following code, and save it as *resizeAndAddLogo.py*:

```python
#! python3
# resizeAndAddLogo.py - Resizes all images in current working directory to fit
# in a 300x300 square, and adds catlogo.png to the lower-right corner.

import os
from PIL import Image

SQUARE_FIT_SIZE = 300
LOGO_FILENAME = 'catlogo.png'

logoIm = Image.open(LOGO_FILENAME)
logoWidth, logoHeight = logoIm.size

    # TODO: Loop over all files in the working directory.

    # TODO: Check if image needs to be resized.

    # TODO: Calculate the new width and height to resize to.

    # TODO: Resize the image.

    # TODO: Add the logo.

    # TODO: Save changes.
```

By setting up the SQUARE_FIT_SIZE ❶ and LOGO_FILENAME ❷ constants at the start of the program, we've made it easy to change the program later. Say the logo that you're adding isn't the cat icon, or say you're reducing the output images' largest dimension to something other than 300 pixels. With these constants at the start of the program, you can just open the code, change those values once, and you're done. (Or you can make it so that the values for these constants are taken from the command line arguments.) Without these constants, you'd instead have to search the code for all instances of 300 and 'catlogo.png' and replace them with the values for your new project. In short, using constants makes your program more generalized.

The logo Image object is returned from Image.open() ❸. For readability, logoWidth and logoHeight are assigned to the values from logoIm.size ❹.

The rest of the program is a skeleton of TODO comments for now.

## Step 2: Loop Over All Files and Open Images

Now you need to find every *.png* file and *.jpg* file in the current working directory. You don't want to add the logo image to the logo image itself, so the program should skip any image with a filename that's the same as LOGO_FILENAME. Add the following to your code:

```
#! python3
# resizeAndAddLogo.py - Resizes all images in current working directory to fit
# in a 300x300 square, and adds catlogo.png to the lower-right corner.

import os
from PIL import Image

--snip--

os.makedirs('withLogo', exist_ok=True)
# Loop over all files in the working directory.
❶ for filename in os.listdir('.'):
❷    if not (filename.endswith('.png') or filename.endswith('.jpg')) \
         or filename == LOGO_FILENAME:
❸        continue    # skip non-image files and the logo file itself

❹    im = Image.open(filename)
     width, height = im.size

--snip--
```

First, the os.makedirs() call creates a *withLogo* folder to store the finished images with logos, instead of overwriting the original image files. The exist_ok=True keyword argument will keep os.makedirs() from raising an exception if *withLogo* already exists. While looping through all the files in the working directory with os.listdir('.') ❶, the long if statement ❷ checks whether each filename doesn't end with *.png* or *.jpg*. If so—or if the file is the logo image itself—then the loop should skip it and use continue ❸ to go to the next file. If filename *does* end with '.png' or '.jpg' (and isn't the logo file), you can open it as an Image object ❹ and set width and height.

## Step 3: Resize the Images

The program should resize the image only if the width or height is larger than SQUARE_FIT_SIZE (300 pixels, in this case), so put all of the resizing code inside an if statement that checks the width and height variables. Add the following code to your program:

```
#! python3
# resizeAndAddLogo.py - Resizes all images in current working directory to fit
# in a 300x300 square, and adds catlogo.png to the lower-right corner.

import os
from PIL import Image

--snip--

    # Check if image needs to be resized.
    if width > SQUARE_FIT_SIZE and height > SQUARE_FIT_SIZE:
        # Calculate the new width and height to resize to.
        if width > height:
          ❶ height = int((SQUARE_FIT_SIZE / width) * height)
            width = SQUARE_FIT_SIZE
        else:
          ❷ width = int((SQUARE_FIT_SIZE / height) * width)
            height = SQUARE_FIT_SIZE

        # Resize the image.
        print('Resizing %s...' % (filename))
      ❸ im = im.resize((width, height))

--snip--
```

If the image does need to be resized, you need to find out whether it is a wide or tall image. If width is greater than height, then the height should be reduced by the same proportion that the width would be reduced ❶. This proportion is the SQUARE_FIT_SIZE value divided by the current width. The new height value is this proportion multiplied by the current height value. Since the division operator returns a float value and resize() requires the dimensions to be integers, remember to convert the result to an integer with the int() function. Finally, the new width value will simply be set to SQUARE_FIT_SIZE.

If the height is greater than or equal to the width (both cases are handled in the else clause), then the same calculation is done, except with the height and width variables swapped ❷.

Once width and height contain the new image dimensions, pass them to the resize() method and store the returned Image object in im ❸.

## Step 4: Add the Logo and Save the Changes

Whether or not the image was resized, the logo should still be pasted to the bottom-right corner. Where exactly the logo should be pasted depends on both the size of the image and the size of the logo. Figure 19-12 shows how to calculate the pasting position. The left coordinate for where to paste the logo will be the image width minus the logo width; the top coordinate for where to paste the logo will be the image height minus the logo height.

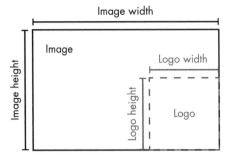

Figure 19-12: The left and top coordinates for placing the logo in the bottom-right corner should be the image width/height minus the logo width/height.

After your code pastes the logo into the image, it should save the modified `Image` object. Add the following to your program:

```
#! python3
# resizeAndAddLogo.py - Resizes all images in current working directory to fit
# in a 300x300 square, and adds catlogo.png to the lower-right corner.

import os
from PIL import Image

--snip--

    # Check if image needs to be resized.
    --snip--

    # Add the logo.
❶ print('Adding logo to %s...' % (filename))
❷ im.paste(logoIm, (width - logoWidth, height - logoHeight), logoIm)

    # Save changes.
❸ im.save(os.path.join('withLogo', filename))
```

The new code prints a message telling the user that the logo is being added ❶, pastes `logoIm` onto `im` at the calculated coordinates ❷, and saves the changes to a filename in the *withLogo* directory ❸. When you run this program with the *zophie.png* file as the only image in the working directory, the output will look like this:

```
Resizing zophie.png...
Adding logo to zophie.png...
```

The image *zophie.png* will be changed to a 225×300-pixel image that looks like Figure 19-13. Remember that the paste() method will not paste the transparency pixels if you do not pass the logoIm for the third argument as well. This program can automatically resize and "logo-ify" hundreds of images in just a couple minutes.

Figure 19-13: The image zophie.png *resized and the logo added (left). If you forget the third argument, the transparent pixels in the logo will be copied as solid white pixels (right).*

### Ideas for Similar Programs

Being able to composite images or modify image sizes in a batch can be useful in many applications. You could write similar programs to do the following:

- Add text or a website URL to images.
- Add timestamps to images.
- Copy or move images into different folders based on their sizes.
- Add a mostly transparent watermark to an image to prevent others from copying it.

## Drawing on Images

If you need to draw lines, rectangles, circles, or other simple shapes on an image, use Pillow's ImageDraw module. Enter the following into the interactive shell:

```
>>> from PIL import Image, ImageDraw
>>> im = Image.new('RGBA', (200, 200), 'white')
>>> draw = ImageDraw.Draw(im)
```

First, we import `Image` and `ImageDraw`. Then we create a new image, in this case, a 200×200 white image, and store the `Image` object in im. We pass the `Image` object to the `ImageDraw.Draw()` function to receive an `ImageDraw` object. This object has several methods for drawing shapes and text onto an `Image` object. Store the `ImageDraw` object in a variable like draw so you can use it easily in the following example.

## Drawing Shapes

The following ImageDraw methods draw various kinds of shapes on the image. The `fill` and `outline` parameters for these methods are optional and will default to white if left unspecified.

### Points

The `point(xy, fill)` method draws individual pixels. The *xy* argument represents a list of the points you want to draw. The list can be a list of x- and y-coordinate tuples, such as [(x, y), (x, y), ...], or a list of x- and y-coordinates without tuples, such as [x1, y1, x2, y2, ...]. The *fill* argument is the color of the points and is either an RGBA tuple or a string of a color name, such as 'red'. The *fill* argument is optional.

### Lines

The `line(xy, fill, width)` method draws a line or series of lines. *xy* is either a list of tuples, such as [(x, y), (x, y), ...], or a list of integers, such as [x1, y1, x2, y2, ...]. Each point is one of the connecting points on the lines you're drawing. The optional *fill* argument is the color of the lines, as an RGBA tuple or color name. The optional *width* argument is the width of the lines and defaults to 1 if left unspecified.

### Rectangles

The `rectangle(xy, fill, outline)` method draws a rectangle. The *xy* argument is a box tuple of the form (*left*, *top*, *right*, *bottom*). The *left* and *top* values specify the x- and y-coordinates of the upper-left corner of the rectangle, while *right* and *bottom* specify the lower-right corner. The optional *fill* argument is the color that will fill the inside of the rectangle. The optional *outline* argument is the color of the rectangle's outline.

### Ellipses

The `ellipse(xy, fill, outline)` method draws an ellipse. If the width and height of the ellipse are identical, this method will draw a circle. The *xy* argument is a box tuple (*left*, *top*, *right*, *bottom*) that represents a box that precisely contains the ellipse. The optional *fill* argument is the color of the inside of the ellipse, and the optional *outline* argument is the color of the ellipse's outline.

## Polygons

The polygon(*xy*, *fill*, *outline*) method draws an arbitrary polygon. The *xy* argument is a list of tuples, such as [(x, y), (x, y), ...], or integers, such as [x1, y1, x2, y2, ...], representing the connecting points of the polygon's sides. The last pair of coordinates will be automatically connected to the first pair. The optional *fill* argument is the color of the inside of the polygon, and the optional *outline* argument is the color of the polygon's outline.

## Drawing Example

Enter the following into the interactive shell:

```
>>> from PIL import Image, ImageDraw
>>> im = Image.new('RGBA', (200, 200), 'white')
>>> draw = ImageDraw.Draw(im)
❶ >>> draw.line([(0, 0), (199, 0), (199, 199), (0, 199), (0, 0)], fill='black')
❷ >>> draw.rectangle((20, 30, 60, 60), fill='blue')
❸ >>> draw.ellipse((120, 30, 160, 60), fill='red')
❹ >>> draw.polygon(((57, 87), (79, 62), (94, 85), (120, 90), (103, 113)),
fill='brown')
❺ >>> for i in range(100, 200, 10):
        draw.line([(i, 0), (200, i - 100)], fill='green')

>>> im.save('drawing.png')
```

After making an Image object for a 200×200 white image, passing it to ImageDraw.Draw() to get an ImageDraw object, and storing the ImageDraw object in draw, you can call drawing methods on draw. Here we make a thin, black outline at the edges of the image ❶, a blue rectangle with its top-left corner at (20, 30) and bottom-right corner at (60, 60) ❷, a red ellipse defined by a box from (120, 30) to (160, 60) ❸, a brown polygon with five points ❹, and a pattern of green lines drawn with a for loop ❺. The resulting *drawing.png* file will look like Figure 19-14.

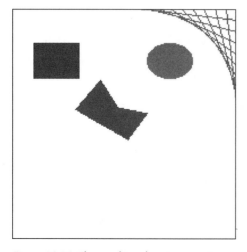

*Figure 19-14: The resulting* drawing.png *image*

There are several other shape-drawing methods for `ImageDraw` objects. The full documentation is available at *https://pillow.readthedocs.io/en/latest /reference/ImageDraw.html*.

### Drawing Text

The `ImageDraw` object also has a `text()` method for drawing text onto an image. The `text()` method takes four arguments: *xy*, *text*, *fill*, and *font*.

- The *xy* argument is a two-integer tuple specifying the upper-left corner of the text box.
- The *text* argument is the string of text you want to write.
- The optional *fill* argument is the color of the text.
- The optional *font* argument is an `ImageFont` object, used to set the typeface and size of the text. This is described in more detail in the next section.

Since it's often hard to know in advance what size a block of text will be in a given font, the `ImageDraw` module also offers a `textsize()` method. Its first argument is the string of text you want to measure, and its second argument is an optional `ImageFont` object. The `textsize()` method will then return a two-integer tuple of the width and height that the text in the given font would be if it were written onto the image. You can use this width and height to help you calculate exactly where you want to put the text on your image.

The first three arguments for `text()` are straightforward. Before we use `text()` to draw text onto an image, let's look at the optional fourth argument, the `ImageFont` object.

Both `text()` and `textsize()` take an optional `ImageFont` object as their final arguments. To create one of these objects, first run the following:

```
>>> from PIL import ImageFont
```

Now that you've imported Pillow's `ImageFont` module, you can call the `ImageFont.truetype()` function, which takes two arguments. The first argument is a string for the font's *TrueType file*—this is the actual font file that lives on your hard drive. A TrueType file has the *.ttf* file extension and can usually be found in the following folders:

- On Windows: *C:\Windows\Fonts*
- On macOS: */Library/Fonts* and */System/Library/Fonts*
- On Linux: */usr/share/fonts/truetype*

You don't actually need to enter these paths as part of the TrueType file string because Python knows to automatically search for fonts in these directories. But Python will display an error if it is unable to find the font you specified.

The second argument to `ImageFont.truetype()` is an integer for the font size in *points* (rather than, say, pixels). Keep in mind that Pillow creates PNG images that are 72 pixels per inch by default, and a point is 1/72 of an inch.

Enter the following into the interactive shell, replacing FONT_FOLDER with the actual folder name your operating system uses:

```
>>> from PIL import Image, ImageDraw, ImageFont
>>> import os
❶ >>> im = Image.new('RGBA', (200, 200), 'white')
❷ >>> draw = ImageDraw.Draw(im)
❸ >>> draw.text((20, 150), 'Hello', fill='purple')
>>> fontsFolder = 'FONT_FOLDER' # e.g. '/Library/Fonts'
❹ >>> arialFont = ImageFont.truetype(os.path.join(fontsFolder, 'arial.ttf'), 32)
❺ >>> draw.text((100, 150), 'Howdy', fill='gray', font=arialFont)
>>> im.save('text.png')
```

After importing Image, ImageDraw, ImageFont, and os, we make an Image object for a new 200×200 white image ❶ and make an ImageDraw object from the Image object ❷. We use text() to draw *Hello* at (20, 150) in purple ❸. We didn't pass the optional fourth argument in this text() call, so the typeface and size of this text aren't customized.

To set a typeface and size, we first store the folder name (like */Library /Fonts*) in fontsFolder. Then we call ImageFont.truetype(), passing it the *.ttf* file for the font we want, followed by an integer font size ❹. Store the Font object you get from ImageFont.truetype() in a variable like arialFont, and then pass the variable to text() in the final keyword argument. The text() call at ❺ draws *Howdy* at (100, 150) in gray in 32-point Arial.

The resulting *text.png* file will look like Figure 19-15.

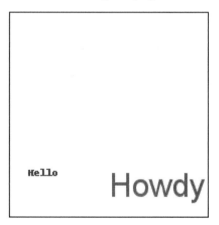

*Figure 19-15: The resulting* text.png *image*

## Summary

Images consist of a collection of pixels, and each pixel has an RGBA value for its color and its addressable by x- and y-coordinates. Two common image formats are JPEG and PNG. The pillow module can handle both of these image formats and others.

When an image is loaded into an Image object, its width and height dimensions are stored as a two-integer tuple in the size attribute. Objects of the Image data type also have methods for common image manipulations: crop(), copy(), paste(), resize(), rotate(), and transpose(). To save the Image object to an image file, call the save() method.

If you want your program to draw shapes onto an image, use ImageDraw methods to draw points, lines, rectangles, ellipses, and polygons. The module also provides methods for drawing text in a typeface and font size of your choosing.

Although advanced (and expensive) applications such as Photoshop provide automatic batch processing features, you can use Python scripts to do many of the same modifications for free. In the previous chapters, you wrote Python programs to deal with plaintext files, spreadsheets, PDFs, and other formats. With the pillow module, you've extended your programming powers to processing images as well!

## Practice Questions

1. What is an RGBA value?

2. How can you get the RGBA value of 'CornflowerBlue' from the Pillow module?

3. What is a box tuple?

4. What function returns an Image object for, say, an image file named *zophie.png*?

5. How can you find out the width and height of an Image object's image?

6. What method would you call to get Image object for a 100×100 image, excluding the lower-left quarter of it?

7. After making changes to an Image object, how could you save it as an image file?

8. What module contains Pillow's shape-drawing code?

9. Image objects do not have drawing methods. What kind of object does? How do you get this kind of object?

## Practice Projects

For practice, write programs that do the following.

### Extending and Fixing the Chapter Project Programs

The *resizeAndAddLogo.py* program in this chapter works with PNG and JPEG files, but Pillow supports many more formats than just these two. Extend *resizeAndAddLogo.py* to process GIF and BMP images as well.

Another small issue is that the program modifies PNG and JPEG files only if their file extensions are set in lowercase. For example, it will process *zophie.png* but not *zophie.PNG*. Change the code so that the file extension check is case insensitive.

Finally, the logo added to the bottom-right corner is meant to be just a small mark, but if the image is about the same size as the logo itself, the result will look like Figure 19-16. Modify *resizeAndAddLogo.py* so that the image must be at least twice the width and height of the logo image before the logo is pasted. Otherwise, it should skip adding the logo.

Figure 19-16: When the image isn't much larger than the logo, the results look ugly.

## Identifying Photo Folders on the Hard Drive

I have a bad habit of transferring files from my digital camera to temporary folders somewhere on the hard drive and then forgetting about these folders. It would be nice to write a program that could scan the entire hard drive and find these leftover "photo folders."

Write a program that goes through every folder on your hard drive and finds potential photo folders. Of course, first you'll have to define what you consider a "photo folder" to be; let's say that it's any folder where more than half of the files are photos. And how do you define what files are photos? First, a photo file must have the file extension *.png* or *.jpg*. Also, photos are large images; a photo file's width and height must both be larger than 500 pixels. This is a safe bet, since most digital camera photos are several thousand pixels in width and height.

As a hint, here's a rough skeleton of what this program might look like:

```
#! python3
# Import modules and write comments to describe this program.

for foldername, subfolders, filenames in os.walk('C:\\'):
    numPhotoFiles = 0
    numNonPhotoFiles = 0
    for filename in filenames:
        # Check if file extension isn't .png or .jpg.
```

```
        if TODO:
            numNonPhotoFiles += 1
            continue    # skip to next filename

        # Open image file using Pillow.

        # Check if width & height are larger than 500.
        if TODO:
            # Image is large enough to be considered a photo.
            numPhotoFiles += 1
        else:
            # Image is too small to be a photo.
            numNonPhotoFiles += 1

    # If more than half of files were photos,
    # print the absolute path of the folder.
    if TODO:
        print(TODO)
```

When the program runs, it should print the absolute path of any photo folders to the screen.

## Custom Seating Cards

Chapter 15 included a practice project to create custom invitations from a list of guests in a plaintext file. As an additional project, use the pillow module to create images for custom seating cards for your guests. For each of the guests listed in the *guests.txt* file from the resources at *https://nostarch.com/automatestuff2/*, generate an image file with the guest name and some flowery decoration. A public domain flower image is also available in the book's resources.

To ensure that each seating card is the same size, add a black rectangle on the edges of the invitation image so that when the image is printed out, there will be a guideline for cutting. The PNG files that Pillow produces are set to 72 pixels per inch, so a 4×5-inch card would require a 288×360-pixel image.

# 20

## CONTROLLING THE KEYBOARD
## AND MOUSE WITH GUI
## AUTOMATION

Knowing various Python modules for editing spreadsheets, downloading files, and launching programs is useful, but sometimes there just aren't any modules for the applications you need to work with. The ultimate tools for automating tasks on your computer are programs you write that directly control the keyboard and mouse. These programs can control other applications by sending them virtual keystrokes and mouse clicks, just as if you were sitting at your computer and interacting with the applications yourself.

This technique is known as *graphical user interface automation*, or *GUI automation* for short. With GUI automation, your programs can do anything that a human user sitting at the computer can do, except spill coffee on the keyboard. Think of GUI automation as programming a robotic arm. You can program the robotic arm to type at your keyboard and move your mouse for you. This technique is particularly useful for tasks that involve a lot of mindless clicking or filling out of forms.

Some companies sell innovative (and pricey) "automation solutions," usually marketed as *robotic process automation (RPA)*. These products are effectively no different than the Python scripts you can make yourself with the pyautogui module, which has functions for simulating mouse movements, button clicks, and mouse wheel scrolls. This chapter covers only a subset of PyAutoGUI's features; you can find the full documentation at *https://pyautogui.readthedocs.io/*.

## Installing the pyautogui Module

The pyautogui module can send virtual keypresses and mouse clicks to Windows, macOS, and Linux. Windows and macOS users can simply use pip to install PyAutoGUI. However, Linux users will first have to install some software that PyAutoGUI depends on. Open a terminal window and enter the following commands:

- `sudo apt-get install scrot`
- `sudo apt-get install python3-tk`
- `sudo apt-get install python3-dev`

To install PyAutoGUI, run `pip install --user pyautogui`. Don't use `sudo` with `pip`; you may install modules to the Python installation that the operating system uses, causing conflicts with any scripts that rely on its original configuration. However, you should use the `sudo` command when installing applications with `apt-get`.

Appendix A has complete information on installing third-party modules. To test whether PyAutoGUI has been installed correctly, run `import pyautogui` from the interactive shell and check for any error messages.

**WARNING**   *Don't save your program as* pyautogui.py. *When you run* `import pyautogui`, *Python will import your program instead of the PyAutoGUI and you'll get error messages like* `AttributeError: module 'pyautogui' has no attribute 'click'`.

## Setting Up Accessibility Apps on macOS

As a security measure, macOS doesn't normally let programs control the mouse or keyboard. To make PyAutoGUI work on macOS, you must set the program running your Python script to be an accessibility application. Without this step, your PyAutoGUI function calls will have no effect.

Whether you run your Python programs from Mu, IDLE, or the Terminal, have that application open. Then open the System Preferences and go to the Accessibility tab. The currently open applications will appear under the "Allow the apps below to control your computer" label. Check

Mu, IDLE, Terminal, or whichever app you use to run your Python scripts. You'll be prompted to enter your password to confirm these changes.

## Staying on Track

Before you jump into a GUI automation, you should know how to escape problems that may arise. Python can move your mouse and type keystrokes at an incredible speed. In fact, it might be too fast for other programs to keep up with. Also, if something goes wrong but your program keeps moving the mouse around, it will be hard to tell what exactly the program is doing or how to recover from the problem. Like the enchanted brooms from Disney's *The Sorcerer's Apprentice*, which kept filling—and then overfilling—Mickey's tub with water, your program could get out of control even though it's following your instructions perfectly. Stopping the program can be difficult if the mouse is moving around on its own, preventing you from clicking the Mu Editor window to close it. Fortunately, there are several ways to prevent or recover from GUI automation problems.

### Pauses and Fail-Safes

If your program has a bug and you're unable to use the keyboard and mouse to shut it down, you can use PyAutoGUI's fail-safe feature. Quickly slide the mouse to one of the four corners of the screen. Every PyAutoGUI function call has a 10th-of-a-second delay after performing its action to give you enough time to move the mouse to a corner. If PyAutoGUI then finds that the mouse cursor is in a corner, it raises the pyautogui.FailSafeException exception. Non-PyAutoGUI instructions will not have this 10th-of-a-second delay.

If you find yourself in a situation where you need to stop your PyAutoGUI program, just slam the mouse toward a corner to stop it.

### Shutting Down Everything by Logging Out

Perhaps the simplest way to stop an out-of-control GUI automation program is to log out, which will shut down all running programs. On Windows and Linux, the logout hotkey is CTRL-ALT-DEL. On macOS, it is ⌘-SHIFT-OPTION-Q. By logging out, you'll lose any unsaved work, but at least you won't have to wait for a full reboot of the computer.

## Controlling Mouse Movement

In this section, you'll learn how to move the mouse and track its position on the screen using PyAutoGUI, but first you need to understand how PyAutoGUI works with coordinates.

The mouse functions of PyAutoGUI use x- and y-coordinates. Figure 20-1 shows the coordinate system for the computer screen; it's similar to the coordinate system used for images, discussed in Chapter 19. The *origin*, where *x* and *y* are both zero, is at the upper-left corner of the screen. The x-coordinates increase going to the right, and the y-coordinates increase going down. All coordinates are positive integers; there are no negative coordinates.

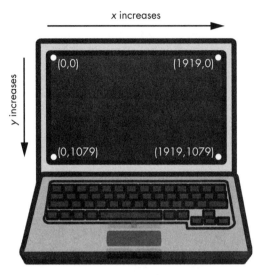

Figure 20-1: The coordinates of a computer screen with 1920×1080 resolution

Your *resolution* is how many pixels wide and tall your screen is. If your screen's resolution is set to 1920×1080, then the coordinate for the upper-left corner will be (0, 0), and the coordinate for the bottom-right corner will be (1919, 1079).

The pyautogui.size() function returns a two-integer tuple of the screen's width and height in pixels. Enter the following into the interactive shell:

```
>>> import pyautogui
>>> wh = pyautogui.size() # Obtain the screen resolution.
>>> wh
Size(width=1920, height=1080)
>>> wh[0]
1920
>>> wh.width
1920
```

The pyautogui.size() function returns (1920, 1080) on a computer with a 1920×1080 resolution; depending on your screen's resolution, your return value may be different. The Size object returned by size() is a named tuple. *Named tuples* have numeric indexes, like regular tuples, and attribute names, like objects: both wh[0] and wh.width evaluate to the width of the screen. (Named tuples are beyond the scope of this book. Just remember that you can use them the same way you use tuples.)

## Moving the Mouse

Now that you understand screen coordinates, let's move the mouse. The pyautogui.moveTo() function will instantly move the mouse cursor to a specified position on the screen. Integer values for the x- and y-coordinates make up the function's first and second arguments, respectively. An optional duration integer or float keyword argument specifies the number of seconds it should take to move the mouse to the destination. If you leave it out, the default is 0 for instantaneous movement. (All of the duration keyword arguments in PyAutoGUI functions are optional.) Enter the following into the interactive shell:

```
>>> import pyautogui
>>> for i in range(10): # Move mouse in a square.
...        pyautogui.moveTo(100, 100, duration=0.25)
...        pyautogui.moveTo(200, 100, duration=0.25)
...        pyautogui.moveTo(200, 200, duration=0.25)
...        pyautogui.moveTo(100, 200, duration=0.25)
```

This example moves the mouse cursor clockwise in a square pattern among the four coordinates provided a total of 10 times. Each movement takes a quarter of a second, as specified by the duration=0.25 keyword argument. If you hadn't passed a third argument to any of the pyautogui.moveTo() calls, the mouse cursor would have instantly teleported from point to point.

The pyautogui.move() function moves the mouse cursor *relative to its current position*. The following example moves the mouse in the same square pattern, except it begins the square from wherever the mouse happens to be on the screen when the code starts running:

```
>>> import pyautogui
>>> for i in range(10):
...        pyautogui.move(100, 0, duration=0.25)     # right
...        pyautogui.move(0, 100, duration=0.25)     # down
...        pyautogui.move(-100, 0, duration=0.25)    # left
...        pyautogui.move(0, -100, duration=0.25)    # up
```

The pyautogui.move() function also takes three arguments: how many pixels to move horizontally to the right, how many pixels to move vertically downward, and (optionally) how long it should take to complete the movement. A negative integer for the first or second argument will cause the mouse to move left or upward, respectively.

## Getting the Mouse Position

You can determine the mouse's current position by calling the pyautogui .position() function, which will return a Point named tuple of the mouse cursor's *x* and *y* positions at the time of the function call. Enter the following into the interactive shell, moving the mouse around after each call:

```
>>> pyautogui.position() # Get current mouse position.
Point(x=311, y=622)
```

```
>>> pyautogui.position()  # Get current mouse position again.
Point(x=377, y=481)
>>> p = pyautogui.position()  # And again.
>>> p
Point(x=1536, y=637)
>>> p[0]  # The x-coordinate is at index 0.
1536
>>> p.x  # The x-coordinate is also in the x attribute.
1536
```

Of course, your return values will vary depending on where your mouse cursor is.

## Controlling Mouse Interaction

Now that you know how to move the mouse and figure out where it is on the screen, you're ready to start clicking, dragging, and scrolling.

### Clicking the Mouse

To send a virtual mouse click to your computer, call the pyautogui.click() method. By default, this click uses the left mouse button and takes place wherever the mouse cursor is currently located. You can pass x- and y-coordinates of the click as optional first and second arguments if you want it to take place somewhere other than the mouse's current position.

If you want to specify which mouse button to use, include the button keyword argument, with a value of 'left', 'middle', or 'right'. For example, pyautogui.click(100, 150, button='left') will click the left mouse button at the coordinates (100, 150), while pyautogui.click(200, 250, button='right') will perform a right-click at (200, 250).

Enter the following into the interactive shell:

```
>>> import pyautogui
>>> pyautogui.click(10, 5)  # Move mouse to (10, 5) and click.
```

You should see the mouse pointer move to near the top-left corner of your screen and click once. A full "click" is defined as pushing a mouse button down and then releasing it back up without moving the cursor. You can also perform a click by calling pyautogui.mouseDown(), which only pushes the mouse button down, and pyautogui.mouseUp(), which only releases the button. These functions have the same arguments as click(), and in fact, the click() function is just a convenient wrapper around these two function calls.

As a further convenience, the pyautogui.doubleClick() function will perform two clicks with the left mouse button, while the pyautogui.rightClick() and pyautogui.middleClick() functions will perform a click with the right and middle mouse buttons, respectively.

## Dragging the Mouse

*Dragging* means moving the mouse while holding down one of the mouse buttons. For example, you can move files between folders by dragging the folder icons, or you can move appointments around in a calendar app.

PyAutoGUI provides the pyautogui.dragTo() and pyautogui.drag() functions to drag the mouse cursor to a new location or a location relative to its current one. The arguments for dragTo() and drag() are the same as moveTo() and move(): the x-coordinate/horizontal movement, the y-coordinate/vertical movement, and an optional duration of time. (macOS does not drag correctly when the mouse moves too quickly, so passing a duration keyword argument is recommended.)

To try these functions, open a graphics-drawing application such as MS Paint on Windows, Paintbrush on macOS, or GNU Paint on Linux. (If you don't have a drawing application, you can use the online one at *https://sumopaint.com/*.) I will use PyAutoGUI to draw in these applications.

With the mouse cursor over the drawing application's canvas and the Pencil or Brush tool selected, enter the following into a new file editor window and save it as *spiralDraw.py*:

```
  import pyautogui, time
❶ time.sleep(5)
❷ pyautogui.click()      # Click to make the window active.
  distance = 300
  change = 20
  while distance > 0:
❸     pyautogui.drag(distance, 0, duration=0.2)    # Move right.
❹     distance = distance - change
❺     pyautogui.drag(0, distance, duration=0.2)    # Move down.
❻     pyautogui.drag(-distance, 0, duration=0.2)   # Move left.
      distance = distance - change
      pyautogui.drag(0, -distance, duration=0.2)   # Move up.
```

When you run this program, there will be a five-second delay ❶ for you to move the mouse cursor over the drawing program's window with the Pencil or Brush tool selected. Then *spiralDraw.py* will take control of the mouse and click to make the drawing program's window active ❷. The *active window* is the window that currently accepts keyboard input, and the actions you take—like typing or, in this case, dragging the mouse—will affect that window. The active window is also known as the *focused* or *foreground window*. Once the drawing program is active, *spiralDraw.py* draws a square spiral pattern like the one on the left of Figure 20-2. While you can also create a square spiral image by using the Pillow module discussed in Chapter 19, creating the image by controlling the mouse to draw it in MS Paint lets you make use of this program's various brush styles, like in Figure 20-2 on the right, as well as other advanced features, like gradients or the fill bucket. You can preselect the brush settings yourself (or have your Python code select these settings) and then run the spiral-drawing program.

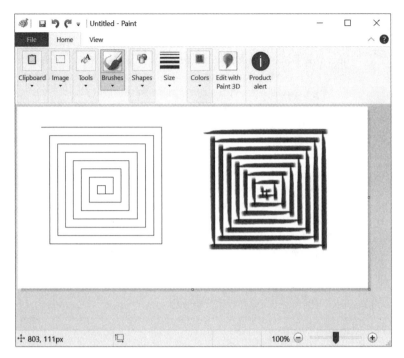

*Figure 20-2: The results from the pyautogui.drag() example, drawn with MS Paint's different brushes*

The distance variable starts at 200, so on the first iteration of the while loop, the first drag() call drags the cursor 200 pixels to the right, taking 0.2 seconds ❸. distance is then decreased to 195 ❹, and the second drag() call drags the cursor 195 pixels down ❺. The third drag() call drags the cursor –195 horizontally (195 to the left) ❻, distance is decreased to 190, and the last drag() call drags the cursor 190 pixels up. On each iteration, the mouse is dragged right, down, left, and up, and distance is slightly smaller than it was in the previous iteration. By looping over this code, you can move the mouse cursor to draw a square spiral.

You could draw this spiral by hand (or rather, by mouse), but you'd have to work slowly to be so precise. PyAutoGUI can do it in a few seconds!

**NOTE** *At the time of this writing, PyAutoGUI can't send mouse clicks or keystrokes to certain programs, such as antivirus software (to prevent viruses from disabling the software) or video games on Windows (which use a different method of receiving mouse and keyboard input). You can check the online documentation at https://pyautogui .readthedocs.io/ to see if these features have been added.*

## Scrolling the Mouse

The final PyAutoGUI mouse function is scroll(), which you pass an integer argument for how many units you want to scroll the mouse up or down. The size of a unit varies for each operating system and application, so you'll have to experiment to see exactly how far it scrolls in your particular situation.

The scrolling takes place at the mouse cursor's current position. Passing a positive integer scrolls up, and passing a negative integer scrolls down. Run the following in Mu Editor's interactive shell while the mouse cursor is over the Mu Editor window:

```
>>> pyautogui.scroll(200)
```

You'll see Mu scroll upward if the mouse cursor is over a text field that can be scrolled up.

## Planning Your Mouse Movements

One of the difficulties of writing a program that will automate clicking the screen is finding the x- and y-coordinates of the things you'd like to click. The pyautogui.mouseInfo() function can help you with this.

The pyautogui.mouseInfo() function is meant to be called from the interactive shell, rather than as part of your program. It launches a small application named MouseInfo that's included with PyAutoGUI. The window for the application looks like Figure 20-3.

Figure 20-3: The MouseInfo application's window

Enter the following into the interactive shell:

```
>>> import pyautogui
>>> pyautogui.mouseInfo()
```

This makes the MouseInfo window appear. This window gives you information about the mouse's cursor current position, as well the color of the pixel underneath the mouse cursor, as a three-integer RGB tuple and as a hex value. The color itself appears in the color box in the window.

To help you record this coordinate or pixel information, you can click one of the eight Copy or Log buttons. The Copy All, Copy XY, Copy RGB, and Copy RGB Hex buttons will copy their respective information to the clipboard. The Log All, Log XY, Log RGB, and Log RGB Hex buttons will write their respective information to the large text field in the window. You can save the text in this log text field by clicking the Save Log button.

By default, the 3 Sec. Button Delay checkbox is checked, causing a three-second delay between clicking a Copy or Log button and the copying or logging taking place. This gives you a short amount of time in which to click the button and then move the mouse into your desired position. It may be easier to uncheck this box, move the mouse into position, and press the F1 to F8 keys to copy or log the mouse position. You can look at the Copy and Log menus at the top of the MouseInfo window to find out which key maps to which buttons.

For example, uncheck the 3 Sec. Button Delay, then move the mouse around the screen while pressing the F6 button, and notice how the x- and y-coordinates of the mouse are recorded in the large text field in the middle of the window. You can later use these coordinates in your PyAutoGUI scripts.

For more information on MouseInfo, review the complete documentation at *https://mouseinfo.readthedocs.io/*.

## Working with the Screen

Your GUI automation programs don't have to click and type blindly. PyAutoGUI has screenshot features that can create an image file based on the current contents of the screen. These functions can also return a Pillow `Image` object of the current screen's appearance. If you've been skipping around in this book, you'll want to read Chapter 17 and install the `pillow` module before continuing with this section.

On Linux computers, the `scrot` program needs to be installed to use the screenshot functions in PyAutoGUI. In a Terminal window, run **sudo apt-get install scrot** to install this program. If you're on Windows or macOS, skip this step and continue with the section.

### Getting a Screenshot

To take screenshots in Python, call the `pyautogui.screenshot()` function. Enter the following into the interactive shell:

```
>>> import pyautogui
>>> im = pyautogui.screenshot()
```

The `im` variable will contain the `Image` object of the screenshot. You can now call methods on the `Image` object in the `im` variable, just like any other `Image` object. Chapter 19 has more information about `Image` objects.

## Analyzing the Screenshot

Say that one of the steps in your GUI automation program is to click a gray button. Before calling the click() method, you could take a screenshot and look at the pixel where the script is about to click. If it's not the same gray as the gray button, then your program knows something is wrong. Maybe the window moved unexpectedly, or maybe a pop-up dialog has blocked the button. At this point, instead of continuing—and possibly wreaking havoc by clicking the wrong thing—your program can "see" that it isn't clicking the right thing and stop itself.

You can obtain the RGB color value of a particular pixel on the screen with the pixel() function. Enter the following into the interactive shell:

```
>>> import pyautogui
>>> pyautogui.pixel((0, 0))
(176, 176, 175)
>>> pyautogui.pixel((50, 200))
(130, 135, 144)
```

Pass pixel() a tuple of coordinates, like (0, 0) or (50, 200), and it'll tell you the color of the pixel at those coordinates in your image. The return value from pixel() is an RGB tuple of three integers for the amount of red, green, and blue in the pixel. (There is no fourth value for alpha, because screenshot images are fully opaque.)

PyAutoGUI's pixelMatchesColor() function will return True if the pixel at the given x- and y-coordinates on the screen matches the given color. The first and second arguments are integers for the x- and y-coordinates, and the third argument is a tuple of three integers for the RGB color the screen pixel must match. Enter the following into the interactive shell:

```
    >>> import pyautogui
❶ >>> pyautogui.pixel((50, 200))
    (130, 135, 144)
❷ >>> pyautogui.pixelMatchesColor(50, 200, (130, 135, 144))
    True
❸ >>> pyautogui.pixelMatchesColor(50, 200, (255, 135, 144))
    False
```

After using pixel() to get an RGB tuple for the color of a pixel at specific coordinates ❶, pass the same coordinates and RGB tuple to pixelMatchesColor() ❷, which should return True. Then change a value in the RGB tuple and call pixelMatchesColor() again for the same coordinates ❸. This should return false. This method can be useful to call whenever your GUI automation programs are about to call click(). Note that the color at the given coordinates must *exactly* match. If it is even slightly different—for example, (255, 255, 254) instead of (255, 255, 255)—then pixelMatchesColor() will return False.

# Image Recognition

But what if you do not know beforehand where PyAutoGUI should click? You can use image recognition instead. Give PyAutoGUI an image of what you want to click, and let it figure out the coordinates.

For example, if you have previously taken a screenshot to capture the image of a Submit button in *submit.png*, the locateOnScreen() function will return the coordinates where that image is found. To see how locateOnScreen() works, try taking a screenshot of a small area on your screen; then save the image and enter the following into the interactive shell, replacing 'submit.png' with the filename of your screenshot:

```
>>> import pyautogui
>>> b = pyautogui.locateOnScreen('submit.png')
>>> b
Box(left=643, top=745, width=70, height=29)
>>> b[0]
643
>>> b.left
643
```

The Box object is a named tuple that locateOnScreen() returns and has the x-coordinate of the left edge, the y-coordinate of the top edge, the width, and the height for the first place on the screen the image was found. If you're trying this on your computer with your own screenshot, your return value will be different from the one shown here.

If the image cannot be found on the screen, locateOnScreen() returns None. Note that the image on the screen must match the provided image perfectly in order to be recognized. If the image is even a pixel off, locateOnScreen() raises an ImageNotFoundException exception. If you've changed your screen resolution, images from previous screenshots might not match the images on your current screen. You can change the scaling in the display settings of your operating system, as shown in Figure 20-4.

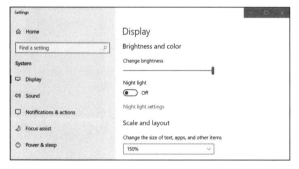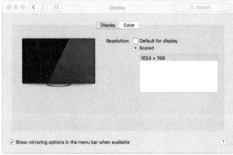

*Figure 20-4: The scale display settings in Windows 10 (left) and macOS (right)*

If the image can be found in several places on the screen, locateAllOnScreen() will return a Generator object. Generators are beyond the scope of this book, but you can pass them to list() to return a list of four-integer tuples. There will be one four-integer tuple for each location where the image is found on the screen. Continue the interactive shell example by entering the following (and replacing 'submit.png' with your own image filename):

```
>>> list(pyautogui.locateAllOnScreen('submit.png'))
[(643, 745, 70, 29), (1007, 801, 70, 29)]
```

Each of the four-integer tuples represents an area on the screen. In the example above, the image appears in two locations. If your image is only found in one area, then using list() and locateAllOnScreen() returns a list containing just one tuple.

Once you have the four-integer tuple for the specific image you want to select, you can click the center of this area by passing the tuple to click(). Enter the following into the interactive shell:

```
>>> pyautogui.click((643, 745, 70, 29))
```

As a shortcut, you can also pass the image filename directly to the click() function:

```
>>> pyautogui.click('submit.png')
```

The moveTo() and dragTo() functions also accept image filename arguments. Remember locateOnScreen() raises an exception if it can't find the image on the screen, so you should call it from inside a try statement:

```
try:
    location = pyautogui.locateOnScreen('submit.png')
except:
    print('Image could not be found.')
```

Without the try and except statements, the uncaught exception would crash your program. Since you can't be sure that your program will always find the image, it's a good idea to use the try and except statements when calling locateOnScreen().

## Getting Window Information

Image recognition is a fragile way to find things on the screen; if a single pixel is a different color, then pyautogui.locateOnScreen() won't find the image. If you need to find where a particular window is on the screen, it's faster and more reliable to use PyAutoGUI's window features.

**NOTE**    *As of version 0.9.46, PyAutoGUI's window features work only on Windows, not on macOS or Linux. These features come from PyAutoGUI's inclusion of the PyGetWindow module.*

## Obtaining the Active Window

The active window on your screen is the window currently in the foreground and accepting keyboard input. If you're currently writing code in the Mu Editor, the Mu Editor's window is the active window. Of all the windows on your screen, only one will be active at a time.

In the interactive shell, call the `pyautogui.getActiveWindow()` function to get a `Window` object (technically a `Win32Window` object when run on Windows).

Once you have that `Window` object, you can retrieve any of the object's attributes, which describe its size, position, and title:

**left, right, top, bottom**   A single integer for the x- or y-coordinate of the window's side

**topleft, topright, bottomleft, bottomright**   A named tuple of two integers for the (x, y) coordinates of the window's corner

**midleft, midright, midleft, midright**   A named tuple of two integers for the (x, y) coordinate of the middle of the window's side

**width, height**   A single integer for one of the window's dimensions, in pixels

**size**   A named tuple of two integers for the (width, height) of the window

**area**   A single integer representing the area of the window, in pixels

**center**   A named tuple of two integers for the (x, y) coordinate of the window's center

**centerx, centery**   A single integer for the x- or y-coordinate of the window's center

**box**   A named tuple of four integers for the (left, top, width, height) measurements of the window

**title**   A string of the text in the title bar at the top of the window

To get the window's position, size, and title information from the `window` object, for example, enter the following into the interactive shell:

```
>>> import pyautogui
>>> fw = pyautogui.getActiveWindow()
>>> fw
Win32Window(hWnd=2034368)
>>> str(fw)
'<Win32Window left="500", top="300", width="2070", height="1208", title="Mu
1.0.1 - test1.py">'
>>> fw.title
'Mu 1.0.1 - test1.py'
>>> fw.size
(2070, 1208)
>>> fw.left, fw.top, fw.right, fw.bottom
(500, 300, 2070, 1208)
>>> fw.topleft
(256, 144)
```

```
>>> fw.area
2500560
>>> pyautogui.click(fw.left + 10, fw.top + 20)
```

You can now use these attributes to calculate precise coordinates within a window. If you know that a button you want to click is always 10 pixels to the right of and 20 pixels down from the window's top-left corner, and the window's top-left corner is at screen coordinates (300, 500), then calling pyautogui.click(310, 520) (or pyautogui.click(fw.left + 10, fw.top + 20) if fw contains the Window object for the window) will click the button. This way, you won't have to rely on the slower, less reliable locateOnScreen() function to find the button for you.

## *Other Ways of Obtaining Windows*

While getActiveWindow() is useful for obtaining the window that is active at the time of the function call, you'll need to use some other function to obtain Window objects for the other windows on the screen.

The following four functions return a list of Window objects. If they're unable to find any windows, they return an empty list:

**pyautogui.getAllWindows()**   Returns a list of Window objects for every visible window on the screen.

**pyautogui.getWindowsAt(x, y)**   Returns a list of Window objects for every visible window that includes the point (x, y).

**pyautogui.getWindowsWithTitle(title)**   Returns a list of Window objects for every visible window that includes the string title in its title bar.

**pyautogui.getActiveWindow()**   Returns the Window object for the window that is currently receiving keyboard focus.

PyAutoGUI also has a pyautogui.getAllTitles() function, which returns a list of strings of every visible window.

## *Manipulating Windows*

Windows attributes can do more than just tell you the size and position of the window. You can also set their values in order to resize or move the window. For example, enter the following into the interactive shell:

```
>>> import pyautogui
>>> fw = pyautogui.getActiveWindow()
❶ >>> fw.width # Gets the current width of the window.
1669
❷ >>> fw.topleft # Gets the current position of the window.
(174, 153)
❸ >>> fw.width = 1000 # Resizes the width.
❹ >>> fw.topleft = (800, 400) # Moves the window.
```

First, we use the `Window` object's attributes to find out information about the window's size ❶ and position ❷. After calling these functions in Mu Editor, the window should move ❹ and become narrower ❸, as in Figure 20-5.

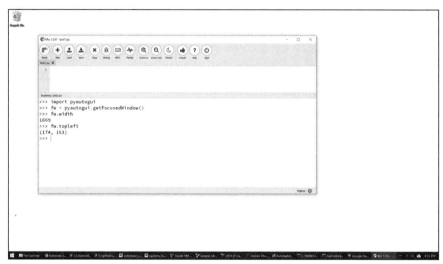

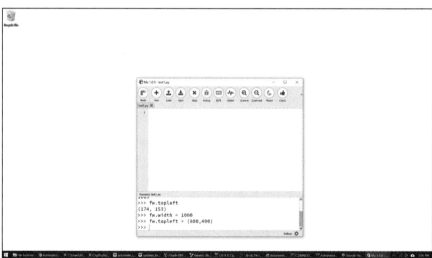

Figure 20-5: The Mu Editor window before (top) and after (bottom) using the `Window` object attributes to move and resize it

You can also find out and change the window's minimized, maximized, and activated states. Try entering the following into the interactive shell:

```
>>> import pyautogui
>>> fw = pyautogui.getActiveWindow()
❶ >>> fw.isMaximized # Returns True if window is maximized.
False
❷ >>> fw.isMinimized # Returns True if window is minimized.
```

```
    False
❸ >>> fw.isActive # Returns True if window is the active window.
    True
❹ >>> fw.maximize() # Maximizes the window.
  >>> fw.isMaximized
    True
❺ >>> fw.restore() # Undoes a minimize/maximize action.
❻ >>> fw.minimize() # Minimizes the window.
  >>> import time
  >>> # Wait 5 seconds while you activate a different window:
❼ >>> time.sleep(5); fw.activate()
❽ >>> fw.close() # This will close the window you're typing in.
```

The isMaximized ❶, isMinimized ❷, and isActive ❸ attributes contain Boolean values that indicate whether the window is currently in that state. The maximize() ❹, minimize() ❻, activate() ❼, and restore() ❺ methods change the window's state. After you maximize or minimize the window with maximize() or minimize(), the restore() method will restore the window to its former size and position.

The close() method ❽ will close a window. Be careful with this method, as it may bypass any message dialogs asking you to save your work before quitting the application.

The complete documentation for PyAutoGUI's window-controlling feature can be found at *https://pyautogui.readthedocs.io/*. You can also use these features separately from PyAutoGUI with the PyGetWindow module, documented at *https://pygetwindow.readthedocs.io/*.

# Controlling the Keyboard

PyAutoGUI also has functions for sending virtual keypresses to your computer, which enables you to fill out forms or enter text into applications.

## Sending a String from the Keyboard

The pyautogui.write() function sends virtual keypresses to the computer. What these keypresses do depends on what window is active and what text field has focus. You may want to first send a mouse click to the text field you want in order to ensure that it has focus.

As a simple example, let's use Python to automatically type the words *Hello, world!* into a file editor window. First, open a new file editor window and position it in the upper-left corner of your screen so that PyAutoGUI will click in the right place to bring it into focus. Next, enter the following into the interactive shell:

```
>>> pyautogui.click(100, 200); pyautogui.write('Hello, world!')
```

Notice how placing two commands on the same line, separated by a semicolon, keeps the interactive shell from prompting you for input between running the two instructions. This prevents you from accidentally bringing a new window into focus between the click() and write() calls, which would mess up the example.

Python will first send a virtual mouse click to the coordinates (100, 200), which should click the file editor window and put it in focus. The write() call will send the text *Hello, world!* to the window, making it look like Figure 20-6. You now have code that can type for you!

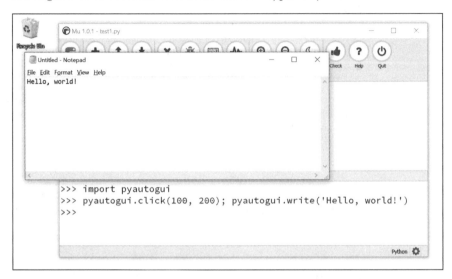

*Figure 20-6: Using PyAutogGUI to click the file editor window and type* Hello, world! *into it*

By default, the write() function will type the full string instantly. However, you can pass an optional second argument to add a short pause between each character. This second argument is an integer or float value of the number of seconds to pause. For example, pyautogui.write('Hello, world!', 0.25) will wait a quarter-second after typing *H*, another quarter-second after *e*, and so on. This gradual typewriter effect may be useful for slower applications that can't process keystrokes fast enough to keep up with PyAutoGUI.

For characters such as *A* or *!*, PyAutoGUI will automatically simulate holding down the SHIFT key as well.

### Key Names

Not all keys are easy to represent with single text characters. For example, how do you represent SHIFT or the left arrow key as a single character? In PyAutoGUI, these keyboard keys are represented by short string values instead: 'esc' for the ESC key or 'enter' for the ENTER key.

Instead of a single string argument, a list of these keyboard key strings can be passed to write(). For example, the following call presses the A key, then the B key, then the left arrow key twice, and finally the X and Y keys:

```
>>> pyautogui.write(['a', 'b', 'left', 'left', 'X', 'Y'])
```

Because pressing the left arrow key moves the keyboard cursor, this will output *XYab*. Table 20-1 lists the PyAutoGUI keyboard key strings that you can pass to write() to simulate pressing any combination of keys.

You can also examine the pyautogui.KEYBOARD_KEYS list to see all possible keyboard key strings that PyAutoGUI will accept. The 'shift' string refers to the left SHIFT key and is equivalent to 'shiftleft'. The same applies for 'ctrl', 'alt', and 'win' strings; they all refer to the left-side key.

**Table 20-1:** PyKeyboard Attributes

| Keyboard key string | Meaning |
| --- | --- |
| 'a', 'b', 'c', 'A', 'B', 'C', '1', '2', '3', '!', '@', '#', and so on | The keys for single characters |
| 'enter' (or 'return' or '\n') | The ENTER key |
| 'esc' | The ESC key |
| 'shiftleft', 'shiftright' | The left and right SHIFT keys |
| 'altleft', 'altright' | The left and right ALT keys |
| 'ctrlleft', 'ctrlright' | The left and right CTRL keys |
| 'tab' (or '\t') | The TAB key |
| 'backspace', 'delete' | The BACKSPACE and DELETE keys |
| 'pageup', 'pagedown' | The PAGE UP and PAGE DOWN keys |
| 'home', 'end' | The HOME and END keys |
| 'up', 'down', 'left', 'right' | The up, down, left, and right arrow keys |
| 'f1', 'f2', 'f3', and so on | The F1 to F12 keys |
| 'volumemute', 'volumedown', 'volumeup' | The mute, volume down, and volume up keys (some keyboards do not have these keys, but your operating system will still be able to understand these simulated keypresses) |
| 'pause' | The PAUSE key |
| 'capslock', 'numlock', 'scrolllock' | The CAPS LOCK, NUM LOCK, and SCROLL LOCK keys |
| 'insert' | The INS or INSERT key |
| 'printscreen' | The PRTSC or PRINT SCREEN key |
| 'winleft', 'winright' | The left and right WIN keys (on Windows) |
| 'command' | The Command (⌘) key (on macOS) |
| 'option' | The OPTION key (on macOS) |

### Pressing and Releasing the Keyboard

Much like the mouseDown() and mouseUp() functions, pyautogui.keyDown() and pyautogui.keyUp() will send virtual keypresses and releases to the computer. They are passed a keyboard key string (see Table 20-1) for their argument. For convenience, PyAutoGUI provides the pyautogui.press() function, which calls both of these functions to simulate a complete keypress.

Run the following code, which will type a dollar sign character (obtained by holding the SHIFT key and pressing 4):

```
>>> pyautogui.keyDown('shift'); pyautogui.press('4'); pyautogui.keyUp('shift')
```

This line presses down SHIFT, presses (and releases) 4, and then releases SHIFT. If you need to type a string into a text field, the write() function is more suitable. But for applications that take single-key commands, the press() function is the simpler approach.

### Hotkey Combinations

A *hotkey* or *shortcut* is a combination of keypresses to invoke some application function. The common hotkey for copying a selection is CTRL-C (on Windows and Linux) or ⌘-C (on macOS). The user presses and holds the CTRL key, then presses the C key, and then releases the C and CTRL keys. To do this with PyAutoGUI's keyDown() and keyUp() functions, you would have to enter the following:

```
pyautogui.keyDown('ctrl')
pyautogui.keyDown('c')
pyautogui.keyUp('c')
pyautogui.keyUp('ctrl')
```

This is rather complicated. Instead, use the pyautogui.hotkey() function, which takes multiple keyboard key string arguments, presses them in order, and releases them in the reverse order. For the CTRL-C example, the code would simply be as follows:

```
pyautogui.hotkey('ctrl', 'c')
```

This function is especially useful for larger hotkey combinations. In Word, the CTRL-ALT-SHIFT-S hotkey combination displays the Style pane. Instead of making eight different function calls (four keyDown() calls and four keyUp() calls), you can just call hotkey('ctrl', 'alt', 'shift', 's').

## Setting Up Your GUI Automation Scripts

GUI automation scripts are a great way to automate the boring stuff, but your scripts can also be finicky. If a window is in the wrong place on a desktop or some pop-up appears unexpectedly, your script could be clicking on the wrong things on the screen. Here are some tips for setting up your GUI automation scripts:

- Use the same screen resolution each time you run the script so that the position of windows doesn't change.
- The application window that your script clicks should be maximized so that its buttons and menus are in the same place each time you run the script.

- Add generous pauses while waiting for content to load; you don't want your script to begin clicking before the application is ready.

- Use `locateOnScreen()` to find buttons and menus to click, rather than relying on XY coordinates. If your script can't find the thing it needs to click, stop the program rather than let it continue blindly clicking.

- Use `getWindowsWithTitle()` to ensure that the application window you think your script is clicking on exists, and use the `activate()` method to put that window in the foreground.

- Use the `logging` module from Chapter 11 to keep a log file of what your script has done. This way, if you have to stop your script halfway through a process, you can change it to pick up from where it left off.

- Add as many checks as you can to your script. Think about how it could fail if an unexpected pop-up window appears or if your computer loses its internet connection.

- You may want to supervise the script when it first begins to ensure that it's working correctly.

You might also want to put a pause at the start of your script so the user can set up the window the script will click on. PyAutoGUI has a `sleep()` function that acts identically to `time.sleep()` (it just frees you from having to also add `import time` to your scripts). There is also a `countdown()` function that prints numbers counting down to give the user a visual indication that the script will continue soon. Enter the following into the interactive shell:

```
>>> import pyautogui
>>> pyautogui.sleep(3) # Pauses the program for 3 seconds.
>>> pyautogui.countdown(10) # Counts down over 10 seconds.
10 9 8 7 6 5 4 3 2 1
>>> print('Starting in ', end=''); pyautogui.countdown(3)
Starting in 3 2 1
```

These tips can help make your GUI automation scripts easier to use and more able to recover from unforeseen circumstances.

# Review of the PyAutoGUI Functions

Since this chapter covered many different functions, here is a quick summary reference:

**moveTo(x, y)**  Moves the mouse cursor to the given *x* and *y* coordinates.

**move(xOffset, yOffset)**  Moves the mouse cursor relative to its current position.

**dragTo(x, y)**  Moves the mouse cursor while the left button is held down.

**drag(xOffset, yOffset)**  Moves the mouse cursor relative to its current position while the left button is held down.

**click(x, y, button)**  Simulates a click (left button by default).

**rightClick()**    Simulates a right-button click.

**middleClick()**    Simulates a middle-button click.

**doubleClick()**    Simulates a double left-button click.

**mouseDown(*x, y, button*)**    Simulates pressing down the given button at the position *x, y*.

**mouseUp(*x, y, button*)**    Simulates releasing the given button at the position *x, y*.

**scroll(*units*)**    Simulates the scroll wheel. A positive argument scrolls up; a negative argument scrolls down.

**write(*message*)**    Types the characters in the given message string.

**write([*key1, key2, key3*])**    Types the given keyboard key strings.

**press(*key*)**    Presses the given keyboard key string.

**keyDown(*key*)**    Simulates pressing down the given keyboard key.

**keyUp(*key*)**    Simulates releasing the given keyboard key.

**hotkey([*key1, key2, key3*])**    Simulates pressing the given keyboard key strings down in order and then releasing them in reverse order.

**screenshot()**    Returns a screenshot as an Image object. (See Chapter 19 for information on Image objects.)

**getActiveWindow(), getAllWindows(), getWindowsAt(), and getWindowsWithTitle()**    These functions return Window objects that can resize and reposition application windows on the desktop.

**getAllTitles()**    Returns a list of strings of the title bar text of every window on the desktop.

---

### CAPTCHAS AND COMPUTER ETHICS

"Completely Automated Public Turing test to tell Computers and Humans Apart" or "captchas" are those small tests that ask you to type the letters in a distorted picture or click on photos of fire hydrants. These are tests that are easy, if annoying, for humans to pass but nearly impossible for software to solve. After reading this chapter, you can see how easy it is to write a script that could, say, sign up for billions of free email accounts or flood users with harassing messages. Captchas mitigate this by requiring a step that only a human can pass.

However not all websites implement captchas, and these can be vulnerable to abuse by unethical programmers. Learning to code is a powerful and exciting skill, and you may be tempted to misuse this power for personal gain or even just to show off. But just as an unlocked door isn't justification for trespass, the responsibility for your programs falls upon you, the programmer. There is nothing clever about circumventing systems to cause harm, invade privacy, or gain unfair advantage. I hope that my efforts in writing this book enable you to become your most productive self, rather than a mercenary one.

# Project: Automatic Form Filler

Of all the boring tasks, filling out forms is the most dreaded of chores. It's only fitting that now, in the final chapter project, you will slay it. Say you have a huge amount of data in a spreadsheet, and you have to tediously retype it into some other application's form interface—with no intern to do it for you. Although some applications will have an Import feature that will allow you to upload a spreadsheet with the information, sometimes it seems that there is no other way than mindlessly clicking and typing for hours on end. You've come this far in this book; you know that *of course* must be a way to automate this boring task.

The form for this project is a Google Docs form that you can find at *https://autbor.com/form*. It looks like Figure 20-7.

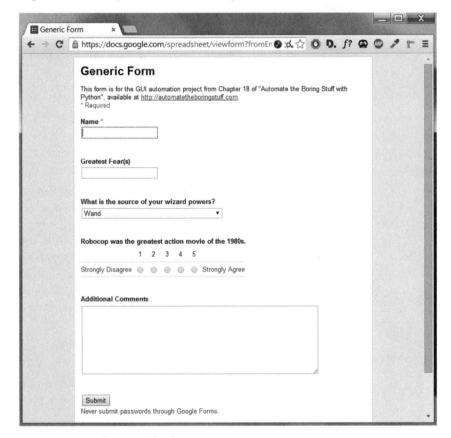

*Figure 20-7: The form used for this project*

At a high level, here's what your program should do:

1. Click the first text field of the form.
2. Move through the form, typing information into each field.
3. Click the Submit button.
4. Repeat the process with the next set of data.

This means your code will need to do the following:

1. Call `pyautogui.click()` to click the form and Submit button.
2. Call `pyautogui.write()` to enter text into the fields.
3. Handle the `KeyboardInterrupt` exception so the user can press CTRL-C to quit.

Open a new file editor window and save it as *formFiller.py*.

## Step 1: Figure Out the Steps

Before writing code, you need to figure out the exact keystrokes and mouse clicks that will fill out the form once. The application launched by calling `pyautogui.mouseInfo()` can help you figure out specific mouse coordinates. You need to know only the coordinates of the first text field. After clicking the first field, you can just press TAB to move focus to the next field. This will save you from having to figure out the x- and y-coordinates to click for every field.

Here are the steps for entering data into the form:

1. Put the keyboard focus on the Name field so that pressing keys types text into the field.
2. Type a name and then press TAB.
3. Type a greatest fear and then press TAB.
4. Press the down arrow key the correct number of times to select the wizard power source: once for *wand*, twice for *amulet*, three times for *crystal ball*, and four times for *money*. Then press TAB. (Note that on macOS, you will have to press the down arrow key one more time for each option. For some browsers, you may need to press ENTER as well.)
5. Press the right arrow key to select the answer to the RoboCop question. Press it once for *2*, twice for *3*, three times for *4*, or four times for *5* or just press the spacebar to select *1* (which is highlighted by default). Then press TAB.
6. Type an additional comment and then press TAB.
7. Press ENTER to "click" the Submit button.
8. After submitting the form, the browser will take you to a page where you will need to follow a link to return to the form page.

Different browsers on different operating systems might work slightly differently from the steps given here, so check that these keystroke combinations work for your computer before running your program.

## Step 2: Set Up Coordinates

Load the example form you downloaded (Figure 20-7) in a browser by going to *https://autbor.com/form*.

Make your source code look like the following:

```
#! python3
# formFiller.py - Automatically fills in the form.

import pyautogui, time

# TODO: Give the user a chance to kill the script.

# TODO: Wait until the form page has loaded.

# TODO: Fill out the Name Field.

# TODO: Fill out the Greatest Fear(s) field.

# TODO: Fill out the Source of Wizard Powers field.

# TODO: Fill out the RoboCop field.

# TODO: Fill out the Additional Comments field.

# TODO: Click Submit.

# TODO: Wait until form page has loaded.

# TODO: Click the Submit another response link.
```

Now you need the data you actually want to enter into this form. In
the real world, this data might come from a spreadsheet, a plaintext file,
or a website, and it would require additional code to load into the program.
But for this project, you'll just hardcode all this data in a variable. Add the
following to your program:

```
#! python3
# formFiller.py - Automatically fills in the form.

--snip--

formData = [{'name': 'Alice', 'fear': 'eavesdroppers', 'source': 'wand',
            'robocop': 4, 'comments': 'Tell Bob I said hi.'},
            {'name': 'Bob', 'fear': 'bees', 'source': 'amulet', 'robocop': 4,
            'comments': 'n/a'},
            {'name': 'Carol', 'fear': 'puppets', 'source': 'crystal ball',
            'robocop': 1, 'comments': 'Please take the puppets out of the
            break room.'},
            {'name': 'Alex Murphy', 'fear': 'ED-209', 'source': 'money',
            'robocop': 5, 'comments': 'Protect the innocent. Serve the public
            trust. Uphold the law.'},
            ]

--snip--
```

The `formData` list contains four dictionaries for four different names. Each dictionary has names of text fields as keys and responses as values. The last bit of setup is to set PyAutoGUI's PAUSE variable to wait half a second after each function call. Also, remind the user to click on the browser to make it the active window. Add the following to your program after the `formData` assignment statement:

```
pyautogui.PAUSE = 0.5
print('Ensure that the browser window is active and the form is loaded!')
```

## Step 3: Start Typing Data

A for loop will iterate over each of the dictionaries in the `formData` list, passing the values in the dictionary to the PyAutoGUI functions that will virtually type in the text fields.

Add the following code to your program:

```
#! python3
# formFiller.py - Automatically fills in the form.

--snip--

for person in formData:
    # Give the user a chance to kill the script.
    print('>>> 5-SECOND PAUSE TO LET USER PRESS CTRL-C <<<')
❶ time.sleep(5)

--snip--
```

As a small safety feature, the script has a five-second pause ❶ that gives the user a chance to hit CTRL-C (or move the mouse cursor to the upper-left corner of the screen to raise the `FailSafeException` exception) to shut the program down in case it's doing something unexpected. After the code that waits to give the page time to load, add the following:

```
#! python3
# formFiller.py - Automatically fills in the form.

--snip--

❶ print('Entering %s info...' % (person['name']))
❷ pyautogui.write(['\t', '\t'])

    # Fill out the Name field.
❸ pyautogui.write(person['name'] + '\t')

    # Fill out the Greatest Fear(s) field.
❹ pyautogui.write(person['fear'] + '\t')

--snip--
```

We add an occasional print() call to display the program's status in its Terminal window to let the user know what's going on ❶.

Since the form has had time to load, call pyautogui.write(['\t', '\t']) to press TAB twice and put the Name field into focus ❷. Then call write() again to enter the string in person['name'] ❸. The '\t' character is added to the end of the string passed to write() to simulate pressing TAB, which moves the keyboard focus to the next field, Greatest Fear(s). Another call to write() will type the string in person['fear'] into this field and then tab to the next field in the form ❹.

## Step 4: Handle Select Lists and Radio Buttons

The drop-down menu for the "wizard powers" question and the radio buttons for the RoboCop field are trickier to handle than the text fields. To click these options with the mouse, you would have to figure out the x- and y-coordinates of each possible option. It's easier to use the keyboard arrow keys to make a selection instead.

Add the following to your program:

```
#! python3
# formFiller.py - Automatically fills in the form.

--snip--

   # Fill out the Source of Wizard Powers field.
❶ if person['source'] == 'wand':
   ❷ pyautogui.write(['down', '\t'] , 0.5)
   elif person['source'] == 'amulet':
      pyautogui.write(['down', 'down', '\t'] , 0.5)
   elif person['source'] == 'crystal ball':
      pyautogui.write(['down', 'down', 'down', '\t'] , 0.5)
   elif person['source'] == 'money':
      pyautogui.write(['down', 'down', 'down', 'down', '\t'] , 0.5)

   # Fill out the RoboCop field.
❸ if person['robocop'] == 1:
   ❹ pyautogui.write([' ', '\t'] , 0.5)
   elif person['robocop'] == 2:
      pyautogui.write(['right', '\t'] , 0.5)
   elif person['robocop'] == 3:
      pyautogui.write(['right', 'right', '\t'] , 0.5)
   elif person['robocop'] == 4:
      pyautogui.write(['right', 'right', 'right', '\t'] , 0.5)
   elif person['robocop'] == 5:
      pyautogui.write(['right', 'right', 'right', 'right', '\t'] , 0.5)

--snip--
```

Once the drop-down menu has focus (remember that you wrote code to simulate pressing TAB after filling out the Greatest Fear(s) field), pressing the down arrow key will move to the next item in the selection list. Depending on the value in person['source'], your program should send a number of down arrow keypresses before tabbing to the next field. If the value at the 'source' key in this user's dictionary is 'wand' ❶, we simulate pressing the down arrow key once (to select *Wand*) and pressing TAB ❷. If the value at the 'source' key is 'amulet', we simulate pressing the down arrow key twice and pressing TAB, and so on for the other possible answers. The 0.5 argument in these write() calls add a half-second pause in between each key so that our program doesn't move too fast for the form.

The radio buttons for the RoboCop question can be selected with the right arrow keys—or, if you want to select the first choice ❸, by just pressing the spacebar ❹.

## Step 5: Submit the Form and Wait

You can fill out the Additional Comments field with the write() function by passing person['comments'] as an argument. You can type an additional '\t' to move the keyboard focus to the next field or the Submit button. Once the Submit button is in focus, calling pyautogui.press('enter') will simulate pressing the ENTER key and submit the form. After submitting the form, your program will wait five seconds for the next page to load.

Once the new page has loaded, it will have a *Submit another response* link that will direct the browser to a new, empty form page. You stored the coordinates of this link as a tuple in submitAnotherLink in step 2, so pass these coordinates to pyautogui.click() to click this link.

With the new form ready to go, the script's outer for loop can continue to the next iteration and enter the next person's information into the form.

Complete your program by adding the following code:

```python
#! python3
# formFiller.py - Automatically fills in the form.

--snip--

    # Fill out the Additional Comments field.
    pyautogui.write(person['comments'] + '\t')

    # "Click" Submit button by pressing Enter.
    time.sleep(0.5) # Wait for the button to activate.
    pyautogui.press('enter')

    # Wait until form page has loaded.
    print('Submitted form.')
    time.sleep(5)

    # Click the Submit another response link.
    pyautogui.click(submitAnotherLink[0], submitAnotherLink[1])
```

Once the main for loop has finished, the program will have plugged in the information for each person. In this example, there are only four people to enter. But if you had *4,000* people, then writing a program to do this would save you a lot of time and typing!

## Displaying Message Boxes

The programs you've been writing so far all tend to use plaintext output (with the print() function) and input (with the input() function). However, PyAutoGUI programs will use your entire desktop as its playground. The text-based window that your program runs in, whether it's Mu or a Terminal window, will probably be lost as your PyAutoGUI program clicks and interacts with other windows. This can make getting input and output from the user hard if the Mu or Terminal windows get hidden under other windows.

To solve this, PyAutoGUI offers pop-up message boxes to provide notifications to the user and receive input from them. There are four message box functions:

**pyautogui.alert(text)**    Displays text and has a single OK button.

pyautogui.confirm(text)    Displays text and has OK and Cancel buttons, returning either 'OK' or 'Cancel' depending on the button clicked.

**pyautogui.prompt(text)**    Displays text and has a text field for the user to type in, which it returns as a string.

**pyautogui.password(text)**    Is the same as prompt(), but displays asterisks so the user can enter sensitive information such as a password.

These functions also have an optional second parameter that accepts a string value to use as the title in the title bar of the message box. The functions won't return until the user has clicked a button on them, so they can also be used to introduce pauses into your PyAutoGUI programs. Enter the following into the interactive shell:

```
>>> import pyautogui
>>> pyautogui.alert('This is a message.', 'Important')
'OK'
>>> pyautogui.confirm('Do you want to continue?') # Click Cancel
'Cancel'
>>> pyautogui.prompt("What is your cat's name?")
'Zophie'
>>> pyautogui.password('What is the password?')
'hunter2'
```

The pop-up message boxes that these lines produce look like Figure 20-8.

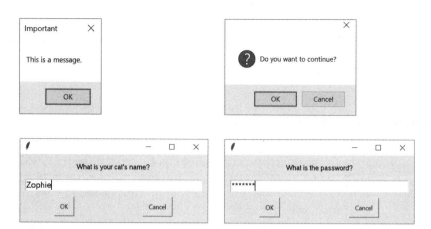

*Figure 20-8: From top left to bottom right, the windows created by* `alert()`, `confirm()`, `prompt()`, *and* `password()`

These functions can be used to provide notifications or ask the user questions while the rest of the program interacts with the computer through the mouse and keyboard. The full online documentation can be found at *https://pymsgbox.readthedocs.io.*

## Summary

GUI automation with the `pyautogui` module allows you to interact with applications on your computer by controlling the mouse and keyboard. While this approach is flexible enough to do anything that a human user can do, the downside is that these programs are fairly blind to what they are clicking or typing. When writing GUI automation programs, try to ensure that they will crash quickly if they're given bad instructions. Crashing is annoying, but it's much better than the program continuing in error.

You can move the mouse cursor around the screen and simulate mouse clicks, keystrokes, and keyboard shortcuts with PyAutoGUI. The `pyautogui` module can also check the colors on the screen, which can provide your GUI automation program with enough of an idea of the screen contents to know whether it has gotten offtrack. You can even give PyAutoGUI a screenshot and let it figure out the coordinates of the area you want to click.

You can combine all of these PyAutoGUI features to automate any mindlessly repetitive task on your computer. In fact, it can be downright hypnotic to watch the mouse cursor move on its own and to see text appear on the screen automatically. Why not spend the time you saved by sitting back and watching your program do all your work for you? There's a certain satisfaction that comes from seeing how your cleverness has saved you from the boring stuff.

## Practice Questions

1. How can you trigger PyAutoGUI's fail-safe to stop a program?
2. What function returns the current `resolution()`?
3. What function returns the coordinates for the mouse cursor's current position?
4. What is the difference between `pyautogui.moveTo()` and `pyautogui.move()`?
5. What functions can be used to drag the mouse?
6. What function call will type out the characters of `"Hello, world!"`?
7. How can you do keypresses for special keys such as the keyboard's left arrow key?
8. How can you save the current contents of the screen to an image file named *screenshot.png*?
9. What code would set a two-second pause after every PyAutoGUI function call?
10. If you want to automate clicks and keystrokes inside a web browser, should you use PyAutoGUI or Selenium?
11. What makes PyAutoGUI error-prone?
12. How can you find the size of every window on the screen that includes the text `Notepad` in its title?
13. How can you make, say, the Firefox browser active and in front of every other window on the screen?

## Practice Projects

For practice, write programs that do the following.

### Looking Busy

Many instant messaging programs determine whether you are idle, or away from your computer, by detecting a lack of mouse movement over some period of time—say, 10 minutes. Maybe you're away from your computer but don't want others to see your instant messenger status go into idle mode. Write a script to nudge your mouse cursor slightly every 10 seconds. The nudge should be small and infrequent enough so that it won't get in the way if you do happen to need to use your computer while the script is running.

### Using the Clipboard to Read a Text Field

While you can send keystrokes to an application's text fields with `pyautogui .write()`, you can't use PyAutoGUI alone to read the text already inside a text field. This is where the Pyperclip module can help. You can use PyAutoGUI to obtain the window for a text editor such as Mu or Notepad, bring it to the front of the screen by clicking on it, click inside the text field, and then send the CTRL-A or ⌘-A hotkey to "select all" and CTRL-C or ⌘-C hotkey to "copy to clipboard." Your Python script can then read the clipboard text by running `import pyperclip` and `pyperclip.paste()`.

Write a program that follows this procedure for copying the text from a window's text fields. Use `pyautogui.getWindowsWithTitle('Notepad')` (or whichever text editor you choose) to obtain a Window object. The `top` and `left` attributes of this Window object can tell you where this window is, while the `activate()` method will ensure it is at the front of the screen. You can then click the main text field of the text editor by adding, say, `100` or `200` pixels to the `top` and `left` attribute values with `pyautogui.click()` to put the keyboard focus there. Call `pyautogui.hotkey('ctrl', 'a')` and `pyautogui.hotkey('ctrl', 'c')` to select all the text and copy it to the clipboard. Finally, call `pyperclip.paste()` to retrieve the text from the clipboard and paste it into your Python program. From there, you can use this string however you want, but just pass it to `print()` for now.

Note that the window functions of PyAutoGUI only work on Windows as of PyAutoGUI version 1.0.0, and not on macOS or Linux.

### Instant Messenger Bot

Google Talk, Skype, Yahoo Messenger, AIM, and other instant messaging applications often use proprietary protocols that make it difficult for others to write Python modules that can interact with these programs. But even these proprietary protocols can't stop you from writing a GUI automation tool.

The Google Talk application has a search bar that lets you enter a username on your friend list and open a messaging window when you press ENTER. The keyboard focus automatically moves to the new window. Other instant messenger applications have similar ways to open new message windows. Write a program that will automatically send out a notification message to a select group of people on your friend list. Your program may have to deal with exceptional cases, such as friends being offline, the chat window appearing at different coordinates on the screen, or confirmation boxes that interrupt your messaging. Your program will have to take screenshots to guide its GUI interaction and adopt ways of detecting when its virtual keystrokes aren't being sent.

**NOTE** *You may want to set up some fake test accounts so that you don't accidentally spam your real friends while writing this program.*

### Game-Playing Bot Tutorial

There is a great tutorial titled "How to Build a Python Bot That Can Play Web Games" that you can find a link to at *https://nostarch.com/automatestuff2/*. This tutorial explains how to create a GUI automation program in Python that plays a Flash game called Sushi Go Round. The game involves clicking the correct ingredient buttons to fill customers' sushi orders. The faster you fill orders without mistakes, the more points you get. This is a perfectly suited task for a GUI automation program—and a way to cheat to a high score! The tutorial covers many of the same topics that this chapter covers but also includes descriptions of PyAutoGUI's basic image recognition features. The source code for this bot is at *https://github.com/asweigart/sushigoroundbot/* and a video of the bot playing the game is at *https://youtu.be/lfk_T6VKhTE*.

# A

## INSTALLING
## THIRD-PARTY MODULES

 Many developers have written their own modules, extending Python's capabilities beyond what is provided by the standard library of modules packaged with Python. The primary way to install third-party modules is to use Python's pip tool. This tool securely downloads and installs Python modules onto your computer from *https://pypi.python.org/*, the website of the Python Software Foundation. PyPI, or the Python Package Index, is a sort of free app store for Python modules.

## The pip Tool

While pip comes automatically installed with Python 3.4 and later on Windows and macOS, you may have to install it separately on Linux. You can see whether pip is already installed on Linux by running which pip3 in a Terminal window. If it's installed, you'll see the location of *pip3* displayed. Otherwise, nothing will display. To install *pip3* on Ubuntu or Debian Linux, open a new Terminal window and enter `sudo apt-get install python3-pip`. To install *pip3* on Fedora Linux, enter `sudo yum install python3-pip` into a Terminal window. You'll need to enter the administrator password for your computer.

The pip tool is run from a *terminal* (also called *command line*) window, not from Python's interactive shell. On Windows, run the "Command Prompt" program from the Start menu. On macOS, run Terminal from Spotlight. On Ubuntu Linux, run Terminal from Ubuntu Dash or press CTRL-ALT-T.

If pip's folder is not listed in the PATH environment variable, you may have to change directories in the terminal window with the `cd` command before running pip. If you need to find out your username, run `echo %USERNAME%` on Windows or `whoami` on macOS and Linux. Then run `cd pip folder`, where pip's folder is *C:\Users\<USERNAME>\AppData\Local\Programs \Python\Python37\Scripts* on Windows. On macOS, it is in */Library/Frameworks /Python.framework/Versions/3.7/bin/*. On Linux, it is in */home/<USERNAME> /.local/bin/*. Then you'll be in the right folder to run the pip tool.

## Installing Third-Party Modules

The executable file for the pip tool is called *pip* on Windows and *pip3* on macOS and Linux. From the command line, you pass it the command `install` followed by the name of the module you want to install. For example, on Windows you would enter `pip install --user MODULE`, where `MODULE` is the name of the module.

Because future changes to these third-party modules may be backward incompatible, I recommend that you install the exact versions used in this book, as given later in this section. You can add `-U MODULE==VERSION` to the end of the module name to install a particular version. Note that there are two equal signs in this command line option. For example, `pip install --user -U send2trash==1.5.0` installs version 1.5.0 of the `send2trash` module.

You can install all of the modules covered in this book by downloading the "requirements" files for your operating system from *https://nostarch.com /automatestuff2/* and running one of the following commands:

- On Windows:

```
pip install --user -r automate-win-requirements.txt --user
```

- On macOS:

```
pip3 install --user -r automate-mac-requirements.txt --user
```

- On Linux:

```
pip3 install --user -r automate-linux-requirements.txt --user
```

The following list contains the third-party modules used in this book along with their versions. You can enter these commands separately if you only want to install a few of these modules on your computer.

- `pip install --user send2trash==1.5.0`
- `pip install --user requests==2.21.0`
- `pip install --user beautifulsoup4==4.7.1`
- `pip install --user selenium==3.141.0`
- `pip install --user openpyxl==2.6.1`
- `pip install --user PyPDF2==1.26.0`
- `pip install --user python-docx==0.8.10` (install python-docx, not docx)
- `pip install --user imapclient==2.1.0`
- `pip install --user pyzmail36==1.0.4`
- `pip install --user twilio`
- `pip install --user ezgmail`
- `pip install --user ezsheets`
- `pip install --user pillow==6.0.0`
- `pip install --user pyobjc-framework-Quartz==5.2` (on macOS only)
- `pip install --user pyobjc-core==5.2` (on macOS only)
- `pip install --user pyobjc==5.2` (on macOS only)
- `pip install --user python3-xlib==0.15` (on Linux only)
- `pip install --user pyautogui`

> **NOTE** *For macOS users: The* `pyobjc` *module can take 20 minutes or longer to install, so don't be alarmed if it takes a while. You should also install the* `pyobjc-core` *module first, which will reduce the overall installation time.*

After installing a module, you can test that it installed successfully by running `import ModuleName` in the interactive shell. If no error messages are displayed, you can assume the module was installed successfully.

If you already have the module installed but would like to upgrade it to the latest version available on PyPI, run `pip install --user -U MODULE` (or `pip3 install --user -U MODULE` on macOS and Linux). The `--user` option installs the module in your home directory. This avoids potential permissions errors you might encounter when trying to install for all users.

The latest versions of the Selenium and OpenPyXL modules tend to have changes that are backward incompatible with the versions used in this book. On the other hand, the Twilio, EZGmail, and EZSheets modules interact with online services, and you might be required to install the latest version of these modules with the `pip install --user -U` command.

**WARNING** *The first edition of this book suggested using the sudo command if you encountered permission errors while running pip: sudo pip install module. This is a bad practice, as it installs modules to the Python installation used by your operating system. Your operating system may run Python scripts to carry out system-related tasks, and if you install modules to this Python installation that conflict with its existing modules, you could create hard-to-fix bugs. Never use sudo when installing Python modules.*

## Installing Modules for the Mu Editor

The Mu editor has its own Python environment, separate from the one that typical Python installations have. To install modules so that you can use them in scripts launched by Mu, you must bring up the Admin Panel by clicking the gear icon in the lower-right corner of the Mu editor. In the window that appears, click the Third Party Packages tab and follow the instructions for installing modules on that tab. The ability to install modules into Mu is still an early feature under development, so these instructions may change.

If you are unable to install modules using the Admin Panel, you can also open a Terminal window and run the pip tool specific to the Mu editor. You'll have to use pip's --target command line option to specify Mu's module folder. On Windows, this folder is *C:\Users\<USERNAME>\AppData\Local\Mu\pkgs*. On macOS, this folder is */Applications/mu-editor.app/Contents/Resources/app_packages*. On Linux, you don't need to enter a --target argument; just run the pip3 command normally.

For example, after you download the requirements file for your operating system from *https://nostarch.com/automatestuff2/*, run the following:

- On Windows:

```
pip install -r automate-win-requirements.txt --target "C:\Users\USERNAME\AppData\Local\Mu\pkgs"
```

- On macOS:

```
pip3 install -r automate-mac-requirements.txt --target /Applications/mu-editor.app/Contents/Resources/app_packages
```

- On Linux:

```
pip3 install --user -r automate-linux-requirements.txt
```

If you want to install only some of the modules, you can run the regular pip (or pip3) command and add the --target argument.

# B

## RUNNING PROGRAMS

If you have a program open in Mu, running it is a simple matter of pressing F5 or clicking the Run button at the top of the window. This is an easy way to run programs while writing them, but opening Mu to run your finished programs can be a burden. There are more convenient ways to execute Python scripts, depending on which operating system you're using.

### Running Programs from the Terminal Window

When you open a terminal window (such as Command Prompt on Windows or Terminal on macOS and Linux), you'll see a mostly blank window into which you can enter text commands. You can run your programs from the terminal, but if you're not used to it, using your computer through a terminal (also called a *command line*) can be intimidating: unlike a graphical user interface, it offers no hints about what you're supposed to do.

To open a terminal window on Windows, click the Start button, enter Command Prompt, and press ENTER. On macOS, click on the Spotlight icon in the upper right, type Terminal, and press ENTER. On Ubuntu Linux, you can press the WIN key to bring up Dash, type Terminal, and press ENTER. The keyboard shortcut CTRL-ALT-T will also open a terminal window on Ubuntu.

Just as the interactive shell has a >>> prompt, the terminal will display a prompt for you to enter commands. On Windows, it will be the full path of the folder you are currently in:

---

C:\Users\Al>**your commands go here**

---

On macOS, the prompt shows your computer's name, a colon, the current working directory (with your home folder represented as ~ for short), and your username, followed by a dollar sign ($):

---

Als-MacBook-Pro:~ al$ **your commands go here**

---

On Ubuntu Linux, the prompt is similar to macOS's, except it begins with the username and an @ sign:

---

al@al-VirtualBox:~$ **your commands go here**

---

It's possible to customize these prompts, but that's beyond the scope of this book.

When you enter a command, like python on Windows or python3 on macOS and Linux, the terminal checks for a program with that name in the folder you're currently in. If it doesn't find it there, it will check the folders listed in the PATH environment variable. You can think of *environment variables* as variables for your entire operating system. They'll contain a few system settings. To see the value stored in the PATH environment variable, run echo %PATH% on Windows and echo $PATH on macOS and Linux. Here's an example on macOS:

---

Als-MacBook-Pro:~ al$ **echo $PATH**
/Library/Frameworks/Python.framework/Versions/3.7/bin:/usr/local/bin:/usr/
bin:/bin:/usr/sbin:/sbin

---

On macOS, the *python3* program file is located in the */Library/Frameworks /Python.framework/Versions/3.7/bin* folder, so you don't have to enter /Library /Frameworks/Python.framework/Versions/3.7/bin/python3 or switch to that folder first to run it; you can enter python3 from any folder, and the terminal will find it in one of the PATH environment variable's folders. Adding a program's folder to the PATH environment variable is a convenient shortcut.

If you want to run a *.py* program, you must enter python (or python3) followed by the *.py* filename. This will run Python, and in turn Python will run the code it finds in that *.py* file. After the Python program finishes, you'll return to the terminal prompt. For example, on Windows, a simple "Hello, world!" program would look like this:

```
Microsoft Windows [Version 10.0.17134.648]
(c) 2018 Microsoft Corporation. All rights reserved.

C:\Users\Al>python hello.py
Hello, world!

C:\Users\Al>
```

Running python (or python3) without any filename will cause Python to launch the interactive shell.

## Running Python Programs on Windows

There are a few other ways you can run Python programs on Windows. Instead of opening a terminal window to run your Python scripts, you can press WIN-R to open the Run Dialog box and enter py C:\*path\to\your \pythonScript.py*, as shown in Figure B-1. The *py.exe* program is installed at *C:\Windows\py.exe*, which is already in the PATH environment variable, and typing the *.exe* file extension is optional when running programs.

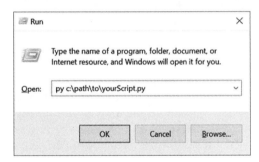

*Figure B-1: The Run dialog on Windows*

The downside of this method is that you must enter the full path to your script. Also, while running your Python script from the dialog box will open a new terminal window to display its output, this window will automatically close when the program ends, and you might miss some output.

You can solve these problems by creating a *batch script*, which is a small text file with the *.bat* file extension that can run multiple terminal commands, much like a shell script in macOS and Linux. You can use a text editor such as Notepad to create these files.

To make a batch file, make a new text file containing a single line, like this:

```
@py.exe C:\path\to\your\pythonScript.py %*
@pause
```

Replace this path with the absolute path to your own program and save this file with a *.bat* file extension (for example, *pythonScript.bat*). The @ sign at the start of each command prevents it from being displayed in the terminal window, and the %* forwards any command line arguments entered after the batch filename to the Python script. The Python script, in turn, reads the command line arguments in the sys.argv list. This batch file will keep you from having to type the full absolute path for the Python program every time you want to run it. In addition, @pause will add "Press any key to continue..." after the end of the Python script to prevent the program's window from disappearing too quickly. I recommend you place all your batch and *.py* files in a single folder that already exists in the PATH environment variable, such as *C:\Users\<USERNAME>*.

With a batch file set up to run your Python script, you don't need to open a terminal window and type the full file path and name of your Python script. Instead, just press WIN-R, enter pythonScript (the full *pythonScript.bat* name isn't necessary), and press ENTER to run your script.

## Running Python Programs on macOS

On macOS, you can create a shell script to run your Python scripts by creating a text file with the *.command* file extension. Create a new file in a text editor such as TextEdit and add the following content:

```
#!/usr/bin/env bash
python3 /path/to/your/pythonScript.py
```

Save this file with the *.command* file extension in your home folder (for example, on my computer it's */Users/al*). In a terminal window, make this shell script executable by running chmod u+x *yourScript*.command. Now you will be able to click the Spotlight icon (or press ⌘-SPACE) and enter *yourScript*.command to run the shell script, which in turn will run your Python script.

## Running Python Programs on Ubuntu Linux

Running your Python scripts in Ubuntu Linux from the Dash menu requires considerable setup. Let's say we have a */home/al/example.py* script (your Python script could be in a different folder with a different filename) that we want to run from Dash. First, use a text editor such as gedit to create a new file with the following content:

```
[Desktop Entry]
Name=example.py
Exec=gnome-terminal -- /home/al/example.sh
Type=Application
Categories=GTK;GNOME;Utility;
```

Save this file to the */home/<al>/.local/share/applications* folder (replacing *al* with your own username) as *example.desktop*. If your text editor doesn't show the *.local* folder (because folders that begin with a period are considered hidden), you may have to save it to your home folder (such as */home/al*) and open a terminal window to move the file with the `mv /home/al/example.desktop /home/al/.local/share/applications` command.

When the *example.desktop* file is in the */home/al/.local/share/applications* folder, you'll be able to press the Windows key on your keyboard to bring up Dash and type *example.py* (or whatever you put for the `Name` field). This opens a new terminal window (specifically, the `gnome-terminal` program) that runs the */home/al/example.sh* shell script, which we'll create next.

In the text editor, create a new file with the following content:

```
#!/usr/bin/env bash
python3 /home/al/example.py
bash
```

Save this file to */home/al/example.sh*. This is a shell script: a script that runs a series of terminal commands. This shell script will run our */home/al/example.py* Python script and then run the bash shell program. Without the bash command on the last line, the terminal window would close as soon as the Python script finishes and you'd miss any text that `print()` function calls put on the screen.

You'll need to add execute permissions to this shell script, so run the following command from a terminal window:

```
al@ubuntu:~$ chmod u+x /home/al/example.sh
```

With the *example.desktop* and *example.sh* files set up, you'll now be able to run the *example.py* script by pressing the Windows key and entering *example.py* (or whatever name you put in the `Name` field of the *example.desktop* file).

## Running Python Programs with Assertions Disabled

You can disable the assert statements in your Python programs to gain a slight performance improvement. When running Python from the terminal, include the -O switch after `python` or `python3` and before the name of the *.py* file. This will run an optimized version of your program that skips the assertion checks.

# C

## ANSWERS TO THE
## PRACTICE QUESTIONS

 This appendix contains the answers to the practice problems at the end of each chapter. I highly recommend that you take the time to work through these problems. Programming is more than memorizing syntax and a list of function names. As when learning a foreign language, the more practice you put into it, the more you will get out of it. There are many websites with practice programming problems as well. You can find a list of these at *https://nostarch.com/automatestuff2/*.

When it comes to the practice projects, there is no one correct program. As long as your program performs what the project asks for, you can consider it correct. However, if you want to see examples of completed projects, they are available in the "Download the files used in the book" link at *https://nostarch.com/automatestuff2/*.

# Chapter 1

1. The operators are +, -, *, and /. The values are 'hello', -88.8, and 5.

2. The variable is spam; the string is 'spam'. Strings always start and end with quotes.

3. The three data types introduced in this chapter are integers, floating-point numbers, and strings.

4. An expression is a combination of values and operators. All expressions evaluate (that is, reduce) to a single value.

5. An expression evaluates to a single value. A statement does not.

6. The bacon variable is set to 20. The bacon + 1 expression does not reassign the value in bacon (that would need an assignment statement: bacon = bacon + 1).

7. Both expressions evaluate to the string 'spamspamspam'.

8. Variable names cannot begin with a number.

9. The int(), float(), and str() functions will evaluate to the integer, floating-point number, and string versions of the value passed to them.

10. The expression causes an error because 99 is an integer, and only strings can be concatenated to other strings with the + operator. The correct way is I have eaten ' + str(99) + ' burritos.'.

# Chapter 2

1. True and False, using capital *T* and *F*, with the rest of the word in lowercase

2. and, or, and not

3. True and True is True.
   True and False is False.
   False and True is False.
   False and False is False.
   True or True is True.
   True or False is True.
   False or True is True.
   False or False is False.
   not True is False.
   not False is True.

4. False
   False
   True
   False
   False
   True

5. ==, !=, <, >, <=, and >=

6. == is the equal to operator that compares two values and evaluates to a Boolean, while = is the assignment operator that stores a value in a variable.

7. A condition is an expression used in a flow control statement that evaluates to a Boolean value.

8. The three blocks are everything inside the if statement and the lines print('bacon') and print('ham').

```
print('eggs')
if spam > 5:
    print('bacon')
else:
    print('ham')
print('spam')
```

9. The code:

```
if spam == 1:
    print('Hello')
elif spam == 2:
    print('Howdy')
else:
    print('Greetings!')
```

10. Press CTRL-C to stop a program stuck in an infinite loop.

11. The break statement will move the execution outside and just after a loop. The continue statement will move the execution to the start of the loop.

12. They all do the same thing. The range(10) call ranges from 0 up to (but not including) 10, range(0, 10) explicitly tells the loop to start at 0, and range(0, 10, 1) explicitly tells the loop to increase the variable by 1 on each iteration.

13. The code:

```
for i in range(1, 11):
    print(i)
```

and:

```
i = 1
while i <= 10:
    print(i)
    i = i + 1
```

14. This function can be called with spam.bacon().

# Chapter 3

1.  Functions reduce the need for duplicate code. This makes programs shorter, easier to read, and easier to update.

2.  The code in a function executes when the function is called, not when the function is defined.

3.  The `def` statement defines (that is, creates) a function.

4.  A function consists of the `def` statement and the code in its `def` clause. A function call is what moves the program execution into the function, and the function call evaluates to the function's return value.

5.  There is one global scope, and a local scope is created whenever a function is called.

6.  When a function returns, the local scope is destroyed, and all the variables in it are forgotten.

7.  A return value is the value that a function call evaluates to. Like any value, a return value can be used as part of an expression.

8.  If there is no return statement for a function, its return value is `None`.

9.  A `global` statement will force a variable in a function to refer to the global variable.

10. The data type of `None` is `NoneType`.

11. That `import` statement imports a module named `areallyourpetsnamederic`. (This isn't a real Python module, by the way.)

12. This function could be called with `spam.bacon()`.

13. Place the line of code that might cause an error in a try clause.

14. The code that could potentially cause an error goes in the try clause. The code that executes if an error happens goes in the except clause.

# Chapter 4

1.  The empty list value, which is a list value that contains no items. This is similar to how `''` is the empty string value.

2.  `spam[2] = 'hello'` (Notice that the third value in a list is at index 2 because the first index is 0.)

3.  `'d'` (Note that `'3' * 2` is the string `'33'`, which is passed to `int()` before being divided by 11. This eventually evaluates to 3. Expressions can be used wherever values are used.)

4.  `'d'` (Negative indexes count from the end.)

5.  `['a', 'b']`

6.  `1`

7.  `[3.14, 'cat', 11, 'cat', True, 99]`

8.  `[3.14, 11, 'cat', True]`

9. The operator for list concatenation is +, while the operator for replication is *. (This is the same as for strings.)

10. While append() will add values only to the end of a list, insert() can add them anywhere in the list.

11. The del statement and the remove() list method are two ways to remove values from a list.

12. Both lists and strings can be passed to len(), have indexes and slices, be used in for loops, be concatenated or replicated, and be used with the in and not in operators.

13. Lists are mutable; they can have values added, removed, or changed. Tuples are immutable; they cannot be changed at all. Also, tuples are written using parentheses, ( and ), while lists use the square brackets, [ and ].

14. (42,) (The trailing comma is mandatory.)

15. The tuple() and list() functions, respectively

16. They contain references to list values.

17. The copy.copy() function will do a shallow copy of a list, while the copy.deepcopy() function will do a deep copy of a list. That is, only copy .deepcopy() will duplicate any lists inside the list.

# Chapter 5

1. Two curly brackets: {}

2. {'foo': 42}

3. The items stored in a dictionary are unordered, while the items in a list are ordered.

4. You get a KeyError error.

5. There is no difference. The in operator checks whether a value exists as a key in the dictionary.

6. The 'cat' in spam checks whether there is a 'cat' key in the dictionary, while 'cat' in spam.values() checks whether there is a value 'cat' for one of the keys in spam.

7. spam.setdefault('color', 'black')

8. pprint.pprint()

# Chapter 6

1. Escape characters represent characters in string values that would otherwise be difficult or impossible to type into code.

2. \n is a newline; \t is a tab.

3. The \\ escape character will represent a backslash character.

4. The single quote in Howl's is fine because you've used double quotes to mark the beginning and end of the string.

5. Multiline strings allow you to use newlines in strings without the \n escape character.

6. The expressions evaluate to the following:
   - `'e'`
   - `'Hello'`
   - `'Hello'`
   - `'lo, world!`

7. The expressions evaluate to the following:
   - `'HELLO'`
   - `True`
   - `'hello'`

8. The expressions evaluate to the following:
   - `['Remember,', 'remember,', 'the', 'fifth', 'of', 'November.']`
   - `'There-can-be-only-one.'`

9. The rjust(), ljust(), and center() string methods, respectively

10. The lstrip() and rstrip() methods remove whitespace from the left and right ends of a string, respectively.

# Chapter 7

1. The re.compile() function returns Regex objects.

2. Raw strings are used so that backslashes do not have to be escaped.

3. The search() method returns Match objects.

4. The group() method returns strings of the matched text.

5. Group 0 is the entire match, group 1 covers the first set of parentheses, and group 2 covers the second set of parentheses.

6. Periods and parentheses can be escaped with a backslash: \., \(, and \).

7. If the regex has no groups, a list of strings is returned. If the regex has groups, a list of tuples of strings is returned.

8. The | character signifies matching "either, or" between two groups.

9. The ? character can either mean "match zero or one of the preceding group" or be used to signify non-greedy matching.

10. The + matches one or more. The * matches zero or more.

11. The {3} matches exactly three instances of the preceding group. The {3,5} matches between three and five instances.

12. The \d, \w, and \s shorthand character classes match a single digit, word, or space character, respectively.

13. The \D, \W, and \S shorthand character classes match a single character that is not a digit, word, or space character, respectively.

14. The .* performs a greedy match, and the .*? performs a non-greedy match.

15. Either [0-9a-z] or [a-z0-9]

16. Passing re.I or re.IGNORECASE as the second argument to re.compile() will make the matching case insensitive.

17. The . character normally matches any character except the newline character. If re.DOTALL is passed as the second argument to re.compile(), then the dot will also match newline characters.

18. The sub() call will return the string 'X drummers, X pipers, five rings, X hens'.

19. The re.VERBOSE argument allows you to add whitespace and comments to the string passed to re.compile().

20. re.compile(r'^\d{1,3}(,\d{3})*$') will create this regex, but other regex strings can produce a similar regular expression.

21. re.compile(r'[A-Z][a-z]*\sWatanabe')

22. re.compile(r'(Alice|Bob|Carol)\s(eats|pets|throws)\s(apples|cats |baseballs)\.', re.IGNORECASE)

# Chapter 8

1. No. PyInputPlus is a third-party module and doesn't come with the Python Standard Library.

2. This optionally makes your code shorter to type: you can type pyip .inputStr() instead of pyinputplus.inputStr().

3. The inputInt() function returns an int value, while the inputFloat() function returns a float value. This is the difference between returning 4 and 4.0.

4. Call pyip.inputint(min=0, max=99).

5. A list of regex strings that are either explicitly allowed or denied

6. The function will raise RetryLimitException.

7. The function returns the value 'hello'.

# Chapter 9

1. Relative paths are relative to the current working directory.

2. Absolute paths start with the root folder, such as / or *C:\*.

3. On Windows, it evaluates to WindowsPath('C:/Users/Al'). On other operating systems, it evaluates to a different kind of Path object but with the same path.

4. The expression 'C:/Users' / 'Al' results in an error, since you can't use the / operator to join two strings.

5. The os.getcwd() function returns the current working directory. The os.chdir() function *changes* the current working directory.

6. The . folder is the current folder, and .. is the parent folder.

7. *C:\bacon\eggs* is the dir name, while *spam.txt* is the base name.

8. The string 'r' for read mode, 'w' for write mode, and 'a' for append mode

9. An existing file opened in write mode is erased and completely overwritten.

10. The read() method returns the file's entire contents as a single string value. The readlines() method returns a list of strings, where each string is a line from the file's contents.

11. A shelf value resembles a dictionary value; it has keys and values, along with keys() and values() methods that work similarly to the dictionary methods of the same names.

# Chapter 10

1. The shutil.copy() function will copy a single file, while shutil.copytree() will copy an entire folder, along with all its contents.

2. The shutil.move() function is used for renaming files as well as moving them.

3. The send2trash functions will move a file or folder to the recycle bin, while shutil functions will permanently delete files and folders.

4. The zipfile.ZipFile() function is equivalent to the open() function; the first argument is the filename, and the second argument is the mode to open the ZIP file in (read, write, or append).

# Chapter 11

1. assert spam >= 10, 'The spam variable is less than 10.'

2. Either assert eggs.lower() != bacon.lower() 'The eggs and bacon variables are the same!' or assert eggs.upper() != bacon.upper(), 'The eggs and bacon variables are the same!'

3. assert False, 'This assertion always triggers.'

4. To be able to call logging.debug(), you must have these two lines at the start of your program:

```
import logging
logging.basicConfig(level=logging.DEBUG, format=' %(asctime)s -
%(levelname)s - %(message)s')
```

5. To be able to send logging messages to a file named *programLog.txt* with `logging.debug()`, you must have these two lines at the start of your program:

```
import logging
>>> logging.basicConfig(filename='programLog.txt', level=logging.DEBUG,
format=' %(asctime)s - %(levelname)s - %(message)s')
```

6. DEBUG, INFO, WARNING, ERROR, and CRITICAL

7. `logging.disable(logging.CRITICAL)`

8. You can disable logging messages without removing the logging function calls. You can selectively disable lower-level logging messages. You can create logging messages. Logging messages provides a timestamp.

9. The Step In button will move the debugger into a function call. The Step Over button will quickly execute the function call without stepping into it. The Step Out button will quickly execute the rest of the code until it steps out of the function it currently is in.

10. After you click Continue, the debugger will stop when it has reached the end of the program or a line with a breakpoint.

11. A breakpoint is a setting on a line of code that causes the debugger to pause when the program execution reaches the line.

12. To set a breakpoint in Mu, click the line number to make a red dot appear next to it.

# Chapter 12

1. The `webbrowser` module has an `open()` method that will launch a web browser to a specific URL, and that's it. The `requests` module can download files and pages from the web. The `BeautifulSoup` module parses HTML. Finally, the `selenium` module can launch and control a browser.

2. The `requests.get()` function returns a `Response` object, which has a text attribute that contains the downloaded content as a string.

3. The `raise_for_status()` method raises an exception if the download had problems and does nothing if the download succeeded.

4. The `status_code` attribute of the `Response` object contains the HTTP status code.

5. After opening the new file on your computer in `'wb'` "write binary" mode, use a for loop that iterates over the `Response` object's `iter_content()` method to write out chunks to the file. Here's an example:

```
saveFile = open('filename.html', 'wb')
for chunk in res.iter_content(100000):
    saveFile.write(chunk)
```

6. F12 brings up the developer tools in Chrome. Pressing CTRL-SHIFT-C (on Windows and Linux) or ⌘-OPTION-C (on OS X) brings up the developer tools in Firefox.

7. Right-click the element in the page and select **Inspect Element** from the menu.

8. `'#main'`

9. `'.highlight'`

10. `'div div'`

11. `'button[value="favorite"]'`

12. `spam.getText()`

13. `linkElem.attrs`

14. The selenium module is imported with `from selenium import webdriver`.

15. The `find_element_*` methods return the first matching element as a `WebElement` object. The `find_elements_*` methods return a list of all matching elements as `WebElement` objects.

16. The `click()` and `send_keys()` methods simulate mouse clicks and keyboard keys, respectively.

17. Calling the `submit()` method on any element within a form submits the form.

18. The `forward()`, `back()`, and `refresh()` WebDriver object methods simulate these browser buttons.

# Chapter 13

1. The `openpyxl.load_workbook()` function returns a `Workbook` object.

2. The `sheetnames` attribute contains a `Worksheet` object.

3. Run `wb['Sheet1']`.

4. Use `wb.active`.

5. `sheet['C5'].value` or `sheet.cell(row=5, column=3).value`

6. `sheet['C5'] = 'Hello'` or `sheet.cell(row=5, column=3).value = 'Hello'`

7. `cell.row` and `cell.column`

8. They hold the highest column and row with values in the sheet, respectively, as integer values.

9. `openpyxl.cell.column_index_from_string('M')`

10. `openpyxl.cell.get_column_letter(14)`

11. `sheet['A1':'F1']`

12. `wb.save('example.xlsx')`

13. A formula is set the same way as any value. Set the cell's value attribute to a string of the formula text. Remember that formulas begin with the = sign.

14. When calling `load_workbook()`, pass `True` for the `data_only` keyword argument.

15. `sheet.row_dimensions[5].height = 100`

16. `sheet.column_dimensions['C'].hidden = True`

17. Freeze panes are rows and columns that will always appear on the screen. They are useful for headers.

18. `openpyxl.chart.Reference()`, `openpyxl.chart.Series()`, `openpyxl.chart.BarChart()`, `chartObj.append(seriesObj)`, and `add_chart()`

# Chapter 14

1. To access Google Sheets, you need a credentials file, a token file for Google Sheets, and a token file for Google Drive.

2. EZSheets has `ezsheets.Spreadsheet` and `ezsheets.Sheet` objects.

3. Call the `downloadAsExcel()` Spreadsheet method.

4. Call the `ezsheets.upload()` function and pass the filename of the Excel file.

5. Access `ss['Students']['B2']`.

6. Call `ezsheets.getColumnLetterOf(999)`.

7. Access the `rowCount` and `columnCount` properties of the `Sheet` object.

8. Call the `delete()` Sheet method. This is only permanent if you pass the `permanent=True` keyword argument.

9. The `createSpreadsheet()` function and `createSheet()` Spreadsheet method will create `Spreadsheet` and `Sheet` objects, respectively.

10. EZSheets will throttle your method calls.

# Chapter 15

1. A `File` object returned from `open()`

2. Read-binary (`'rb'`) for `PdfFileReader()` and write-binary (`'wb'`) for `PdfFileWriter()`

3. Calling `getPage(4)` will return a `Page` object for page 5, since page 0 is the first page.

4. The `numPages` variable stores an integer of the number of pages in the `PdfFileReader` object.

5. Call `decrypt('swordfish')`.

6. The `rotateClockwise()` and `rotateCounterClockwise()` methods. The degrees to rotate is passed as an integer argument.

7. `docx.Document('demo.docx')`

8. A document contains multiple paragraphs. A paragraph begins on a new line and contains multiple runs. Runs are contiguous groups of characters within a paragraph.

9. Use `doc.paragraphs`.

10. A `Run` object has these variables (*not* a Paragraph).

11. `True` always makes the `Run` object bolded and `False` makes it always not bolded, no matter what the style's bold setting is. `None` will make the `Run` object just use the style's bold setting.

12. Call the `docx.Document()` function.

13. `doc.add_paragraph('Hello there!')`

14. The integers 0, 1, 2, 3, and 4

# Chapter 16

1. In Excel, spreadsheets can have values of data types other than strings; cells can have different fonts, sizes, or color settings; cells can have varying widths and heights; adjacent cells can be merged; and you can embed images and charts.

2. You pass a `File` object, obtained from a call to `open()`.

3. `File` objects need to be opened in read-binary (`'rb'`) for `reader` objects and write-binary (`'wb'`) for `writer` objects.

4. The `writerow()` method

5. The `delimiter` argument changes the string used to separate cells in a row. The `lineterminator` argument changes the string used to separate rows.

6. `json.loads()`

7. `json.dumps()`

# Chapter 17

1. A reference moment that many date and time programs use. The moment is January 1, 1970, UTC.

2. `time.time()`

3. `time.sleep(5)`

4. It returns the closest integer to the argument passed. For example, `round(2.4)` returns 2.

5. A datetime object represents a specific moment in time. A `timedelta` object represents a duration of time.

6. Run `datetime.datetime(2019, 1, 7).weekday()`, which returns 0. This means Monday, as the datetime module uses 0 for Monday, 1 for Tuesday, and so on up to 6 for Sunday.

7. `threadObj = threading.Thread(target=spam)`

   `threadObj.start()`

8. Make sure that code running in one thread does not read or write the same variables as code running in another thread.

# Chapter 18

1. SMTP and IMAP, respectively
2. `smtplib.SMTP()`, `smtpObj.ehlo()`, `smptObj.starttls()`, and `smtpObj.login()`
3. `imapclient.IMAPClient()` and `imapObj.login()`
4. A list of strings of IMAP keywords, such as `'BEFORE <date>'`, `'FROM <string>'`, or `'SEEN'`
5. Assign the variable `imaplib._MAXLINE` a large integer value, such as `10000000`.
6. The `pyzmail` module reads downloaded emails.
7. The *credentials.json* and *token.json* files tell the EZGmail module which Google account to use when accessing Gmail.
8. A message represents a single email, while a back-and-forth conversation involving multiple emails is a thread.
9. Include the `'has:attachment'` text in the string you pass to `search()`.
10. You will need the Twilio account SID number, the authentication token number, and your Twilio phone number.

# Chapter 19

1. An RGBA value is a tuple of 4 integers, each ranging from 0 to 255. The four integers correspond to the amount of red, green, blue, and alpha (transparency) in the color.
2. A function call to `ImageColor.getcolor('CornflowerBlue', 'RGBA')` will return `(100, 149, 237, 255)`, the RGBA value for that color.
3. A box tuple is a tuple value of four integers: the left-edge x-coordinate, the top-edge y-coordinate, the width, and the height, respectively.
4. `Image.open('zophie.png')`
5. `imageObj.size` is a tuple of two integers, the width and the height.
6. `imageObj.crop((0, 50, 50, 50))`. Notice that you are passing a box tuple to `crop()`, not four separate integer arguments.
7. Call the `imageObj.save('new_filename.png')` method of the `Image` object.
8. The `ImageDraw` module contains code to draw on images.
9. `ImageDraw` objects have shape-drawing methods such as `point()`, `line()`, or `rectangle()`. They are returned by passing the `Image` object to the `ImageDraw.Draw()` function.

# Chapter 20

1. Move the mouse to the upper-left corner of the screen, that is, the (0, 0) coordinates.
2. `pyautogui.size()` returns a tuple with two integers, for the width and height of the screen.

3. `pyautogui.position()` returns a tuple with two integers, for the x- and y-coordinates of the mouse cursor.

4. The `moveTo()` function moves the mouse to absolute coordinates on the screen, while the `move()` function moves the mouse relative to the mouse's current position.

5. `pyautogui.dragTo()` and `pyautogui.drag()`

6. `pyautogui.typewrite('Hello, world!')`

7. Either pass a list of keyboard key strings to `pyautogui.write()` (such as `'left'`) or pass a single keyboard key string to `pyautogui.press()`.

8. `pyautogui.screenshot('screenshot.png')`

9. `pyautogui.PAUSE = 2`

10. You should use Selenium for controlling a web browser instead of PyAutoGUI.

11. PyAutoGUI clicks and types blindly and cannot easily find out if it's clicking and typing into the correct windows. Unexpected pop-up windows or errors can throw the script off track and require you to shut it down.

12. Call the `pyautogui.getWindowsWithTitle('Notepad')` function.

13. Run `w = pyatuogui.getWindowsWithTitle('Firefox')`, and then run `w.activate()`.

# INDEX

logout() method (imaplib module), 425, 433
loops, 35
lower() string method, 135–136
lstrip() string method, 142

# M

macOS
    backslash vs. forward slash, 202–203
    installing Python, xxxiii, xxxiv
    installing third-party modules, 507–510
    launchd, 408
    opening files with default applications, 409
    open program, 410
    pip tool on, 507–508
    running Python programs, 514
    starting IDLE, xxxv
    starting Mu, xxxv
    terminal window, 511–512
Magic 8 Ball example program, 92
makedirs() function (os module), 207
*mapIt.py* with the webbrowser Module project, 268–270
matching
    greedy, 171, 176
    non-greedy, 171, 176
math operators, 4
    addition (+), 5
    division (/), 5
    exponent (**), 5
    integer division/floored quotient (//), 5
    modulus/remainder (%), 5
    multiplication (*), 5
    order of operations, 5
maximize() function (PyAutoGUI), 489
merge_cells() method (OpenPyXL), 322
methods. *See also names of individual methods*
    chaining calls, 457
    defined, 88
    dictionary, 114–117
    list, 88–92
    string, 134–143
Microsoft Windows. *See* Windows (operating system)
middleClick() function (PyAutoGUI), 478
minimize() function (PyAutoGUI), 489

mkdir() method (pathlib module), 208
modules. *See also names of individual modules*
    importing, 47
    third-party, installing, 507–510
modulus/remainder (%) operator, 5
Monty Python, xxx, 10
mouse
    clicking, 478
    determining position, 477–478
    dragging, 479–480
    moving, 477
    scrolling, 480–481
mouseDown() function (PyAutoGUI), 478
mouseUp() function (PyAutoGUI), 478
move() function
    PyAutoGUI, 477
    shutil module, 233
moveTo() function (PyAutoGUI), 477
moving files/folders, 233
multi-clipboard
    Multi-Clipboard Automatic Messages project, 144–146
    Updatable Multi-Clipboard, 226–228
multiline comments, 132
multiline strings, 131–132
multiple assignment trick, 85
multiplication operator (*), 5
Multiplication Quiz project, 196–197
Multithreaded XKCD Downloader project, 403–406
multithreading
    concurrency issues, 403
    join() method, 405
    overview, 401
    passing arguments to threads, 402
    start() method, 401
    Thread() function, 401
mutable data types, 94

# N

NameError exception, 14
namelist() method (zipfile module), 238
negative character classes, 174
negative indexes, 80
nested dictionaries and lists, 118, 125
nested for loops, 357, 456
new() method (Image module), 452
None value, 61

strings (*continued*)
    lstrip() method, 142
    multiline, 131
    partition() method, 140
    raw, 131
    replication, 8
    rjust() method, 140–142
    rstrip() method, 142
    slicing, 132
    split() method, 138
    startswith() method, 138
    strip() method, 142
    upper() method, 135
str() function, 15
strip() string method, 142
strptime() function, 399, 400
sub() method (re module), 178
submit() method (Selenium), 295
subtraction operator (-), 5
Sudoku, xxx
summary() function (EZGmail), 418
Super Stopwatch project, 392–394
syntax error
    can't assign to keyword, 23
    EOL while scanning string literal, 7
    invalid syntax, 6
sys module
    argv variable, 145
    exit() function, 402

## T

Tag objects, 281
tags, HTML, 274
Task Scheduler, 408
text() method (Pillow), 468, 469
textsize() method (Pillow), 468
threading module
    join() method, 405
    poll() method, 407
    start() method, 402
    Thread object, 401, 404–406
throwaway code, xxviii
tic-tac-toe, 120
timedelta() function (datetime module),
    396, 400
time module
    overview, 390
    sleep() function, 391
    Super Stopwatch project, 392–394
    time() function, 390, 400

TLS encryption, 422–423
top-level domain, 182
total_seconds() method (datetime
    module), 400
transpose() method (Pillow), 458
triple quotes (''''), 131, 132
True Boolean value, 22
truetype() function (Pillow), 468
truth tables, 25–26
"truthy" values, 43
try statement, 71
tuple data type, 96
tuple() function, 97
Twilio, 439
twilio module
    create() method, 439, 440
    TwilioRestClient() function, 439, 440
type() function, 96

## U

Ubuntu
    installing Python, xxxiv
    installing third-party modules,
        507–510
    pip tool on, 507–508
    Popen() function, 407, 409
    running Python programs, 514–515
    starting IDLE, xxxv
    starting Mu, xxxv
    terminal window, 511–512
unary operators, 26
underscore (_), 11, 173
Unicode encodings, 273
Unix epoch, 390
unlink() function (os module), 234
unmerge_cells() method
        (OpenPyXL), 322
unread() function (EZGmail), 418
updateColumn() function (EZSheets), 340
updateColumns() function
        (EZSheets), 340
updateRow() function (EZSheets), 340
updateRows() function (EZSheets), 340
Updatable Multi-Clipboard, 226–228
Updating a Spreadsheet project,
        315–317
upload() function (EZSheets), 339
upper() string method, 135–136
UTC (Coordinated Universal
        Time), 390

*Automate the Boring Stuff with Python, 2nd Edition* is set in New Baskerville, Futura, Dogma, and TheSansMono Condensed. This book was printed and bound at Sheridan Books, Inc. in Chelsea, Michigan. The paper is 60# Finch Offset, which is certified by the Forest Stewardship Council (FSC).

The book uses a layflat binding, in which the pages are bound together with a cold-set, flexible glue and the first and last pages of the resulting book block are attached to the cover. The cover is not actually glued to the book's spine, and when open, the book lies flat and the spine doesn't crack.

# RESOURCES

Visit *https://nostarch.com/automatestuff2/* for resources, errata, and more information.